MATERIAL and ENERGY BALANCES

for Engineers and Environmentalists

Advances in Chemical and Process Engineering – Vol. 1

MATERIAL and ENERGY BALANCES

for Engineers and Environmentalists

Colin Oloman

University of British Columbia, Canada

ICP

Imperial College Press

Published by

Imperial College Press
57 Shelton Street
Covent Garden
London WC2H 9HE

Distributed by

World Scientific Publishing Co. Pte. Ltd.
5 Toh Tuck Link, Singapore 596224
USA office: 27 Warren Street, Suite 401-402, Hackensack, NJ 07601
UK office: 57 Shelton Street, Covent Garden, London WC2H 9HE

British Library Cataloguing-in-Publication Data
A catalogue record for this book is available from the British Library.

Cartoons: Jeremy Blumel, Colin Oloman
Layout: Susan Belford, Borislav Arnaoudov, Eli Koleva

First published as **Ol's Notes on Material and Energy Balances** in 2003 by Colin Oloman.
Second edition: 2005

MATERIAL AND ENERGY BALANCES FOR ENGINEERS AND ENVIRONMENTALISTS
Advances in Chemical and Process Engineering — Vol. 1

ISBN-13 978-1-84816-368-3
ISBN-10 1-84816-368-1
ISBN-13 978-1-84816-369-0 (pbk)
ISBN-10 1-84816-369-X (pbk)

Typeset by Stallion Press
Email: enquiries@stallionpress.com

Printed in Singapore by Mainland Press Pte Ltd.

Acknowledgements

Thanks to the undergraduate students
in the Department of Chemical and Biological Engineering
at the University of British Columbia for motivating me to write this book.

My gratitude also goes to colleagues, students and friends who reviewed
early drafts of the text and made helpful suggestions for its improvement,
notably Bruce Bowen, Dusko Posarac, Chad Bennington, Joel Bert, Sheldon Duff,
Paul Watkinson, Karyn Ho, Alvaro Reyes and Bob Harvey.

Contents

Symbol	Meaning	Typical Units	
A, B, C	constants in the Antoine equation	–, K, K	
A_{hex}	area of heat transfer surface	m^2	
A_i	interface area	m^2	
a, b, c, d	empirical constants	various units	
ã	acceleration	ms^{-2}	[m/s^2]
a	activity of a species	–	
$C^{\#}$	number of components in a system	–	
C_p	heat capacity at constant pressure	$kJ.kmol^{-1}.K^{-1}$	[kJ/kmol.K]
C_v	heat capacity at constant volume	$kJ.kmol^{-1}.K^{-1}$	[kJ/kmol.K]
$C_{p,m}$	mean heat capacity at constant pressure	$kJ.kmol^{-1}.K^{-1}$	[kJ/kmol.K]
$C_{v,m}$	mean heat capacity at constant volume	$kJ.kmol^{-1}.K^{-1}$	[kJ/kmol.K]
c	velocity of light in vacuum	ms^{-1}	[m/s]
D	liquid–liquid distribution coefficient	–	
D of F	degrees of freedom of an M&E balance problem	–	
d	diameter of dispersed bubbles, drops or particles	m	
E	energy	kJ	
E_i	interfacial energy	kJ	
E_p	potential energy	kJ	
E_k	kinetic energy	kJ	
E_v	electric voltage	V	
e	volume fraction of dispersed phase	–	
F	Faraday's number	$kCkmol^{-1}$	[kC/kmol]
$F^{\#}$	number of degrees of freedom of the intensive properties that define the thermodynamic state of a system	–	
F_e	rate of energy degradation by friction of flowing fluid	kW	
f	force	kN	
G^o	standard free energy of reaction	$kJ.kmol^{-1}$	[kJ/kmol]
GEP	gross economic potential	$\$US.y^{-1}$	[$US/y]
g	gravitational constant	$m.s^{-2}$	[m/s^2]
H	enthalpy, w.r.t. a reference state	kJ	
$\dot{H}$	stream enthalpy flow, w.r.t. compounds at reference state	kW	[kJ/s]
$\dot{H}^*$	stream enthalpy flow, w.r.t. elements at reference state	kW	[kJ/s]
H_{mix}	enthalpy (heat) of mixing	kJ	
h	specific enthalpy, w.r.t. compounds at reference state	$kJ.kmol^{-1}$	[kJ/kmol]
h*	specific enthalpy, w.r.t. elements at reference state	$kJ.kmol^{-1}$	[kJ/kmol]
$h^o_{f,Tref}$	standard enthalpy (heat) of formation at T_{ref}	$kJ.kmol^{-1}$	[kJ/kmol]
h^o_c	standard enthalpy (heat) of combustion	$kJ.kmol^{-1}$	[kJ/kmol]
$h_{c,gross}$	gross heat of combustion (–ve higher heating value)	$kJ.kmol^{-1}$	[kJ/kmol]
$h_{c,net}$	net heat of combustion (–ve lower heating value)	$kJ.kmol^{-1}$	[kJ/kmol]

Symbol	Meaning	Typical units	
h_m	specific enthalpy (heat) of fusion	$kJ.kmol^{-1}$	[kJ/kmol]
h_p	specific enthalpy (heat) of a phase change	$kJ.kmol^{-1}$	[kJ/kmol]
h_v	specific enthalpy (heat) of vaporisation	$kJ.kmol^{-1}$	[kJ/kmol]
h_{rxn}	specific enthalpy (heat) of reaction	$kJ.kmol^{-1}$	[kJ/kmol]
I	number of process streams	–	
I′	current	kA	
J	number of process species	–	
K	number of process units	–	
K_A	adsorption constant	kPa^{-1}	[1/kPa]
K_{DIV}	divider flow ratio	–	
K_{eq}	reaction equilibrium constant	–	
K_H	Henry's constant	kPa	
k	chemical reaction rate constant	s^{-1}	[1/s]
L′	vertical height of fluid column	m	
L	number of chemical reactions	–	
M	molar mass (molecular weight)	$kg.kmol^{-1}$	[kg/kmol]
M_m	mean molar mass (molecular weight)	$kg.kmol^{-1}$	[kg/kmol]
m	mass	kg	
$\dot{m}$	mass flow rate	$kg.s^{-1}$	[kg/s]
$\bar{m}$	total stream mass flow rate	$kg.s^{-1}$	[kg/s]
NEP	net economic potential	$\$US.y^{-1}$	[$US/y]
n	number of moles	kmol	
$\dot{n}$	mole flow rate	$kmol.s^{-1}$	[kmol/s]
$\bar{n}$	total stream mole flow rate	$kmol.s^{-1}$	[kmol/s]
n_w	amount of water from combustion reaction	$kmol(kmolfuel)^{-1}$	[kmol/kmolfuel]
P	total pressure	kPa(abs)	
P_c	critical pressure	kPa(abs)	
ΔP	pressure difference	kPa	
p	partial pressure	kPa(abs)	
p*	vapour pressure	kPa(abs)	
p_w	partial pressure of water vapour	kPa(abs)	
p_w^*	vapour pressure of water	kPa(abs)	
Q	net heat input to the system	kJ	
$\dot{Q}$	rate of net heat transfer into the system	kW	[kJ/s]
R	universal gas constant	$kJ.kmol^{-1}.K^{-1}$	[kJ/kmol.K]
ROI	return on investment	$\%.y^{-1}$	[%/y]
r	ratio of heat capacities = C_p / C_v	–	
rh	relative humidity	–	
SG	specific gravity	–	

Symbol	Meaning	Typical units	
S	selectivity	–	
S'	solubility	$kg.kg^{-1}$	[kg/kg]
s	split fraction	–	
T	temperature	K	
T_B	bubble-point temperature	K	
T_c	critical temperature	K	
T_D	dew-point temperature	K	
T_{ref}	reference temperature	K	
ΔT	temperature difference	K	
t	time	s, h	
U	internal energy, w.r.t. a reference state	kJ	
U_{hex}	heat transfer coefficient	$kW.m^{-2}.K^{-1}$	[kW/m^2. K]
u	specific internal energy, w.r.t. a reference state	$kJ.kmol^{-1}$	[kJ/kmol]
$\tilde{u}$	velocity	$m.s^{-1}$	[m/s]
V	volume	m^3	
$\dot{V}$	volume flow rate	$m^3.s^{-1}$	[m^3/s]
V_R	reactor volume	m^3	
v	specific volume	$m^3.kg^{-1}$	[m^3/kg]
v_m	molar volume	$m^3.kmol^{-1}$	[m^3/kmol]
v'	partial volume	m^3	
$\tilde{v}$	volume fraction	–	
W	net work output from (i.e. work done by) system	kJ	
$\dot{W}$	rate of net work transfer out of system	kW	[kJ/h]
w	mass fraction	–	
w_q	quality	–	
X	conversion of reactant	–	
X_{act}	actual conversion of reactant	–	
X_{eq}	equilibrium conversion of reactant	–	
x	mole fraction (in liquid phase)	–	
Y	yield	–	
y	mole fraction (in gas phase)	–	
Z	value of h, u or v in a gas-liquid mixture	various	
z	compressibility factor	–	

Greek Letters

Symbol	Meaning	Typical units	
$\alpha, \beta, \chi, \delta$	reaction stoichiometric coefficients	–	
ν	general reaction stoichiometric coefficient	–	
ε	extent of reaction for the specified reaction	kmol	
θ	coefficient of thermal expansion	K^{-1}	[1/K]

Symbol	Meaning	Typical units	
σ	coefficient of compressibility	kPa^{-1}	[1/kPa]
ρ	density	$kg.m^{-3}$	[kg/m^3]
ρ_m	density of mixture	$kg.m^{-3}$	[kg/m^3]
ρ_{ref}	density of reference substance	$kg.m^{-3}$	[kg/m^3]
$\Pi^{\#}$	number of phases in a system at equilibrium	–	
Σ	sum of values	–	
γ	surface tension	$kN.m^{-1}$	[kN/m]
Γ	surface concentration	$kmol.m^{-2}$	[$kmol/m^2$]

Subscripts

g	gas phase
l	liquid phase
s	solid phase
i	process stream counter
j	process species counter
k	process unit counter
in	input to a system
out	output from a system

Superscripts

o	standard state

– = dimensionless

Abbreviations		Typical Units
ACC	(final amount – initial amount) of a specified quantity in a system	kg, kmol, kJ
IN	input of a specified quantity to a system	kg, kmol, kJ
OUT	output of a specified quantity from a system	kg, kmol, kJ
GEN	generation of a specified quantity in a system	kg, kmol, kJ
CON	consumption of a specified quantity in a system	kg, kmol, kJ
Rate ACC	rate of change of ACC, w.r.t. time	kg/s, kmol/s, kW, kg/h, kmol/h, kJ/h
Rate IN	rate of change of IN, w.r.t. time	kg/s, kmol/s, kW, kg/h, kmol/h, kJ/h
Rate OUT	rate of change of OUT, w.r.t. time	kg/s, kmol/s, kW, kg/h, kmol/h, kJ/h
Rate GEN	rate of change of GEN, w.r.t. time	kg/s, kmol/s, kW, kg/h, kmol/h, kJ/h
Rate CON	rate of change of CON, w.r.t. time	kg/s, kmol/s, kW, kg/h, kmol/h, kJ/h

Arrays: Array dimensions "i" , "j" and "k" designate respectively the streams, species and process units in equations, flowsheets and stream tables.

$ACC = IN - OUT + GEN - CON$	General balance
Rate ACC = Rate IN – Rate OUT + Rate GEN – Rate CON	Differential general balance
Abs pressure = Atm pressure + gauge pressure	Superatmospheric
Abs pressure = Atm pressure – vacuum	Subatmospheric
$P = \rho g L'$	Column of fluid
$K = 273.15 + {}^{\circ}C$	Kelvin
${}^{\circ}R = 459.67 + {}^{\circ}F$	Rankine
${}^{\circ}F = 32 + 1.8\,{}^{\circ}C$	Fahrenheit
$p(j) = y(j)P$	Partial pressure
$v'(j) = y(j)V$	Partial volume
$\Sigma[p(j)] = P$	Dalton's law
$\Sigma[v'(j)] = V$	Amagat's law
$F^{\#} = C^{\#} - P^{\#} + 2$	Phase rule
$PV = nRT$	Ideal gas law
$PV = znRT$	Non-ideal gas
$v = v_o(1 + \beta(T - T_o) - \delta(P - P_o))$	Liquid or solid
$\rho = m/V$	Density
$\rho = MP/RT$	Ideal gas density
$M_m = \Sigma[M(j)y(j)]$	Mean molar mass
$\rho_m = 1/\Sigma[w(j)/\rho(j)]$	Ideal mix density
$SG = \rho/\rho_{ref}$	Specific gravity
$V_m = Mv = M/\rho$	Molar volume
$w(j) = m(j)/\Sigma[m(j)]$	Mass fraction
$x(j) = n(j)/\Sigma[n(j)]$	Mole fraction
$\Sigma[w(j)] = 1 \quad \Sigma[x(j)] = 1 \quad \Sigma[y(j)] = 1$	Stream composition
$x(j) = w(j)/M(j)/\Sigma[w(j)/M(j)]$	Mass to mole fraction
$w(j) = x(j)M(j)/\Sigma[x(j)M(j)]$	Mole to mass fraction
Mass % = weight % = wt % = 100w(j)	Mass percent
Mole % = 100x(j)	Mole percent
$p^*(j) = \exp[A(j) - B(j)/(T + C(j))]$	Antoine equation
$\ln(p^*_1/p^*_2) = (h_v/R)(1/T_2 - 1/T_1)$	Clausius-Clapeyron equation
$p(j) = x(j)p^*(j)$	Raoult's law
$p(j) = K_H x(j)$	Henry's law

Equation	Name
$1 = \Sigma[x(j)p^*(j)]/P$ (Ideal mixture liquid phase)	Bubble-point
$1 = P\Sigma[y(j)/p^*(j)]$ (Ideal mixture liquid phase)	Dew-point
$\Sigma[v(j)M(j)] = 0$	Stoichiometry
$E_i = \gamma A_i$	Surface energy
$X(j) = \dfrac{\text{amount of j converted to all products}}{\text{amount of j introduced to reaction}}$	Conversion
$\varepsilon(\ell) = \dfrac{\text{amount of species j converted in reaction } l}{\text{stoichiometric coefficient of j in reaction } l}$	Extent of reaction
$S(j, q) = \dfrac{\text{amount of j converted to product q}}{\text{amount of j converted to all products}}$	Selectivity
$Y(j, q) = \dfrac{\text{amount of j converted to product q}}{\text{amount of j introduced to reaction}}$	Yield
$Y(j, q) = X(j)S(j, q)$	Yield
$K_{eq} = (a_C^{\chi} a_D^{\delta})/(a_A^{\alpha} a_B^{\beta})$	Equilibrium constant
$a(j) \approx x(j)$	Activity (solids)
$a(j) \approx [J]/[1\ \text{kmol/m}^3]$	Activity (liquids)
$a(j) \approx p(j)/101.3\ \text{kPa}$	Activity (gases)
$X_{act}/X_{eq} < 1$	Conversion
$C_p = C_v + R$	Heat capacity
$C_p = a + bT + cT^2 + dT^3$	Heat capacity
$C_{p,m} = \int_{T_{ref}}^{T} C_p\, dT/(T - T_{ref})$ [No phase change, no reaction]	Mean heat capacity
$H = U + PV$	Enthalpy
$h_{f,Tref}(\text{gas}) = h_{f,Tref}(\text{solid}) + h_{m,Tref} + h_{v,Tref}$	Heat of formation
$h_{f,Tref}(\text{gas}) = h_{f,Tref}(\text{liquid}) + h_{v,Tref}$	Heat of formation
$A_i = (6e/d)V$	Interface area
$Z_m = w_q Z_g + (1 - w_q)Z_l$ [Z = h, u or v]	Quality
$rh = p_w / p^*_w$	Relative humidity
GEP = Value of products – Value of feeds	Gross economic potential
NEP = GEP – cost (utilities + labour + maintenance + interest)	Net economic potential
ROI = Annual NEP/Capital cost	Return on investment

$s(i,j) = \dot{n}(i,j)/\dot{n}(1,j) = \dot{m}(i,j)/\dot{m}(1,j)$		Split-fraction
$0 = \Sigma[\overline{m}(i)in] - \Sigma[\overline{m}(i)out]$		Steady-state mass balance
$e_M = 100\Sigma[\overline{m}(i)out]/\Sigma[\overline{m}(i)in]$		Mass balance closure
D of F = no. of unknowns – no. of independent equations		Specification
$E_{final} - E_{initial} = Q - W$	[Closed system]	Energy balance
$E_{final} - E_{initial} = E_{input} - E_{output} + Q - W - (PV_{out} - PV_{in})$	[Open system]	Energy balance
$dE/dt = \dot{E}_{input} - \dot{E}_{output} + \dot{Q} - \dot{W} - (P\dot{V}_{out} - P\dot{V}_{in})$	[Open system]	Energy balance
$dE/dt = 0 = [\dot{H} + \dot{E}_k + \dot{E}_p]_{in} - [\dot{H} + \dot{E}_k + \dot{E}_p]_{out} + \dot{Q} - \dot{W}$	[Open system]	Energy balance (steady-state)
$0 = \dot{H}_{input} - \dot{H}_{ouptut} + \dot{Q} - \dot{W}$	[Open system]	Enthalpy balance (steady-state)
$\Gamma(j) = aK_A p(j)/(1 + K_A p(j))$		Adsorption isotherm
$D(j) = x(j)_A / x(j)_B$		Distribution coefficient
$S'(j) = m(j)/m_s$		Solubility
Separation efficiency of component j $=$ $\dfrac{\text{amount of j in specified outlet stream}}{\text{amount of j in the inlet stream}}$		Separator
$\dot{m}(1,j)/\overline{m}(1) = \dot{m}(2,j)/\overline{m}(2) = \dot{m}(3,j)/\overline{m}(3)$, etc.		Divider composition
$K_{DIV}(i) = \overline{n}(i)/\overline{n}(1) = \overline{m}(i)/\overline{m}(1)$		Divider ratio
$C_v = [\partial u/\partial T]_v$	[No phase change, no reaction]	Heat capacity constant volume
$C_p = [\partial h/\partial T]_p$	[No phase change, no reaction]	Heat capacity constant pressure
$\dot{H}^*(i) = \Sigma[\dot{n}(i,j)h^*(j)]$	[Stream enthalpy]	Ideal mixture
$\dot{H}^*(i) = \Sigma[\dot{n}(i,j)h^*(j)] + \dot{H}_{mix}$	[Stream enthalpy]	Non-ideal mixture
$h^* \approx \int_{Tref}^{T(i)} C_p dT + h^o_{f,\,Tref}$	[Respect the phase]	Specific enthalpy, w.r.t. elements
$h^* \approx C_{p.m}[T(i) - T_{ref}] + h^o_{f,\,Tref}$	[Respect the phase]	Specific enthalpy, w.r.t. elements
$h \approx \int_{Tref}^{T(i)} C_p dT + h_{p,\,Tref}$	[Respect the phase]	Specific enthalpy, w.r.t. compounds
$h \approx C_{p.m}[T(i) - T_{ref}] + h_{p,\,Tref}$	[Respect the phase]	Specific enthalpy, w.r.t. compounds
$\dot{Q} = U_{hex}A_{hex}\Delta T$	[Thermal duty]	Heat exchanger
$\dot{W} = -(P_1\dot{V}_1)(r/(r-1))[(P_2/P_1)^{((r-1)/r)} - 1]$	[Power]	Adiabatic gas compressor

$T_2 = T_1[P_2/P_1]^{((r-1)/r)}$	[Temperature]	Adiabatic gas compressor
$\dot{W} = -(P_1\dot{V}_1)[1n(P_2/P_1)]$	[Power]	Isothermal compressor
$\dot{W} = -\dot{V}_1(P_2 - P_1)$	[Power]	Liquid pump
$\dot{W} = E_V I'$	[Power]	Battery, fuel-cell, resistor
$e_E = 100[\Sigma\dot{H}^*(i)out + \dot{W}]/[\Sigma\dot{H}^*(i)in + \dot{Q}]$		Energy balance closure
$T_{out} = T_{ref} + [(\dot{H}^*(i)in + \dot{Q} - \dot{W})/\dot{n}(in) - h^o_{f,Tref}]/C_{p,m}$	[No phase change]	Heat exchanger
$0 = \Sigma[\dot{H}(i)]in - \Sigma[\dot{H}(i)]out + \dot{Q} - \dot{W} - \dot{n}_{fuel}h_c$		Combustion
$h_{c,net} = h_{c, gross} + n_w h_{v,H2O}$		Heat of combustion
$d(m)/dt = (\dot{m})_{in} - (\dot{m})_{out} + (\dot{m})_{gen} - (\dot{m})_{con}$	[Unsteady-state]	Mass balance
$d(n)/dt = (\dot{n})_{in} - (\dot{n})_{out} + (\dot{n})_{gen} - (\dot{n})_{con}$	[Unsteady-state]	Mole balance
$dU/dt = \dot{H}_{in} - \dot{H}_{out} + \dot{Q} - \dot{W}$	[Unsteady-state]	Enthalpy balance
$0 = 0.5[(\tilde{u}_{out})^2 - (\tilde{u}_{in})^2] + g[L'_{out} - L'_{in}] + [P_{out} - P_{in}]/\rho$		Mechanical energy balance
$0 = [\dot{E}_k + \dot{E}_p]_{in} - [\dot{E}_k + \dot{E}_p]_{out} - \int_{P_{in}}^{P_{out}} \dot{V}dP - \dot{F}_e - \dot{W}$		Bernoulli equation

Absorption	Transfer of species between two bulk phases.
Acceleration	Rate of change of velocity with respect to time.
Adiabatic	Zero transfer of heat across the system envelope.
Accumulation	Change in the amount of a material and/or energy inside the system envelope.
Adsorption	Concentration of species at an interface.
Atom	Smallest particle of an element, identifiable as that element by its nucleus.
Atom balance	Material balance in which the quantity is atoms or moles of specified elements.
Batch	Operation in a closed system.
Blower	Gas compressor operating up to ca. 400 kPa(abs).
Boiler	Apparatus in which liquid is vaporised (boiled), usually by transfer of heat from combustion.
Bubble-point	Condition at which first bubbles of vapour form in a liquid.
Capital cost	Total cost for the design and installation of a plant.
Carbon cycle	Geochemical cycle of carbon through carbon dioxide, biomass and carbonates.
Catalyst	Substance that speeds a reaction without itself suffering a net conversion.
Closed system	System with zero transfer of material across the boundary.
Closure	Ratio of output to input of mass or energy for an open system. Steady-state closure = 1.
Colloid	Mixture with super-molecular particles or drops below ca. 1 micron dispersed in liquid.
Combustion	Thermochemical reaction of a fuel with an oxidant (the oxidant is usually oxygen from air).
Component	Constituent of a mixture identified as a phase or as a species.
Compressibility factor	Multiplier (range approx. 0.2 to 4) that adapts the ideal gas law for non-ideal gases.
Compressor	Apparatus used to pump gas (to super-atmospheric pressure).
Concentration	Quantitative measure of the composition of a mixture.
Condensation	Change of phase from vapour (gas) to liquid or solid.
Condensed phase	Solid or liquid phase (usually related to a corresponding vapour).
Continuous	Operation in an open system.
Controlled mass	Closed system. Zero transfer of material across system boundary.
Controlled volume	Open system. Finite transfer of material across the system boundary.
Convergence	Tending to a single condition, usually the solution of a non-linear problem.
Conversion	Fraction of a reactant species consumed by chemical reaction.
Critical pressure	Pressure required to liquefy a substance at its critical temperature.
Critical temperature	Temperature above which a substance cannot be liquefied at any pressure.
Degrees of freedom	Number of variables that can be independently manipulated to influence a system.
Density	Ratio of mass to volume.
Dew-point	Condition at which first drops of liquid form from a vapour.
Differential equation	Relation between the rates of change of variables with respect to other variables.
Dispersion	Mixture in which one (or more) phase is dispersed as bubbles, drops or particles.
Divider	Process unit that divides an input stream to multiple outputs of the same composition.
Dry-bulb	Dry thermometer bulb.

Electrolysis	Decomposition of a substance at complementary electrodes by an electric current (usually DC).
Emulsion	Mixture of liquids with a dispersed liquid phase (usually in drops < ca. 1 mm).
Energy	Measure of the potential to do work.
Energy balance	Quantitative account of amounts of energy within and across a system.
Enthalpy	Internal energy plus pressure–volume energy.
Enthalpy balance	Energy balance with the stream energy replaced by stream enthalpy.
Entropy	Capacity factor for isothermally unavailable energy. Measure of system disorder.
Equation of state	Mathematical relation between intensive variables that uniquely fix the thermodynamic state of a system. A single-phase, single-component system requires 3 intensive variables — typically P, T and v.
Equilibrium	Condition of balance. Zero gradient in chemical potential.
Evaporation	Change of phase from solid or liquid to vapour.
Extensive property	A property that depends on the amount of material.
Extent of reaction	Ratio of moles of a species consumed or produced in a chemical reaction to its stoichiometric coefficient.
Fan	Apparatus used to pump gas (usually within about 5 kPa of atmospheric pressure).
Figure of merit	Quantitative measure of the value of an object, design, process, etc.
1st law of thermodynamics	Mathematical statement of the conservation of energy in non-nuclear processes.
Flash split	Separation of a mixture into liquid and vapour by dropping the pressure.
Flowsheet	Diagram of process units and their functions, with their interconnecting process streams.
Flow work	Net energy needed as pressure–volume work to move material (usually a fluid) in and out of a system.
Fluid	Gas or liquid. Easily deformed when subject to differential pressure.
Foam	Mixture of gas with liquid, in which the continuous liquid phase exists as bubble films.
Force	Product of mass with acceleration.
Fuel cell	Spontaneous electrochemical cell that generates DC electrical power with catalytic electrodes.
Furnace	Apparatus in which a fuel is burned (usually in air) to generate heat.
Gas	A compressible state of matter that cannot hold its shape without support. Material in the gas phase above its critical temperature.
Gauge pressure	Pressure measured relative to the local atmospheric pressure.
Greenhouse effect	Heating of Earth's atmosphere due to absorption of IR radiation by greenhouse gases.
Greenhouse gases	Polyatomic gases (e.g. CO_2, H_2O, CH_4) that absorb IR radiation more effectively than air (N_2 and O_2).
Gross economic potential	Value of products minus value of feeds.
Heat	Energy transfer to a system that is driven by a temperature difference.
Heat capacity	Specific heat input to raise the temperature of a substance without changing its phase.
Heat exchanger	Process unit that transfers heat from one input stream to a second input stream.

Heat of combustion[1]	Heat input required to carry out a combustion reaction under isothermal conditions.
Heat of formation[1]	Heat input required to form a compound from its elements under isothermal conditions.
Heat of fusion[1]	Heat input to melt a substance under isothermal conditions.
Heat of mixing[1]	Heat input required to form a mixture from its components under isothermal conditions.
Heat of phase change[1]	Heat input required to change the phase of a substance under isothermal conditions.
Heat of reaction[1]	Heat input required to carry out a (chemical) reaction under isothermal conditions.
Heat of vaporisation[1]	Heat input to vaporise a substance under isothermal conditions.
Ideal gas	Gas with zero interaction between molecules.
Ideal mixture	Mixture with zero interaction between components (at the molecular level).
Independent equation	Equation not obtained by combination of other equations in the set.
Infinite dilution	Concentration corresponding to an infinite ratio of solvent to solute.
Intensive property	A property that is independent of the amount of material.
Interaction	Condition where the level of one variable affects the response of another variable(s).
Isothermal	Constant temperature.
Iteration	Repeating a calculation with recycled values of key variables, usually to convergence.
Irreversible reaction	Reaction that goes in only the forward (L to R) direction.
Kilomole	1000 moles. Molar mass expressed in kilograms.
Kinetic energy	Energy of a mass due to its motion (velocity).
Latent heat	Heat input to change the phase of matter without changing composition or temperature.
Linear equation	Equation in which each independent variable appears in a separate term and with exponent = 1.
Liquid	A (relatively) incompressible state of matter that cannot hold its shape without support.
Mass	Measure of the quantity of material.
Mass fraction	Ratio of mass of a component to the total mass of a mixture.
Material	Matter, physical substance, stuff.
Material balance	Quantitative account of amounts of material within and across a system.
Mixer	Process unit that combines multiple input streams to a single output stream.
Mixture	Combination of components in a single or multi-phase system without chemical reaction.
Model	Representation of a physical system in equations that predict the system behaviour.
Molarity	Measure of concentration expressed as kilomole solute per cubic meter solution.
Mole	Amount of material in Avogadro's number (6.023E23) of "molecules".
Mole balance	Material balance in which the quantity is moles of specified substance(s).
Mole fraction	Ratio of moles of a component to the total moles of a mixture.
Momentum	Product of mass and velocity.
Net economic potential	Gross economic potential minus operating costs.
Non-condensable	Gas above its critical temperature.
Non-ideal mixture	Mixture in which the components interact (properties of the mixture are not a linear combination of those of the components).
Non-linear	Relation between variables that is not linear (e.g. involving products, exponents, etc.).

[1] The heats (a.k.a. enthalpies) of combustion, formation, fusion, mixing, phase change and reaction involve the condition of constant pressure, which is usually one standard atmosphere.

Non-linear equation	Equation in which one or more independent variables are multiplied by other variables, and/or exponentiated and/or are arguments in algebraic functions (e.g. exp, log, tan, etc.).
Non-process element	Element that is not necessary (i.e. is superfluous) for operation of a process.
Normal boiling point	Boiling point temperature of a pure substance under 101.3 kPa(abs) pressure.
Normality	Measure on concentration expressed as kiloequivalents solute per cubic meter solution.
N.T.P.	Normal temperature and pressure [various, but usually ca. 293 K, 101.3 kPa(abs)].
Open system	System with finite transfer of material across the boundary.
Optimisation	Finding conditions for the best (usually a maximum or minimum) value of an objective.
Path function	Value dependent on the path between two states. [Heat and work are both path functions]
Partial pressure	Pressure exerted by a component of a gas mixture if it alone were at the temperature and volume of the mixture.
Partial volume	Volume occupied by a component of a gas mixture if it alone were present at the total pressure and temperature of the mixture.
Phase	State of matter identified as solid, liquid or gas.
Phase split	Distribution of species between the phases in a multi-phase system.
Photovoltaic cell	Device for converting radiant energy (e.g. sunlight) to electricity.
Potential energy	Energy of a mass due to its position in a potential field (e.g. gravitational).
Power	Rate of transfer of energy with respect to time.
Pressure	Force per unit area.
Process	Operation that changes the condition of material or energy.
Process rate	Specific rate w.r.t. time, determined by intrinsic kinetics and/or mass or energy transfer.
Process unit	Apparatus that operates on material or energy input(s) to produce modified output(s).
Pump	Process unit that changes the pressure of a process stream.
Purge	Removing undesired substances by bleeding material from recycle loops.
Quantity	Property that can be measured.
Quality	Mass fraction of vapour in vapour–liquid mixture.
Reactor	Process unit that transforms molecular or atomic species by chemical or nuclear reaction.
Reaction rate	Rate of consumption of a reactant or generation of a product with respect to time.
Recycle	Return to a previous stage of a cyclic process (e.g. of material, from downstream to upstream process units, via a "recycle loop").
Relaxation	Procedure used to speed convergence of iterative calculations.
Return on investment	Ratio of net economic potential to capital cost.
Reversible reaction	Reaction that can go in both the forward (L to R) and reverse (R to L) directions.
R.T.P.	Room temperature and pressure. [ca. 293 K, 101.3 kPa(abs)]
Rule of thumb	Approximation (based on experience) adequate for practical calculations in most cases.
Selectivity	Fraction of a reactant converted to a specified product.
Sensible heat	Heat to change the temperature of matter without changing composition or phase.

Separator Process unit that separates an input stream to multiple output streams of different composition.
Sequential modular Method of calculating material and/or energy balances taking process units in sequence.
Shaft work Work transfer from a system, excluding pressure–volume work (flow work) on/by material entering/leaving the system.
Simultaneous equations A set of independent equations embracing multiple variables.
Sink Consumption term in the general balance equation.
Slurry Mixture in which particles of solid exceeding about 1 micron diameter are dispersed in a liquid.
Solid State of matter that retains its shape without support.
Solubility Maximum concentration of a solute in its solution (under specified conditions).
Solute Constituent of a solution dissolved in the solvent.
Solution Mixture in which the components are (uniformly) distributed on a molecular scale.
Solvent Basic constituent of a solution, usually present in the largest molar amount.
Source Generation term in the general balance equation.
Sparge Blow gas into a liquid (usually through an orifice to produce small bubbles).
Species Constituent identified by a specific atomic or molecular configuration.
Specification Determination of the degrees of freedom of a system (esp. in process design).
Specific gravity Ratio of density of substance to density of a reference substance (usually liquid water or air).
Specific heat Ratio of heat capacity of substance to heat capacity of liquid water.
Standard state Equilibrium state of a substance under 101.3 kPa(abs) pressure (and usually 298 K).
State Condition of a material that fixes its thermodynamic properties.
State function Property depending on the state of a substance and independent of the path to that state.
Steady-state Condition of a system with zero rate of accumulation. Invariant with respect to time.
Steam table Table of the thermodynamic properties of water in its gas and liquid phases.
Stoichiometry Quantitative relation between substances consumed and generated in chemical reaction.
Stream table Tabular quantitative summary of a material and energy balance on an open system.
Strip Remove material by desorption or redistribution into a carrier phase.
S.T.P. Standard temperature and pressure [273.15 K, 101.3 kPa(abs)].
Surface energy Excess internal energy of a surface relative to an equal amount of bulk material.
Sublimation Change of phase from solid to vapour.
System Region of space defined by a (real or imaginary) closed envelope (envelope = system boundary).
Temperature Measure of potential to transfer heat.
Threshold effect Sudden change in behaviour of a system when process variable(s) cross a critical value.

Transient Varying over time. Unsteady-state.

Triple-point Condition of a pure substance at which the solid, liquid and gas co-exist at equilibrium.

Turbine Rotating machine that generates energy from pressure drop in a flowing fluid.

Unsteady-state Condition of a system in which rate of accumulation is not zero. Variant with respect to time. Transient.

Vacuum Sub-atmospheric pressure (absolute vacuum = zero absolute pressure).

Vacuum pump Apparatus used to pump gas from a system under vacuum.

Vaporisation Change of phase from solid or liquid to gas (a.k.a. vapour).

Vapour Material in the gas phase below its critical temperature (i.e. condensable).

Vapour pressure Pressure exerted by a solid or liquid in equilibrium with its own vapour.

Velocity Rate of change of distance with respect to time.

Volume fraction Ratio of partial volume of a component to the total volume of a mixture.

Weight Force exerted on a mass by a gravitational field.
[Weight = (mass)(acceleration of gravity) = mg]

Wet-bulb Thermometer bulb wetted by liquid (usually wrapped in a wetted porous material).

Work Force times distance. Energy transfer from a system, other than as heat or internal energy of transferred material.

Yield Product of reactant conversion and the corresponding selectivity for a given species. Fraction of reactant converted to a specific product (usually the desired product).

An Introduction

A material and energy balance is essentially a quantitative account of the redistribution of material and/or energy that occurs when anything happens. This basic tool of process engineering can be used to solve many practical problems. The list below shows that such problems range from those of interest to engineers and scientists in their daily work (see *Refs. 1–2*) to those of relevance to all people concerned with the sustainability of life on planet Earth (see *Refs. 3–5*).

USES OF MATERIAL AND ENERGY BALANCES

- Modelling and design of industrial chemical[1] processes
- Life-cycle analysis and industrial ecology
- Development of medical technology
- Modelling the global environment, geochemical cycles and climate change
- Management of resources (energy, food and water) for the population of planet Earth
- Design for space stations and interplanetary travel

This text focuses on the use of material and energy balances in the modelling and design of industrial chemical processes, but includes examples that illustrate the application of material and energy balances to chemical processes in other fields, such as those listed above.

[1] The term "chemical process" in this context means any change resulting in the redistribution of material and/or energy in a system, and embraces processes that involve *biochemical, electrochemical, photochemical, physico-chemical or thermochemical phenomena*. As used in this text the term "chemical process" excludes changes that involve the inter-conversion of mass and energy via nuclear reactions.

PREREQUISITES

The principles involved in material and energy balance calculations are those that you probably learned in high school and/or the first year of university. These principles include, for example:

- The law of conservation
- Dimensions and units of physical quantities
- Chemical reactions and stoichiometry
- The ideal gas law
- Multi-phase equilibrium
- The first law of thermodynamics
- Basic thermochemistry and thermophysics
- Simple algebra and a little calculus
- Computer calculations with spreadsheets

These basics are reviewed in Chapters 1 and 2 then combined and extended in Chapters 3 to 7 to a formalised procedure for treating chemical processes, which is a key skill in the practice of chemical, biological and environmental engineering.

DEALING WITH COMPLEX PROBLEMS

Material and energy balances provide a tool for students in many fields to grasp and to quantify the factors that determine the behaviour of both man-made and natural systems. Chemical processes (as defined broadly above) are complex and typically involve many interacting factors in non-linear relationships. In this respect, the material and energy balance techniques used by engineers to model industrial processes can be used as both an educational and a predictive tool to deal with complex natural phenomena such as, the global geochemical cycles, population dynamics and the thermal "greenhouse" effect on planet Earth.

The constraints of material and energy in natural systems appear poorly understood by our politicians, as well as by many economists and practitioners of the social sciences (see *Refs. 6–7*). Perhaps this text will help to provide the needed conceptual basis for dealing with the problems of material and energy flow that are basic to the progress and sustainability of our civilisation.

FURTHER READING

[1] T. M. Duncan and J. A. Reimer, *Chemical Engineering Design and Analysis,* Cambridge University Press, Cambridge, 1998.

[2] D. T. Allen and K. S. Rosselot, *Pollution Prevention for Chemical Processes,* John Wiley & Sons, New York, 1997.

[3] J. A. Fava *et al.* (Eds.), *A Technical Framework for Life-Cycle Assessment,* Society of Environmental Toxicology and Chemistry, 1994.

[4] M. Wackernagel and W. Rees, *Our Ecological Footprint — Reducing the Human Impact on Earth*, New Society Publishers, Gabriola Island, 1996.

[5] A. Ford, *Modeling the Environment,* Island Press, Washington DC, 1999.

[6] T. Homer-Dixon, *The Ingenuity Gap,* Alfred Knopf, New York, 2000.

[7] R. Wright, *A Short History of Progress,* ANANSI, Toronto, 2004.

CHAPTER ONE

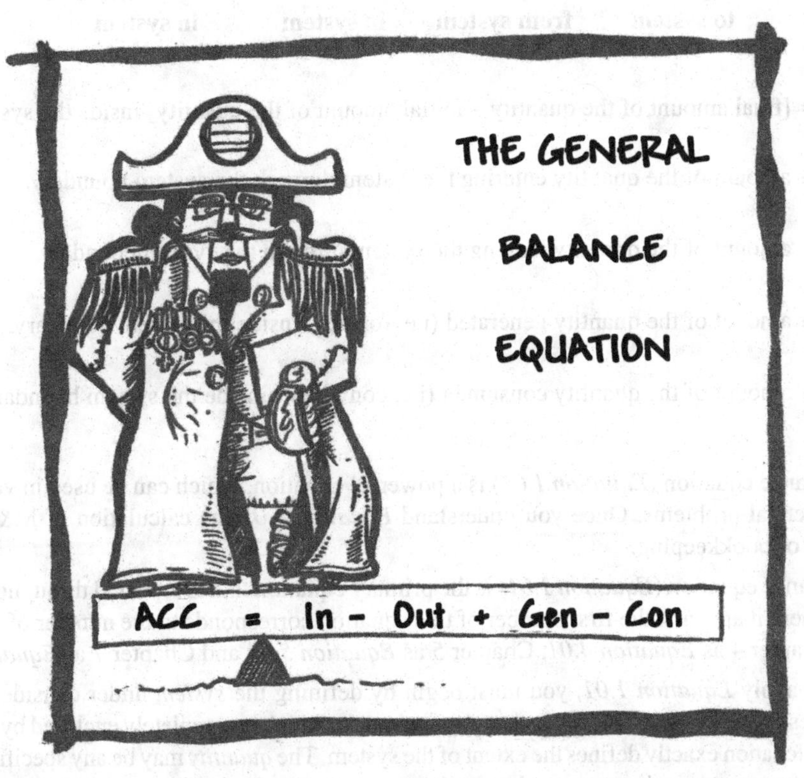
THE GENERAL
BALANCE
EQUATION
ACC = In - Out + Gen - Con

THE GENERAL BALANCE

All material and energy (M&E) balance calculations are based on our experience that matter and energy may change its form, but it cannot appear from nor disappear to nothing. This observation is expressed mathematically in *Equation 1.01*, the *General Balance Equation.*

For a defined *system* and a specified *quantity:*

Accumulation in system	=	**Input to system**	–	**Output from system**	+	**Generation in system**	–	**Consumption in system**	*Equation 1.01*

where:

Accumulation in system = [final amount of the quantity – initial amount of the quantity] inside the system boundary.

Input to system = amount of the quantity entering the system through the system boundary. (input)

Output from system = amount of the quantity leaving the system through the system boundary. (output)

Generation in system = amount of the quantity generated (i.e. formed) inside the system boundary. (source)

Consumption in system = amount of the quantity consumed (i.e. converted) inside the system boundary. (sink)

The general balance equation (*Equation 1.01*) is a powerful equation, which can be used in various ways to solve many practical problems. Once you understand *Equation 1.01* the calculation of M&E balances is simply a matter of bookkeeping.

The general balance equation (*Equation 1.01*) is the primary equation that is repeated throughout this text. In each chapter where it appears, the first number of this equation corresponds to the number of the chapter, so that it enters Chapter 4 as *Equation 4.01*, Chapter 5 as *Equation 5.01* and Chapter 7 as *Equation 7.01*.

Every time you apply *Equation 1.01*, you must begin by defining the *system* under consideration and the *quantity* of interest in the system. The *system* is a physical space, which is completely enclosed by a hypothetical envelope whose location exactly defines the extent of the system. The *quantity* may be any specified measurable (extensive[1]) property, such as the mass, moles,[2] volume or energy content of one or more components of the system. When *Equation 1.01* is applied to energy balances, the quantity also includes energy transfer across the system envelope as heat and/or work (see Chapter 5).

---

[1] An extensive property is a property whose value is proportional to the amount of material.

[2] A *mole* is the amount of a substance containing the same number of "*elementary particles*" as there are atoms in 0.012 kg of carbon 12 (i.e. Avogadro's number = 6.028E23 particles). The number of moles in a given quantity of a molecular species (or element) is its mass divided by its molar mass, i.e. n = m / M. For molecular and ionic compounds in chemical processes (e.g. H_2O, H_2, O_2, CH_4, NaCl) the "*elementary particles*" are molecules. For unassociated elements and for ions the "*elementary particles*" are atoms (e.g. C, Na^+, Cl^-) or charged groups of atoms (e.g. NH_4^+, ClO^-). Refer to a basic chemistry text for more comprehensive information of atoms, ions, molecules and chemical bonding.

When you use the general balance equation it is best to begin with a conceptual diagram of the system, which clearly shows the complete system boundary plus the inputs and outputs of the system. *Figure 1.01* is an example of such a diagram. Note that:

- *Figure 1.01* is a two-dimensional representation of a real three-dimensional system
- The system has a closed boundary
- The system may have singular or multiple inputs and/or outputs

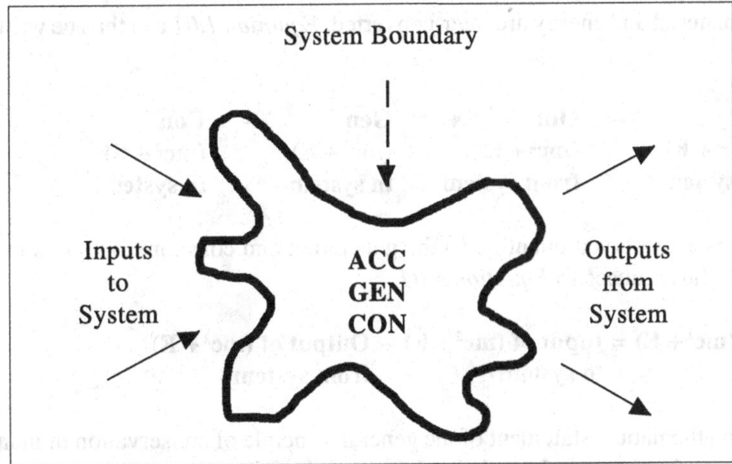

Figure 1.01. Conceptual diagram of a system.

CONSERVED AND CHANGING QUANTITIES

The general balance equation (*Equation 1.01*) applies to any extensive quantity. However, when you use *Equation 1.01*, it is helpful to distinguish between quantities that are conserved and quantities that are not conserved. A conserved quantity is a quantity whose total amount is maintained constant and is understood to obey the *principle of conservation*, which states that such a quantity can be neither created nor destroyed. Alternatively, a conserved quantity is one whose total amount remains constant in an isolated system,[3] regardless of what changes occur inside the system.

For quantities that are conserved, the generation and consumption terms in *Equation 1.01* are both zero, whereas for quantities that are not conserved, either or both of the generation and consumption terms is not zero.

[3] An isolated system is a hypothetical system that has zero interaction with its surroundings, i.e. zero transfer of material, heat, work, radiation, etc. across the boundary.

MATERIAL AND ENERGY BALANCES

In general, material and energy are equivalent through *Equation 1.02.*

$$E = mc^2 \qquad \textit{Equation 1.02}$$

where:
- E = energy — J
- m = mass (a.k.a. the rest mass) — kg
- c = velocity of light in vacuum = 3E8 — ms^{-1}

In a system where material and energy are inter-converted, *Equation 1.01* can thus be written for the quantity (mc^2 + E) as:

Acc	=	**In**	–	**Out**	+	**Gen**	–	**Con**	*Equation 1.03*
(mc^2 + E)		**(mc^2 + E)**		**(mc^2 + E)**		**(mc^2 + E)**		**(mc^2 + E)**	
in system		**to system**		**from system**		**in system**		**in system**	

The sum (mc^2 + E) is a conserved quantity, so the generation and consumption terms in *Equation 1.03* are zero. *Equation 1.03* thus reduces to *Equation 1.04.*

Accumulation of (mc^2 + E) = Input of (mc^2 + E) – Output of (mc^2 + E) *Equation 1.04*
in system **to system** **from system**

Equation 1.04 is a mathematical statement of the general principle of conservation of matter and energy that applies to all systems within human knowledge.[4]

Excluding processes in nuclear engineering, nuclear physics, nuclear weapons and radiochemistry, systems on earth do not involve significant inter-conversion of mass and energy. In such non-nuclear systems, individual atomic species (i.e. elements) as well as the total mass and total energy are *independently* conserved quantities.

For individual atomic species in non-nuclear systems, *Equation 1.01* then reduces to *Equation 1.05.*

Accumulation of atoms = Input of atoms – Output of atoms *Equation 1.05*
in system **to system** **from system**

Also, *Equation 1.04* can be written as two separate and independent equations, one for total mass (*Equation 1.06*) and another for total energy (*Equation 1.07*).

Accumulation of mass = Input of mass – Output of mass *Equation 1.06*
in system **to system** **from system**

---

[4] Modern cosmology may have more to say about the principle of conservation (see *Ref. 8*).

Accumulation of energy in system = **Input of energy to system** − **Output of energy from system** *Equation 1.07*

Equation 1.05, for the conservation of elements ≡ ATOM BALANCE
Equation 1.06, for the conservation of total mass ≡ MASS BALANCE
Equation 1.07, for the conservation of total energy ≡ ENERGY BALANCE

The word "total" is used here to ensure that *Equations 1.06* and *1.07* are not applied to the mass or energy associated with only part of the system (e.g. a single species or form of energy) — which may not be a conserved quantity.

Equations 1.05, 1.06 and *1.07* are **special cases of *Equation 1.01*** that apply to conserved quantities. The strength of *Equation 1.01* is that it can also be applied to quantities that are not conserved. The generation and consumption terms of *Equation 1.01* then account for changes in the amounts of non-conserved quantities in the system (see *Ref. 7*).

Some of the quantities that can be the subject of *Equation 1.01*, applied to non-nuclear proceses, are:

Elements : Conserved — *Equation 1.05* — Atom balance
Total mass : Conserved — *Equation 1.06* — Total mass balance
Total energy : Conserved — *Equation 1.07* — Total energy balance
Momentum : Conserved — Momentum balance
Moles : Conserved in systems without chemical reaction.
Not necessarily conserved in systems with chemical reaction
Enthalpy : Not necessarily conserved
Entropy : Not necessarily conserved
Volume : Not necessarily conserved

Examples 1.01 and *1.02* show how *Equation 1.01* is applied respectively to a conserved quantity and to a non-conserved quantity.

EXAMPLE 1.01 Material balance on water in a reservoir.

A water reservoir initially contains 100 tonne of water. Over a period of time 60 tonne of water flow into the reservoir, 20 tonne of water flow out of the reservoir and 3 tonne of water are lost from the reservoir by evaporation. Water is not involved in any chemical reactions in the reservoir.

Problem:
What is the amount of water in the reservoir at the end of the period of time? [Answer in tonnes of water]

Solution:
Define the system = the reservoir
Specify the quantity = mass of water [conserved quantity]

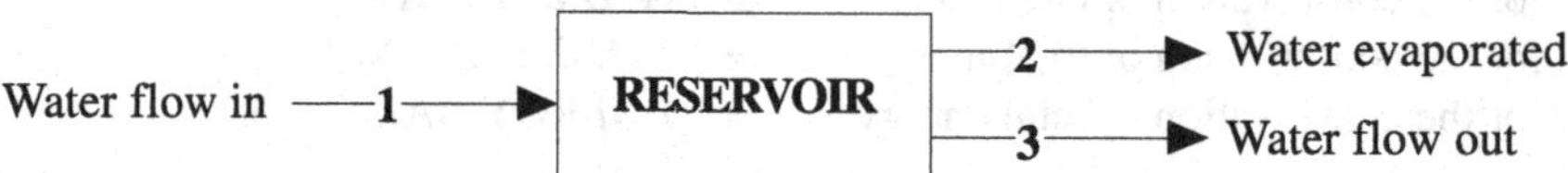

Write the balance equation:

ACC of water in system	=	IN of water to system	–	OUT of water from system	+	GEN of water in system	–	CON of water in system

Interpret the terms:

Accumulation of water in system = final mass of water in system – initial mass of water in system = unknown = X

Input of water to system	= 60	tonne
Output of water from system	= 20 (flow) + 3 (evaporation) = 23	tonne
Generation of water in system	= 0 (water is conserved)	tonne
Consumption of water in system	= 0 (water is conserved)	tonne

Substitute values for each term into the general balance equation

$$X = 60 - (20 + 3) + 0 - 0$$

Solve the balance equation for the unknown.

$$X = 37 \text{ tonne}$$

Final mass of water in system

= Initial mass of water in system + X = 100 + 37 = **137 tonne**

EXAMPLE 1.02 Material balance on the population of a country.

A country had a population of 10 million people in 1900 AD. Over the period from 1900 to 2000 AD, 4 million people immigrated into the country, 2 million people emigrated from the country, 5 million people were born in the country and 3 million people died in the country.

Problem:
What is the population of the country in the year 2000 AD?

Solution:
Define the system = country
Specify the quantity = number of people
[a non-conserved quantity]

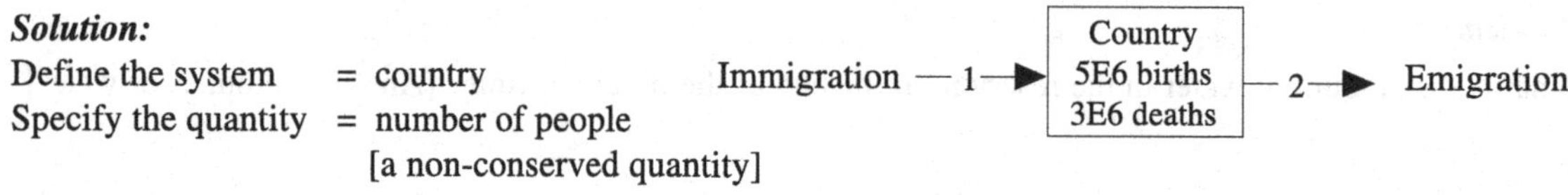

Write the balance equation:

ACC of people in system	=	IN of people to system	–	OUT of people from system	+	GEN of people in system	–	CON of people in system

Interpret the terms:

Accumulation = Final no. of people in system – Initial no.of people in system = unknown = X

Input of people to system	= immigration	= 4 million people
Output of people from system	= emigration	= 2 million people
Generation of people in system	= born	= 5 million people
Consumption of people in system	= died	= 3 million people

Substitute values into the general balance equation

$X = 4 - 2 + 5 - 3 = 4$ million people

Final number of people in 2000 AD:

Initial no. of people in 1900 + X = 10 + 4 = **<u>14 million people</u>**

Example 1.03 shows a slightly more difficult case of a non-conserved quantity involving chemical reaction stoichiometry.

EXAMPLE 1.03 Material balance on carbon dioxide from an internal combustion engine.

An automobile driven by an internal combustion engine burns 10 kmol of gasoline[5] consisting of 100% octane (C_8H_{18}) and converts it completely to carbon dioxide and water vapour by a combustion reaction, whose stoichiometric equation is:

$$2C_8H_{18} + 25O_2 \rightarrow 16\,CO_2 + 18H_2O \qquad \textit{Reaction 1}$$

All CO_2 and H_2O produced in the reaction is discharged to the atmosphere via the engine's exhaust pipe (i.e. zero accumulation).

Assume CO_2 content of input combustion air = zero

Problem: What is the amount of carbon dioxide discharged to the atmosphere from 10 kmol octane? [Answer in kg of CO_2]

[5] kmol ≡ kilomole ≡ molar mass (molecular weight) expressed as kilograms.
1 kmol C_8H_{18} = (12.01)(8) + (1.008)(18) = 114.22 kg = 162 litre = 43 US gallons, i.e. about 3 tanks of gasoline (a.k.a. petrol) for a family car.

Solution: First note that the amount (mass or moles) of CO_2 is not a conserved quantity. Chemical reactions involve the consumption of reactant species and the generation of product species, so that no species appearing in the stoichiometric equation is conserved when a reaction occurs.

Define the system = automobile internal combustion engine

Specify the quantity = mol (kmol) of CO_2 [mole quantities give the simplest calculations for chemical reactions]

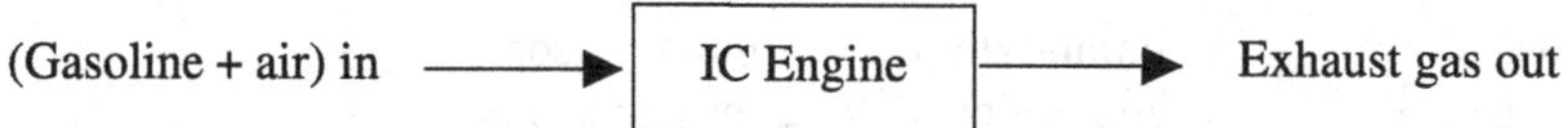

Write the balance equation:

ACC of CO_2 in system	=	IN of CO_2 to system	−	OUT of CO_2 from system	+	GEN of CO_2 in system	−	CON of CO_2 in system

Interpret the terms:

Accumulation of CO_2 in system	= 0	kmol	(all CO_2 is discharged)
Input of CO_2 to system	= 0	kmol	(zero CO_2 enters engine)
Output of CO_2 from system	= unknown = X	kmol	
Generation of CO_2 in system	= (16/2)(consumption of C_8H_{18} in system) = (16/2)(10 kmol)		
	= 80	kmol	(from stoichiometry of *Reaction 1*)
Consumption of CO_2 in system	= 0	kmol	(CO_2 is a product, not a reactant)

Substitute values for each term into the general balance equation.

$$0 = 0 - X + 80 - 0$$

Solve the balance equation for the unknown "X".

X = 80 kmol CO_2 discharged to atmosphere
≡ (80 kmol)(((1)(12.01) + (2)(16.00)) kg/kmol) = **<u>3521 kg CO_2</u>**

FORMS OF THE GENERAL BALANCE EQUATION

There are several forms of the general balance equation (*Equation 1.01*), with each form suited to a specific class of problem.

First, for convenience, *Equation 1.01* is usually written in shorthand as:

ACC = IN − OUT + GEN − CON (see *Equation 1.01*)

where:

ACC = Accumulation of specified quantity in system = (Final amount − Initial amount) of quantity in system.
IN = Input of specified quantity to system (input)
OUT = Output of specified quantity from system (output)
GEN = Generation of specified quantity in system (source)
CON = Consumption of specified quantity in system (sink)

As written above *Equation 1.01* is an *integral balance* suited to deal with the *integrated (total) amounts* of the specified quantity in each of the five terms over a period of time from the initial to the final state of the system. The integral balance equation can be used for any system with any process but is limited to seeing only the total changes occurring between the initial and final states of the system. Calculations with the integral balance require the solution of algebraic equations.

When *Equation 1.01* is differentiated with respect to (w.r.t.) time it becomes a *differential balance* equation (*Equation 1.08*), with each term representing a RATE of change of the specified quantity with respect to time, i.e.

Rate ACC = Rate IN − Rate OUT + Rate GEN − Rate CON *Equation 1.08*

where:

Rate ACC = Rate of accumulation of specified quantity in system, w.r.t. time
Rate IN = Rate of input of specified quantity to system, w.r.t. time
Rate OUT = Rate of output of specified quantity from system, w.r.t. time
Rate GEN = Rate of generation of specified quantity in system, w.r.t. time
Rate CON = Rate of consumption of specified quantity in system, w.r.t. time

Equation 1.08 is suited to deal with both closed systems and open systems involving unsteady-state processes, as well as with open systems involving steady-state processes. These terms are defined below in "Class of Problem".

The differential balance equation is more versatile than the integral balance equation because it can trace changes in the system over time. Application of the differential balance to unsteady-state processes requires knowledge of calculus to solve differential equations. For steady-state processes, the differential equation(s) can be simplified and solved as algebraic equation(s).

CLASS OF PROBLEM

Practical problems are classified according to the type of system and the nature of the process occurring in the system, as follows:

Closed system [Controlled mass]
Zero material* is transferred in or out of the system [during the time period of interest].
i.e. in the material balance equation: IN = OUT = 0
A process occurring in a closed system is called a BATCH process.

Open system [Controlled volume]
Material is transferred in and/or out of the system.
i.e. in the material balance equation: IN $\neq$ 0 and/or OUT $\neq$ 0
A process occurring in an open system is called a CONTINUOUS process.

Steady-state process
A process in which all conditions are invariant with time.
i.e. at steady-state: Rate ACC = 0 for all quantities.

Unsteady-state process
A process in which one or more conditions vary with time [these are *transient* conditions], i.e. at unsteady-state: Rate ACC $\neq$ 0 for one or more quantities.

* *Energy can be transferred in and/or out of both closed and open systems.*

From these generalisations we can write balance equations for special cases that occur frequently in practical problems.

Differential material[6] balance on a closed system:	**Rate ACC = Rate GEN – Rate CON**	*Equation 1.09*
Differential total mass balance on a closed system:	**Rate ACC = 0**	*Equation 1.10*
Differential material balance on an open system at steady-state:	**0 = Rate IN – Rate OUT + Rate GEN – Rate CON**	*Equation 1.11*
Differential total mass balance on an open system:	**Rate ACC = Rate IN – Rate OUT**	*Equation 1.12*
Differential total mass balance on open system at steady-state:	**0 = Rate IN – Rate OUT**	*Equation 1.13*
Differential total energy balance on a closed or an open system:	**Rate ACC = Rate IN – Rate OUT**	*Equation 1.14*
Differential total energy balance at steady-state:	**0 = Rate IN – Rate OUT**	*Equation 1.15*

[6] Note that "material" and "mass" are not synonymous terms. Material may change its form, but the total mass is constant (in non-nuclear processes).

Equations 1.09 to *1.15* can be applied either as differential balances or as integral balances with the terms integrated over time.

A powerful feature of *Equation 1.01* is your freedom to define both the *system* and the *quantity* to suit the problem at hand. For example, if you are working on global warming, the *system* could be a raindrop, a tree, a cloud, an ocean or the whole earth with its atmosphere, while the *quantity* may be the mass (or moles) of carbon dioxide, the mass (or moles) water vapour or the energy content associated with the system. If you are working on a slow release drug, the *system* could be a single cell, an organ or the whole human body, while the *quantity* may be the mass of the drug or its metabolic product, etc.

For an industrial chemical process (the focus of this text[7]) the *system* could be defined as, for example:

- a single molecule
- a microelement of fluid
- a bubble of gas, drop of liquid or particle of solid
- a section of a process unit (e.g. a tube in a heat exchanger, a plate in a distillation column, etc.)
- a complete process unit (e.g. a heat exchanger, an evaporator, a reactor, etc.)
- part of a process plant including several process units and piping
- a complete process plant
- a complete process plant plus the surrounding environment

In an industrial chemical process the *quantity* of interest may be, for example:

- total mass or total moles of all species
- mass or moles of any single species
- mass or atoms of any single element
- total energy

Other quantities that may be used in process calculations are, for example:

- total mass plus energy of all species (nuclear processes)
- one form of energy (e.g. internal energy, kinetic energy, potential energy)
- volume of one or more components
- enthalpy
- entropy [not treated in this text]
- momentum [not treated in this text]

The choice of both the *system* and the *quantity* are major decisions that you must make when beginning an M&E balance calculation. These choices are usually obvious but sometimes they are not. You should always give these choices careful consideration because they will determine the degree of difficulty you encounter in solving the problem.

---

[7] The techniques described here for an industrial chemical process can be applied to any system in which you are interested.

Examples 1.04 and *1.05* show that selection of the *system* and the *quantity* can affect the ease of solution of M&E balance problems. *Example 1.04* is a multi-stage balance problem that can be solved either by stage-wise calculations or by an overall balance. Solution A shows the stage-wise approach while solution B shows the more efficient solution obtained by defining an overall system that includes all the process steps inside a single envelope. In *Example 1.05* you can see how a relatively complex multi-stage problem involving chemical reactions can be reduced to one overall balance on the quantity of a single element (carbon). Such atom balances (on one or more elements) are handy for resolving many material balance problems. However there are other more complex problems represented in Chapter 4 whose solution requires stage-wise balances on each chemical species over every process unit.

EXAMPLE 1.04 Material balance on a two-stage separation process.

A 100 kg initial mixture of oil and water contains 10 kg oil + 90 kg water. The mixture is separated in two steps.[8]

Step 1. 8 kg oil with 2 kg water is removed by decantation. **Step 2.** 1 kg oil with 70 kg water is removed by evaporation.

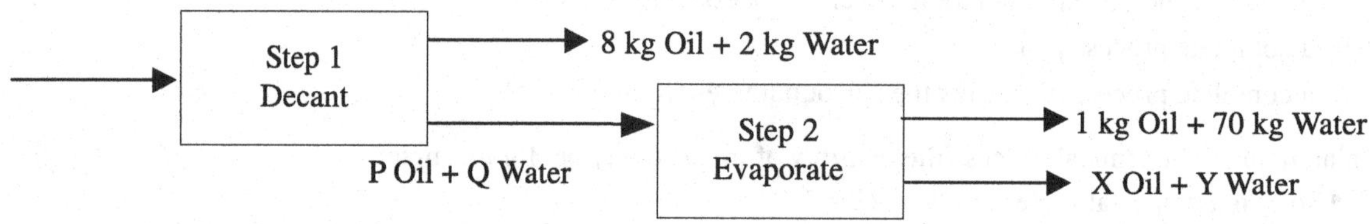

Problem: Find the amounts of oil and water in the final mixture (i.e. X kg oil + Y kg water).

Solution A: [Stepwise balances]

Step 1: Define the system = Decanter (an open system)
Specify the quantities = Mass of oil, mass of water

Write the integral balance equation: **ACC = IN – OUT + GEN – CON**

For the oil:				**For the water:**			
ACC	= Final oil in system – Initial oil in system = 0			ACC	= Final water in system – Initial water in system = 0		
IN	= 10	kg		IN	= 90	kg	
OUT	= 8 + P	kg		OUT	= 2 + Q	kg	
GEN	= 0	kg	[Oil is conserved]	GEN	= 0	kg	[Water is conserved]
CON	= 0	kg		CON	= 0	kg	

Mass balance on oil: ACC = 0 = 10 – (8 + P) + 0 – 0 P = 2 kg oil
Mass balance on water: ACC = 0 = 90 – (2 + Q) + 0 – 0 Q = 88 kg water

[8] Assumes the decanter and the evaporator are empty before and after the operation; i.e. accumulation = 0.

Step 2: Define the system = Evaporator (an open system)
Specify the quantities = Mass of oil, mass of water

Write the integral balance equation: **ACC = IN – OUT + GEN – CON**

For the oil:

ACC = Final oil in system – Initial oil in system = 0
IN = 2 kg
OUT = 1 + X kg
GEN = 0 kg [Oil is conserved]
CON = 0 kg

For the water:

ACC = Final water in system – Initial water in system = 0
IN = 88 kg
OUT = 70 + Y kg
GEN = 0 kg [Water is conserved]
CON = 0 kg

Mass balance on oil: ACC = 0 = 2 – (1 + X) + 0 – 0 **X = 1 kg oil**
Mass balance on water: ACC = 0 = 88 – (70 + Y) + 0 – 0 **Y = 18 kg water**

Solution B: [Overall balance]

Define the system = Decanter + evaporator (i.e. overall process)
Specify the quantities = Mass of oil, mass of water

Write the integral balance equation: **ACC = IN – OUT + GEN – CON**

For the oil:

ACC = Final oil in system – Initial oil in system = 0
IN = 10 kg
OUT = 8 + 1 + X kg
GEN = 0 kg [Oil is conserved]
CON = 0 kg

For the water:

ACC = Final water in system – Initial water in system = 0
IN = 90 kg
OUT = 2 + 70 + Y kg
GEN = 0 kg [Water is conserved]
CON = 0 kg

Mass balance on oil: ACC = 0 = 10 – (8 + 1 + X) + 0 – 0 **X = 1 kg oil**
Mass balance on water: ACC = 0 = 90 – (2 + 70 + Y) + 0 – 0 **Y = 18 kg water**

Example 1.04 shows that (when the intermediate conditions are not required) the overall balance gives the most efficient solution.

EXAMPLE 1.05 Material balance on carbon dioxide from a direct methanol fuel cell.

Methanol (CH_3OH) is used as the fuel in an automobile engine based on a direct methanol fuel cell. The fuel supply for one tank of methanol is produced from 10 kmol[9] of natural gas (methane, CH_4) by the following process:

1. Eight kmol of CH_4 reacts completely with water (steam) to produce CO and H_2 by the stoichiometry of *Reaction 1*.

$CH_4 + H_2O \rightarrow CO + 3H_2$ *Reaction 1*

[9] 10 kmol $CH_4 \equiv$ (10 kmol)(16.04 kg/kmol) = 160.4 kg CH_4.

2. Two kmol of CH_4 is burned in *Reaction 2* to supply heat to drive *Reaction 1.*

$CH_4 + 2O_2 \rightarrow CO_2 + 2H_2O$ (CO_2 to atmosphere) *Reaction 2*

3. One-third of the H_2 from *Reaction 1* is converted to water by *Reaction 3.*

$2H_2 + O_2 \rightarrow 2H_2O$ *Reaction 3*

4. The remaining CO and H_2 are converted to CH_3OH by *Reaction 4.*

$CO + 2H_2 \rightarrow CH_3OH$ *Reaction 4*

5. The methanol is subsequently completely converted to CO_2 and H_2O by the net *Reaction 5*, which occurs in the fuel cell to produce power for the automobile.

$CH_3OH + 1.5\,O_2 \rightarrow CO_2 + 2H_2O$ (CO_2 to atmosphere) *Reaction 5*

Problem: Find the total amount of CO_2 released to the atmosphere from these operations, starting with the 10 kmol of CH_4. Assume zero accumulation of material in all the process steps. [kg CO_2]

Solution: We can define each of the 5 reaction steps as a separate system and trace the production of CO_2 through the sequence of operations, or we can define any combination of steps as the system (provided it is enclosed by a complete envelope). In this case it is most efficient to define the system to include all 5 reaction steps and carry out an overall atom balance on the open system.

Define the system = the overall process of 5 reaction steps [closed envelope, shown by broken line].
Specify the quantity = moles (kmol) of the element carbon [contained in various compounds].
Write the integral balance equation: **ACC = IN – OUT + GEN – CON**

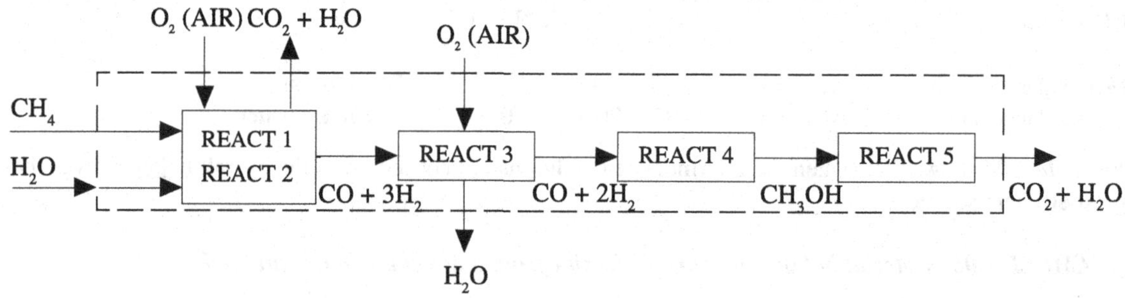

ACC = Final C in system – Initial C in system = 0 (i.e. zero accumulation)
IN = (10 kmol CH_4)(1 kmol C/kmol CH_4) = 10 kmol C
OUT = (X kmol CO_2)(1 kmol C/kmol CO_2) = X kmol C
GEN = 0 [carbon is conserved]
CON = 0 [carbon is conserved]

$0 = 10 - X + 0 - 0$
$X = 10$ kmol $CO_2 \equiv$ (10 kmol)(44 kg/kmol) = **<u>440 kg CO_2 total to atmosphere</u>**

Example 1.06 shows the solution of a material balance using a differential balance on an open system and also illustrates how a problem involving chemical reaction can be solved either by mole balances from the reaction stoichiometry or by atom balances on selected elements. You should note here that for a single reaction the atom balance method is simpler than the mole balance because the atom balance does not need to use the reaction stoichiometric equation. In more complex cases, such as those involving reactions with incomplete conversion and/or where the specified elements appear in several products, the mole balance is often the prefered method of solution.

EXAMPLE 1.06 Material balance on water vapour from a jet engine.

Water vapour trails from jet planes are known to initiate cloud formation and are one suspected cause of global climate change.

A jet engine burns kerosene fuel ($C_{14}H_{30}$) at the steady rate of 1584 kg/h.

The fuel undergoes complete combustion to CO_2 and H_2O.

Problem: Calculate the flow of water vapour in the engine exhaust stream. [kg/h]

Solution A: (Mole balance) Define the system = jet engine (an open system)
Specify the quantity = moles (kmol) of water

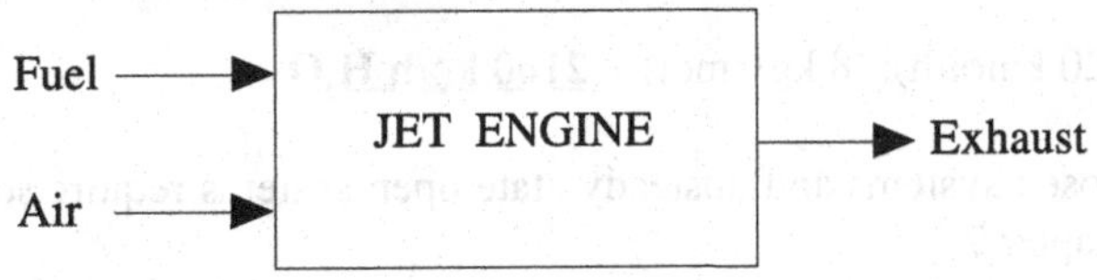

Write the reaction stoichiometry:

$$2C_{14}H_{30} + 43O_2 \rightarrow 28CO_2 + 30H_2O \qquad \textit{Reaction 1}$$

Let X = flow of water in engine exhaust. [kmol/h]

Write the differential material balance:

Rate ACC = Rate IN – Rate OUT + Rate GEN – Rate CON

Rate ACC = 0 (steady-state)
Rate IN = 0
Rate OUT = X
Rate GEN = (15 kmol H_2O/kmol $C_{14}H_{30}$)(1584 kg $C_{14}H_{30}$/h)/(198 kg $C_{14}H_{30}$/kmol $C_{14}H_{30}$) = 120 kmol/h H_2O
Rate CON = 0

0 = 0 – X + 120 – 0   X = 120 kmol H_2O /h = (120 kmol/h)(18 kg/kmol) = **2160 kg/h H_2O**

NOTE: Nitrogen (N_2) in the air is assumed to pass unchanged through the combustion process. In reality a small fraction of the nitrogen is converted to nitrogen oxides, NO and NO_2.

Solution B: (Atom balance) Define the system = jet engine
Specify the quantity = moles (kmol) of hydrogen atoms i.e: hydrogen as H_1

Let X = flow of water in engine exhaust. [kmol/h]

Write the differential material balance:

Rate ACC = Rate IN – Rate OUT + Rate GEN – Rate CON

Rate ACC = 0 (steady-state)
Rate IN = (1584 kg/h $C_{14}H_{30}$)/(198 kg $C_{14}H_{30}$/kmol $C_{14}H_{30}$)(30 kmol H_1/kmol $C_{14}H_{30}$) = 240 kmol/h H_1
Rate OUT = (X kmol/h H_2O)(2 kmol H_1 / kmol H_2O) = 2X kmol/h H_1
Rate GEN = Rate CON = 0 (atoms are conserved)

0 = 240 – 2X + 0 – 0
X = 120 kmol H_2O /h = (120 kmol/h)(18 kg/kmol) = **2160 kg/h H_2O**

Differential balances on closed systems and unsteady-state open systems require solution by calculus and are not considered until Chapter 7.

SUMMARY

[1] The general balance equation (GBE) is the key to material and energy (M&E) balance problems.

The GBE can be used in integral or in differential form.

For a defined *system* and a specified *quantity:*

Integral form of GBE

ACC = IN – OUT + GEN – CON (see *Equation 1.01*)

Differential form of GBE

Rate ACC = Rate IN – Rate OUT + Rate GEN – Rate CON (see *Equation 1.08*)

[2] Always begin work with the GBE by defining the system (a closed envelope) and specifying the quantity of interest. Take care here to avoid ambiguity. *Ambiguity is the mother of confusion.*

[3] The choice of both *system* and *quantity* in the GBE determines the ease of solution of M&E balance problems. The *system* can range from a sub-micron feature to a whole planet and beyond, or from a single unit in a multi-stage sequence to the overall operation. The *quantity* may be any specified measurable (extensive) property, such as the mass, moles, volume or energy of one or more components of the system.

[4] The GBE applies to both conserved and non-conserved quantities.
For conserved quantities: GEN = CON = 0
For non-conserved quantities: GEN and/or CON $\neq$ 0

[5] In non-nuclear processes the individual atomic species (elements) as well as both total mass and total energy are independently conserved quantities, for which the GBE's simplify to:

Atom (element) balance:

Atoms ACC = Atoms IN – Atoms OUT (see *Equation 1.05*)

Total mass balance:

Mass ACC = Mass IN – Mass OUT (see *Equation 1.06*)

Total energy balance:

Energy ACC = Energy IN – Energy OUT (see *Equation 1.07*)

[6] M&E balance problems are broadly classified by system and process as:

Closed system:	Material IN = Material OUT = 0	[Controlled mass]
Open system:	Material IN and/or Material OUT ≠ 0	[Controlled volume]
Batch process:	Process occurring in a closed system.	
Continuous process:	Process occurring in an open system.	
Steady-state process:	Invariant with time. Rate ACC = 0 for all quantities	
Unsteady-state process:	Variant with time. RateACC ≠ 0 for one or more quantities	

[7] Energy may be transferred in and/or out of both closed and open systems.

[8] The calculation of an M&E balance can sometimes be simplified by choosing a conserved quantity and/or by making an overall balance on a multi-unit sequence. Always check these options before starting a more detailed analysis.

[9] *Examples 1.01* to *1.06* are simple material balance problems that illustrate the concepts of the GBE. Subsequent chapters will show how the GBE is used to solve more complex material and energy problems, with emphasis on practical chemical processing.

FURTHER READING

[1] R. M. Felder and R. W. Rousseau, *Elementary Principles of Chemical Processes*, John Wiley & Sons, New York, 2000.

[2] D. M. Himmelblau, *Basic Principles and Calculations in Chemical Engineering*, Prentice Hall, Englewood Cliffs, 1989.

[3] O. A. Hougen, K. M. Watson and R. A. Ragatz, *Chemical Process Principles*, John Wiley, New York, 1954.

[4] P. M. Doran, *Bioprocess Engineering Principles*, Academic Press, San Diego, 1995.

[5] G. V. Reklaitis, *Material and Energy Balances*, John Wiley & Sons, New York, 1983.

[6] R. K. Sinnot, *Chemical Engineering Design*, Butterworth-Heinmann, Oxford, 1999.

[7] S. I. Sandler, *Chemical and Engineering Thermodynamics*, John Wiley & Sons, New York, 1999.

[8] S. W. Hawking, *The Theory of Everything — The Origin and Fate of the Universe*, New Millenium Press, Beverly Hills, 2002.

MATERIAL TRANSFER ACROSS A SYSTEM BOUNDARY

CHAPTER TWO

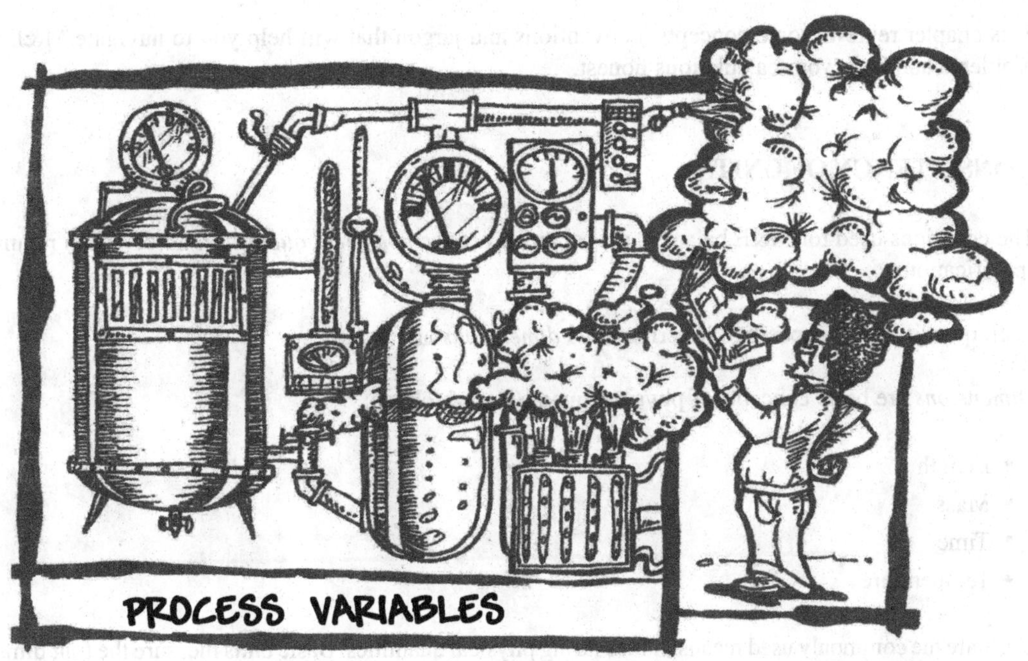

PROCESS VARIABLES AND THEIR RELATIONSHIPS

UNITS AND CALCULATIONS

Material and energy balances are essentially mathematical models that are used to predict the performance of chemical processes. The calculation of M&E balances involves setting up and solving equations related to subjects such as equations of state, chemical stoichiometry, thermochemistry, phase equilibria, etc. in the context of the general balance equation (*Equation 1.01*). These calculations may range from simple arithmetic to complex manipulations with algebraic and differential equations.

This chapter reviews some concepts, conventions and jargon that will help you to navigate M&E balance problems and keep your calculations honest.

CONSISTENCY OF UNITS

The equations used for M&E balances are relations between *physical quantities,* all of which require exact specification.

Each quantity can be specified in terms of its *dimensions* and *units.*

Dimensions are basic concepts of physical quantities, which are:

- Length
- Mass
- Time
- Temperature

Units are the commonly used means of measuring physical quantities. Basic units measure the four dimensions. Derived units measure combinations of the basic units.

Table 2.01 summarises the basic and derived units that are useful for M&E balances. This table also shows units in both the International Metric System [SI] and the old British Imperial System (otherwise slightly modified and called the American Engineering System, now used in the USA). Apart from some example problems in this chapter, this text will deal only in the SI system of units.

You should check that the units "balance" at every step of a calculation, using the rules of *Table 2.02*, as shown in *Examples 2.01* and *2.02*.

Always specify the units of your final result.

Table 2.01. Units used in material and energy balance calculations.

Quantity	SI Unit	Symbol	Combination	Imperial/American Unit	Symbol	Combination
Basic Units						
Length	metre	m	–	foot	ft	–
Mass	kilogram	kg	–	pound (mass)[1]	lb	–
Time	second	s	–	second	s	–
Thermodynamic temp.	Kelvin	K	–	Rankine	°R	–
Current	Ampere	A	–	Ampere	A	–
Amount of substance	mole	mol	–	pound mole	lbmol	–
Force	derived unit, see below	–	–	pound (force)	lb_f	–
Derived Units						
Force	Newton	N	$kg.m.s^{-2}$	basic unit, see above	lb_f	basic unit
Energy	Joule	J	$kg.m^2.s^{-2}$	British thermal unit	BTU	778 $ft.lb_f$
Power	Watt = $J.s^{-1}$	W	$kg.m^2.s^{-3}$	horsepower	HP	550 $ft.lb_f.s^{-1}$
Pressure	Pascal = Nm^{-2}	Pa	$kg.m^{-1}.s^{-2}$	pound per square inch	psi	$lb_f.ft^{-2}/144$
Electric charge	Coulomb	C	A.s	Coulomb	C	A.s
Frequency	Hertz (cycle/s)	Hz	s^{-1}	cycle/s	Hz	s^{-1}
Commonly Used Compound Units						
Customary temp.	Celsius	°C	K -273.15	Fahrenheit	°F	°R – 459.67
Volume	cubic metre	–	m^3	cubic foot	–	ft^3
Density	kilogram per cubic metre	–	$kg.m^{-3}$	pound per cubic foot	–	$lb.ft^{-3}$
Velocity	metre per second	–	$m.s^{-1}$	foot per second	–	$ft.s^{-1}$
Acceleration	metre per sec.squared	–	$m.s^{-2}$	foot per sec.squared	–	$ft.s^{-2}$
Heat capacity	Joule per mole.Kelvin = kJ per kmol.Kelvin	–	$kg.m^2.s^{-2}.mol^{-1}.K^{-1}$	BTU per pound.mole Fahrenheit	–	778 $ft.lb_f.lbmol^{-1}F^{-1}$
Heat of formation	Joule per mole = kJ/kmol	–	$kg.m^2.s^{-2}.mol^{-1}$	BTU per pound mole	–	778 $ft.lb_f.lbmol^{-1}$
Amount of substance	kilomole	kmol	1000 mol	ton[2] mole	tmol	2240 lbmol
Molar mass	–	–	$kg.kmol^{-1}$	–	–	$lb.lbmol^{-1}$
Energy	kiloJoule	kJ	1000 $kg.m^2.s^{-2}$	–	–	–
Power	kiloWatt = $kJ.s^{-1}$	kW	1000 $kg.m^2.s^{-3}$	–	–	–
Pressure	kiloPascal = $kN.m^{-2}$	kPa	1000 $kg.m^{-1}.s^{-2}$	–	–	–

In M&E balance calculations a quantity is meaningless without its units. Every time you write an equation you should specify the units of each variable in that equation. You should then check the equation for dimensional consistency by ensuring that each term in the equation has the same units as those terms to which it is added, subtracted or equated.

All correct equations are dimensionally consistent, and any equation that is not dimensionally consistent is incorrect. All dimensionally consistent equations are not necessarily correct, but when you write an equation from memory there is a good chance that if it is dimensionally consistent then it is correct. A check of dimensional consistency is a simple method to eliminate common errors in your calculations.

1 The pound mass is sometimes written as lb_m.

2 One "long ton" = 2240 lb, one "short ton" = 2000 lb, one "metric" ton = 2204.6 lb = 1000 kg.

Table 2.02. Rules for manipulating units.

1. Treat units like algebraic symbols.
2. Add, subtract or equate terms only when they have the same units as each other.
3. Identical units can be cancelled from numerator and denominator in products and dividends.
4. Logarithm, exponential and trigonometric functions have dimensionless arguments and yield dimensionless results.

For example:

X = A + B − C	The terms A, B, C and X all have the same units.
X = AB	The term AB has the same units as X.
X = AB + CDE	The terms AB, CDE and X all have the same units.
X = ln(Y)	X and Y are both dimensionless.[3]
X = exp(Y)	X and Y are both dimensionless.
X = (2A(exp(B)) + C(log(D)))/(P − RS)	B and D are dimensionless. Units of A = units of C. Units of P = units of RS. Units of X = units of A/P, etc.

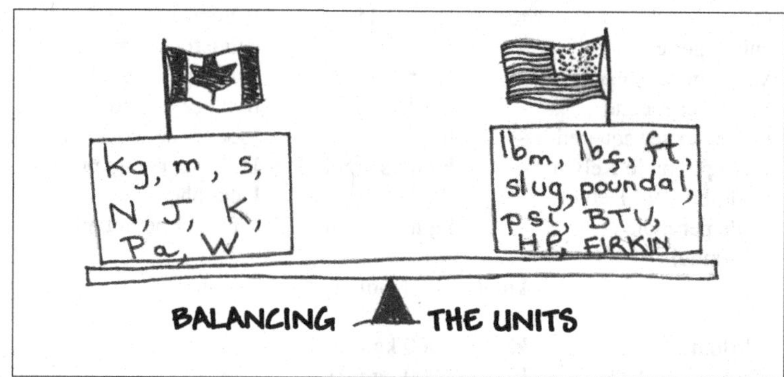

[3] Some equations apparently call for you to exponentiate or take the logarithm of a dimensional variable [e.g. log (p*) in the Antoine equation (*Equation 2.32*)]. In such cases the argument of the function should strictly be made dimensionless by division with unit value of the dimensioned variable.

Use of units to check equations

EXAMPLE 2.01 Checking the dimensional consistency of equations.

The relation between the force "f" required to move an object of mass "m" with an acceleration "ã" is one of the equations (a), (b) and (c) below:

(a) $f = m + \tilde{a}$
(b) $f = m/\tilde{a}$
(c) $f = m\tilde{a}$

Where:

f	= force	N
m	= mass	kg
ã	= acceleration	ms^{-2}

Problem: Which of these three equations is dimensionally consistent?
Solution: Substitute the units for every variable in each equation.

	LHS	RHS	
(a)	$kg.m.s^{-2}$	$= kg + ms^{-2}$	Inconsistent units. Incorrect equation.
(b)	$kg.m.s^{-2}$	$= kg/m.s^{-2} = kg.m^{-1}.s^{2}$	Inconsistent units. Incorrect equation.
(c)	$kg.m.s^{-2}$	$= kg.m.s^{-2}$	Consistent units. Correct equation.

Equation (c) is dimensionally consistent (you will probably recognise it as Newton's second law of motion). Note that if the RHS of equation (c) was multiplied by a dimensionless constant (e.g. $f = 2m\tilde{a}$), then it will be dimensionally consistent but incorrect. A check on dimensional consistency does not guarantee that an equation is correct but it does eliminate many common mistakes.

EXAMPLE 2.02 Finding the units for variables in a simple equation.

The pressure at the base of a vertical column of fluid is given by the (correct) equation:

$$P = \rho g L'$$

where:

P	= pressure	
ρ	= fluid density	$kg.m^{-3}$
g	= gravitation constant	ms^{-2}
L´	= height of column of fluid	m

Problem: What are the units of P?
Solution: Substitute units for each variable.

$$P = (kg.m^{-3})\,(m.s^{-2})\,(m)$$

Cancel m^2 from numerator and denominator.

$$P = kg.m^{-1}.s^{-2} = kg.m.s^{-2}\,m^{-2} = Nm^{-2} = Pa \text{ (i.e. Pascal)}$$

EXAMPLE 2.03 Finding the units of variables in a complicated equation.

The temperature of the fluid leaving a continuous flow heat exchanger is given by the (correct) equation:

$$T_2 = \frac{\{z + \exp[(U_{hex}A_{hex}/(m_cC_c))(z+1)]\}T_1 + \{\exp[(U_{hex}A_{hex}/(m_cC_c))(z+1)] - 1\}zT_3}{(z+1)\exp[(U_{hex}A_{hex}/(m_cC_c))(z+1)]}$$

where:

A_{hex} = heat transfer area	m^2	T_2	= hot fluid outlet temperature	?
C_c = cold fluid heat capacity	$kJ.kg^{-1}.K^{-1}$	T_1	= hot fluid inlet temperature	?
C_h = hot fluid heat capacity	$kJ.kg^{-1}.K^{-1}$	T_3	= cold fluid inlet temperature	K
m_c = cold fluid flow rate	?	U_{hex}	= heat transfer coefficient	?
m_h = hot fluid flow rate	$kg.s^{-1}$	z	$= m_cC_c / (m_hC_h)$	?

Problem: What are the units of z, T_1, T_2, m_c and U?

Solution:

Units of z = dimensionless

Units of T_1 = units of T_2 = units of T_3 = K

Units of $(m_cC_c / (m_hC_h))$ = units of z = dimensionless

Units of m_c = units of (m_hC_h / C_c)

= $(kg.s^{-1})(kJ.kg^{-1}.K^{-1}) / (kJ.kg^{-1}.K^{-1})$

= $kg.s^{-1}$

Units of $(UA_{hex}/(m_cC_c))$ = dimensionless

Units of U = units of (m_cC_c / A_{hex})

= $kg.s^{-1}.\ kJ.kg^{-1}\ .\ K^{-1}/m^2$

= $kJ.s^{-1}.m^{-2}.K^{-1}$

= $kW.m^{-2}.K^{-1} \equiv kW/(m^2.K)$

CONVERSION OF UNITS

Sometimes you need to convert a value in the SI system to its equivalent in the Imperial/American system, or vice-versa. An easy way to make such a conversion is with a table of conversion factors. *Table 2.03* lists some common conversion factors. Extensive tables of unit conversion factors are in the handbooks and texts listed at the end of this chapter.

You can also work out any unit conversions from the equivalents in *Table 2.03*, as shown in *Example 2.04*.

Table 2.03. Unit conversion factors.

Quantity	Value	Units		Value	Units
Mass	1	kg	*Equivalent to:*	2.2046	lb_m
Length	1	m		3.2808	ft
Temperature	1	K		1.8	R°
Amount of substance	1	kmol		2.2046	lbmol
Energy	1	kJ		0.9486	BTU
Power	1	kW		1.341	HP
Force	1	kN		224.8	lb_f
Pressure	1	kPa		0.145	psi
Volume	1	m^3		264.17	US gall
				220.83	Imp gall

EXAMPLE 2.04 Converting units from SI to Imperial/American.

Problem A: Convert a density of 5.000 $kg.m^{-3}$ to its equivalent in units of $lb_m\ ft^{-3}$.
Solution: Substitute conversion factors from *Table 2.03*.

$$5\ kg.m^{-3} \equiv 5\ kg\ (2.2\ lb_m/kg)\ /\ m\ (3.28\ ft/m)^3$$

Cancel the SI units from numerator and denominator of the RHS, and do the arithmetic.

$$5\ kg.m^{-3} \equiv 0.312\ lb_m\ ft^{-3}\ (lb_m \equiv \text{pound mass})$$

Problem B: Convert a heat transfer coefficient of 1.00 $kW.m^{-2}.K^{-1}$ to its equivalent in units of $BTU.hr^{-1}.ft^{-2}.°F^{-1}$.
Solution: Substitute conversion factors from *Table 2.03*.

$$1\ kW.m^{-2}.K^{-1} \equiv \frac{1\ kJ\ (0.9486\ BTU/kJ)}{(s\ (1/3600\ s/h))(m\ (3.28\ ft/m))^2\ (K\ (1.8°F/K))}$$

Cancel SI units from numerator and denominator and do the arithmetic.

$$1.00\ kW.m^{-2}.K^{-1} \equiv 176\ BTU.h^{-1}.ft^{-2}.°F^{-1}$$

SIGNIFICANT FIGURES IN CALCULATIONS

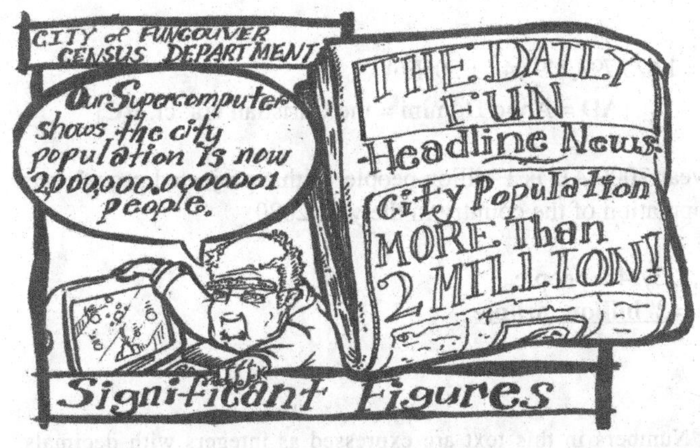

The number of significant figures in a value is the number of digits in the argument when expressed in scientific notation, i.e. x.xxxx Ey ≡ x.xxxx (10^y) shows the 5 significant figures "x.xxxx". Electronic calculators and computers produce up to 10 significant figures but you will rarely do an engineering calculation with a result justified to more than 5 significant figures. Often, 1 to 3 significant figures are satisfactory.

The final results of your calculations should be given with a number of significant figures corresponding to the accuracy of measurements and calculations. Extra significant figures are misleading and inappropriate. They also clutter your work and waste the time of others who have to read it. When you use a calculator or a computer, a good procedure is to carry extra significant figures to the end of the calculation, then round off the final "displayed" result(s) to an appropriate number of significant figures.

In deciding the number of significant figures you should consider the sources of your data, the equations used to process them and the needs of your audience, as illustrated in *Example 2.05 A, B and C*.

EXAMPLE 2.05 Specifying significant figures from measurements and calculations.

Problem A: In a laboratory test the volume of 2.81 ± 0.005 kg of nitrogen gas at 300.0 ± 0.1 K and 122 ± 1 kPa(abs) was measured as 2.05 ± 0.005 m^3. Calculate the value of the universal gas constant.

Solution: By the ideal gas law, PV = nRT, solve for R and substitute values.

R = PV/nT = (122 ± 1) kPa * (2.05 ± 0.005) m^3 / [((2.81 ± 0.005)/28.013) kmol * (300 ± 0.1) K]
= 8.310855635 ± (1/122 + 0.005/2.05 + 0.005/2.81 + 0.1/300)(8.310855635)
= 8.310855635 ± 0.105950432
= **8.31** ± 0.11 $kJ.kmol^{-1}.K^{-1}$

Problem B: The capital cost of a wood pulp mill can be estimated by the correlation:

$C = 17E6\ (P)^{0.6} \quad \pm 20\%$

where: C = capital cost; accurate to ± 20%; $US (2000 AD)
P = pulp production rate metric ton per day

Calculate the capital cost of a pulp mill producing 2000 metric ton per day of pulp.

Solution: Substitute P = 2000 ton/day
$C = 17E6\ (P)^{0.6} = 17E6\ (2000)^{0.6}$ = 1,625,799,249.64 (± 306E6)
i.e. C = **1.6E9 $US** (2000 AD) [AD = Anno Domini = the Christian era. cf. BC]

Problem C: The population of a country in year 2000 AD is 1 billion people, with a projected growth rate of 4% per year. Calculate the population of the country in the year 2020.

Solution: Pop = $1E9\ (1 + 0.04)^{20}$ = 2,191,123,143 people
i.e. Population in year 2020 AD = **2 billion people**

Table 2.04. Prefixes in SI system

Value	Prefix	Abbreviation
1E9	giga	G
1E6	mega	M
1E3	kilo	k
1E-3	milli	m
1E-6	micro	μ
1E-9	nano	n

Note: Numbers in this text are expressed as integers with decimals, e.g. "x.xx" or in scientific format, i.e: x.xxEy ≡ $(x.xx)(10^y)$.

PROCESS VARIABLES

PRESSURE

Pressure is the normal force per unit area. The basic unit of pressure in the SI system is the Pascal (Pa), but since this is a small quantity, ordinary pressures are usually measured in kiloPascals (kPa). Some measuring devices, such as the barometer, display the *absolute pressure*, but others, such as the open ended manometer and the Bourdon gauge display the *gauge pressure*, which is the pressure relative to the local atmospheric pressure. The relation between these scales is given in *Equations 2.01* and *2.02*.

For super-atmospheric pressures: **Absolute pressure = atmospheric pressure + gauge pressure** *Equation 2.01*

For sub-atmospheric pressures: **Absolute pressure = atmospheric pressure – vacuum** *Equation 2.02*

Many calculations with pressure require the absolute pressure (e.g. the gas law PV = nRT). When a *pressure difference* (ΔP) is required either the absolute or the relative scales can be used consistently to calculate ΔP. *Example 2.06* shows calculations with absolute and relative pressures.

The atmospheric pressure features in many problems with pressure. Equivalents between standard atmospheric pressure and other common measures of pressure are given in *Table 2.05*. The pressure equivalent to a column of fluid comes from *Equation 2.03*.

$$\mathbf{P = \rho g L'}$$ *Equation 2.03*

where: P = pressure Pa
ρ = fluid density $kg.m^{-3}$
L′ = vertical height of fluid column m

Table 2.05. Absolute pressure equivalents.

Standard atmosphere	Bar	kPa	psi*	mm (mercury at 0 °C)	m (water at 4 °C)
1.000	0.989	101.3	14.696	760.0	10.333

*psi ≡ pounds (force) per square inch (see *Table 2.01*).
psia = psi (absolute) psig = psi (gauge)

To avoid ambiguity you should always indicate whether pressures are absolute or relative, e.g. 448 kPa(abs), 50 psig, etc. Also note that atmospheric pressure varies with place and time and is rarely exactly one standard atmosphere. For accurate work the local atmospheric pressure should be established, for example, by a barometer measurement. In solving problems where the atmospheric pressure is not given it is usual to assume the value of one standard atmosphere. However, this assumption will give a substantial error for locations far above sea level, e.g. the atmospheric pressure near the top of Mt. Everest is about 30 kPa(abs).

TEMPERATURE

The temperature of a body is a measure of its potential to transfer heat to other bodies. There are two temperature scales in both the SI and the Imperial/American systems. The relations between these scales are given in *Equations 2.04, 2.05* and *2.06.*

	SI	**Imperial/American**
Thermodynamic (absolute) temperature	Kelvin (K)	Rankine (°R)
Customary (relative) temperature	Celsius (°C)	Fahrenheit (°F)

K = 273.15 + °C *Equation 2.04*

°R = 459.67 + °F *Equation 2.05*

°F = 32 + 1.8 °C *Equation 2.06*

Common temperature measurement devices, such as thermometers, thermocouples and thermistors usually display temperature on the customary scale. However, many calculations require the use of temperature on the thermodynamic scale (e.g. the gas law PV = nRT). When a temperature *difference* (ΔT) is required, either the customary or the thermodynamic scale can be used consistently to calculate ΔT. *Example 2.06* shows calculations with absolute and relative pressures and temperatures.

EXAMPLE 2.06 Using pressure and temperature measurements for process calculations.

Figure 2.01 shows a pipe with a set of pressure and temperature gauges used to measure a continuous steady flow of nitrogen gas, which is simultaneously being heated from an external heat source.

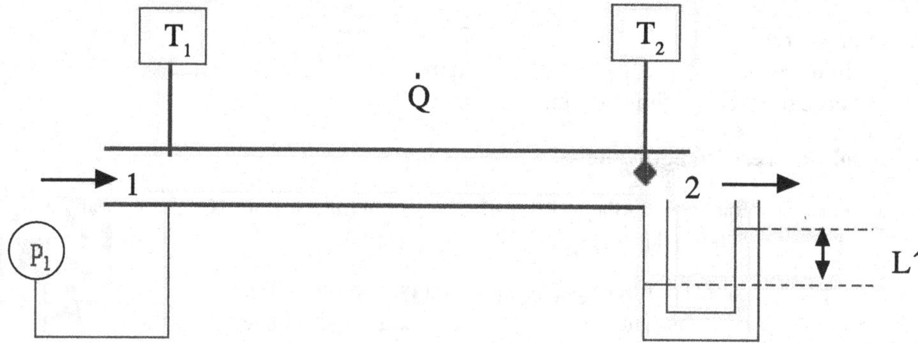

Figure 2.01. Gas flow measurement.

The flow of nitrogen is approximated by:[4] $\dot{m} \approx A(\rho_{av}\Delta P)^{0.5}$

The rate of heat transfer to the flowing nitrogen is given by: $\dot{Q} = \dot{m}C_p\Delta T$

[4] This equation applies only for a narrow range of conditions.

where:

A = cross section area of tube = 0.002 m^2
C_p = heat capacity of nitrogen = 1.05 $kJ.kg^{-1}K^{-1}$
L' = difference in mercury levels in the open manometer = 105 mm Hg
$\dot{m}$ = mass flow rate of nitrogen
ρ_{av} = density of nitrogen at the average temperature and pressure
ΔP = $P_1 - P_2$ = pressure drop along the pipe
$\dot{Q}$ = rate of heat transfer into the flowing nitrogen
ΔT = $T_2 - T_1$ = temperature rise along the pipe
P_1 = inlet nitrogen pressure measured on a Bourdon gauge = 20 kPa(gauge)
T_1 = inlet nitrogen temperature measured by a thermometer = 20°C
T_2 = outlet nitrogen temperature measured by a thermocouple = 158°F

Problem: Calculate the velocity of the nitrogen gas at the pipe outlet (point 2) and the rate of heat transfer to the nitrogen.

Solution:

Note 1 Units are not specified for several of the variables, but they can be found by dimensional consistency.
N_2 outlet pressure = P_2 = (105 mm Hg (gauge))(101.3 kPa/760 mm Hg) = 14.0 kPa(gauge)
Average pressure = $(P_1 + P_2) / 2$ = (20 kPa(gauge) + 14 kPa(gauge)) / 2 = 17 kPa(gauge)
N_2 outlet temperature = (158°F – 32°F) / 1.8 (°F/°C) = 70°C
Average temperature = $(T_1 + T_2) / 2$ = (20°C + 70°C) / 2 = 45°C

Note 2 The outlet values calculated above are the relative pressure and temperature.
Ideal gas density (*Equation 2.14*) = (M/R)(P/T) where M = 28.01 kg/kmol, R = 8.314 kJ/(kmol K)
Density of N_2 at average P and T = (28.01 kg/kmol/8.314 kJ/(kmol.K))(17 + 101.3) kPa(abs)/(45 + 273) K
= 1.25 kg/m^3

Note 3 The P and T used in the ideal gas equation must be absolute values.
Atmospheric pressure is taken from *Table 2.05* as 101.3 kPa(abs).
Pressure difference = ΔP = (20 kPa(gauge) – 14 kPa(gauge)) = (121.3 kPa(abs) – 115.3 kPa(abs)) = 6 kPa
Flow of N_2 = $\dot{m} = A(\rho_{av}\Delta P)^{0.5}$ = (0.002 m^2)(1.25 kg/m^3)(1000 Pa/kPa)(6 kPa)$^{0.5}$ = 0.173 kg/s

Note 4 The pressure difference ΔP can be calculated either from both relative pressures or both absolute pressures.

Note 5 From *Table 2.01* Pa = kg/(m.s^2), thus the units of $\dot{m} = m^2(kg/m^3)(kg/(m.s^2))^{0.5}$ = kg/s
Volumetric flow of N_2 at outlet = $\dot{V} = \dot{n}RT_2/P_2 = (\dot{m}/M)RT_2 / P_2$ Ideal gas, *Equation 2.13*
= (0.173 kg/s/28.01 kg/kmol)(8.314 kJ/(kmol.K)(70 + 273)K/(14 + 101.3)kPa(abs) = 0.153 m^3/s

Note 6 The P and T used in the ideal gas equation must be absolute values.

Velocity of N_2 at outlet	$= \dot{V}/A = (0.158\ m^3/s) / (0.002\ m^2) =$ **76 m/s**
Temperature difference	$= \Delta T = (70°C - 20°C) = 50C° = (343\ K - 293\ K) = 50\ K$
Rate of heat transfer into N_2	$= \dot{Q} = \dot{m}\ C_p \Delta T = (0.173\ kg/s)(1.05\ kJ/(kg.K))(50\ K)$
	$= 9.1\ kJ/s =$ **9.1 kW**

Note 7 The temperature difference ΔT can be calculated either from both relative temperatures or both absolute temperatures.

Partial pressure and partial volume

The *partial pressure* of a component of a gas mixture is the pressure that component would exert if it alone occupied the full volume of the mixture at the temperature of the mixture.

The *partial volume* of a component of a gas mixture is the volume that component would occupy if it alone were present at the total pressure and temperature of the mixture.

The partial pressure and partial volume of a component of an ideal gas mixture are both proportional to its mole fraction, i.e.:

$$\mathbf{p(j) = y(j)P} \quad \text{(fixed temperature)} \qquad \textit{Equation 2.07}$$

$$\mathbf{v'(j) = y(j)V} \quad \text{(fixed temperature)} \qquad \textit{Equation 2.08}$$

Also, since $\Sigma\,[y(j)] = 1$, then:

$$\mathbf{\Sigma\,[p(j)] = P} \quad \text{(Dalton's law of partial pressures)} \qquad \textit{Equation 2.09}$$

$$\mathbf{\Sigma\,[v'(j)] = V} \quad \text{(Amagat's law of partial volumes)} \qquad \textit{Equation 2.10}$$

where:

p(j)	=	partial pressure of component	"j"	kPa(abs)
v′(j)	=	partial volume of component	"j"	m^3
y(j)	=	mole fraction of component	"j" in ideal gas mixture	–
P	=	total pressure of mixture		kPa(abs)
V	=	total volume of mixture		m^3

Phase

Many substances can exist as one or more of the phases: solid (S), liquid (L) or gas (G). The phase of a substance at equilibrium depends on its conditions of temperature and pressure. For example, the phase diagram of water in *Figure 2.02* shows that at equilibrium water is:

- a single phase solid at 200 K,150 kPa(abs) region (S)
- a single phase liquid at 300 K,150 kPa(abs) region (L)
- a single phase gas at 600 K,150 kPa(abs) region (G)
- a two phase solid–liquid along the line between (L) and (S)
- a two phase solid–gas along line between (S) and (G)
- a two phase liquid–gas along line between (L) and (G)
- a three phase solid–liquid–gas at the *triple-point*

  The triple-point is the unique condition at which the solid, liquid and gas co-exist at equilibrium.

A vapour is a gas below its critical temperature and in this condition can be condensed by a sufficient increase in the pressure.

Note that the phase relations of *Figure 2.02* apply only at equilibrium. It is possible to obtain water (and many other substances) in non-equilibrium states, corresponding to conditions such as supercooled liquid water (e.g. P = 101.3 kPa(abs), T = 272 K), superheated liquid water (e.g. P = 101.3 kPa(abs), T = 374 K) and supercooled water vapour (e.g. P = 101.3 kPa(abs), T = 372 K). These non-equilibrium states are thermodynamically unstable but do exist under some circumstances.

The physical properties of any substance depend on its phase, so the description of any process stream should include the stream phase(s).

Depending on the circumstances, determining the phase of an unseen material can be easy, or it can be difficult. In the case of pure substances (i.e. a single molecular species), when you know the temperature and pressure you can look up the phase in a table of thermodynamic properties or in a phase diagram (see *Refs. 1–5*). Alternatively you can find the phase from equations representing the inter-phase equilibria as shown in *Example 2.07*.

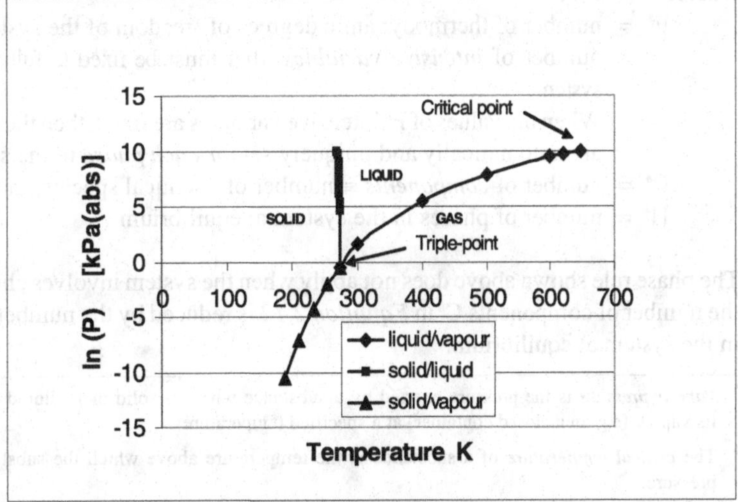

Figure 2.02. Phase diagram for water.

EXAMPLE 2.07 Finding the phase of water by using the Antoine equation.

The *vapour pressure*[5] of pure water is given by:

$$p^* = \exp[16.5362 - 3985.44 / (T - 38.9974)] \qquad \textit{Equation 2.11}$$

where: p^* = vapour pressure kPa(abs)
T = temperature K

Problem: Is pure water a liquid or a gas at 374 K, 100 kPa(abs)?
Solution: Substitute T = 374 K into *Equation 2.11*.
$p^* = \exp[16.5362 - 3985.44 / (374 - 38.9974)] = 104$ kPa(abs)

Since p^* at 374 K is greater than 100 kPa(abs) [i.e. 374 K, 100 kPa lies below the G/L equilibrium line] then water in its equilibrium state at 374 K, 100 kPa(abs) is in the GAS phase.

In this case the gas phase is a *vapour* because the temperature is below the *critical temperature*[6] of water, which is 647 K, and the water (gas) can be condensed to water (liquid) by increasing the pressure.

Phase rule

When you are trying to determine the phase of a material it is sometimes helpful to use the phase rule, as shown in *Example 2.08*.

$$F^\# = C^\# - \Pi^\# + 2 \quad \text{(Phase rule — applies to non-reactive systems)} \qquad \textit{Equation 2.12}$$

where:

$F^\#$ = number of thermodynamic degrees of freedom of the system
= number of *intensive variables*[7] that must be fixed to fully-specify the state of each phase in the system.
When the values of $F^\#$ intensive variables are fixed, then the values of all other intensive variables are automatically and uniquely set *for each phase* in the system

$C^\#$ = number of *components* ≡ number of chemical species, in the system

$\Pi^\#$ = number of phases in the system at equilibrium

The phase rule shown above does not apply when the system involves chemical reactions. In reactive systems the number of components $C^\#$ in *Equation 2.12* is reduced by the number of independent reactions that occur in the system at equilibrium.

---

5 *Vapour pressure* is the pressure exerted by a substance when its solid or its liquid phase is in contact and at equilibrium with its vapour (e.g. in a closed container) at a specified temperature.

6 The *critical temperature* of a substance is the temperature above which the substance cannot be liquefied by increasing the pressure.

7 *Intensive variables* are the properties of each phase that are independent of the amount of material. e.g. P, T, v, u, h, x. Note that for a multi-phase system at equilibrium the values of P and T are uniform throughout the phases, but the values of v, u, h, x, etc. are different in each phase.

EXAMPLE 2.08 Using the phase rule to find the phase of water in a multi-phase system.

Problem A: How many intensive variables are needed to define the equilibrium state of pure liquid water?

Solution: From the phase rule *Equation 2.12:*

$F^{\#} = C^{\#} - \Pi^{\#} + 2 = 1$ component – 1 phase + 2 = **2 intensive variables**

Values of any two intensive variables (e.g. [T, P] or [T, v] or [P, h], etc.) will fully-specify the state of pure liquid water.

Problem B: How many intensive variables are needed to define the equilibrium state of each phase of a mixture of liquid water and gas water?

Solution: From the phase rule *Equation 2.12:*

$F^{\#} = C^{\#} - \Pi^{\#} + 2 = 1$ component – 2 phases + 2 = **1 intensive variable**

The value of any one intensive variable (e.g. P or T or v or h, etc.) will fully-specify the states of the liquid and of the gas.

Note however that the value of one more system variable, such as the specific volume or specific enthalpy, of the 2 phase mixture is needed to fix the distribution of mass between the two phases (i.e. to fix the phase split).

Problem C: The equilibrium state of a mixture of ethanol and water is determined by fixing ONLY the temperature of the mixture and the mole fraction of ethanol in one phase. How many phases are present in the system at equilibrium?

Solution: From the phase rule *Equation 2.12:* $\Pi^{\#} = C^{\#} + 2 - F^{\#}$

$C^{\#} = 2$ (i.e. ethanol, water)
$F^{\#} = 2$ (i.e. T, x)
$\Pi^{\#} = 2 + 2 - 2 =$ **2 phases**

The phase relations for mixtures of substances are outlined later in this chapter under "Multi-Species, Multi-Phase Equilibria".

EQUATIONS OF STATE

An *equation of state*[8] is a relation between the specific volume of a substance, its pressure and its temperature (i.e. v, P and T).

IDEAL GAS

An *ideal gas* is a hypothetical gas in which the individual molecules do not interact with each other. No real gas is truly ideal, but many gases are nearly ideal and can be modelled by the ideal gas *equation of state*, which is called the "ideal gas law":

PV = nRT (Ideal gas law) *Equation 2.13*

where:				
P = pressure in gas	kPa(abs)	V = volume of gas	m^3
n = m / M = amount of gas	kmol	R = universal gas constant = 8.314	$kJ.kmol^{-1}.K^{-1}$
T = temperature of gas	K	M = molar mass (molecular weight) of gas	$kg.kmol^{-1}$
m = mass of gas	kg		

As a rule of thumb, the ideal gas law predicts values within 1% of reality when RT/P > 5 m^3/kmol for diatomic gases, or RT/P > 20 m^3/kmol for other polyatomic gases. The deviation of a real gas from the ideal gas law increases as the conditions approach the *critical state* of the gas, which is defined by its *critical temperature* (T_c) and *critical pressure* (P_c).[9] *Table 2.06* lists the critical temperature and pressure of some common substances. Deviations from ideality are measured by the gas compressibility factor "z", which ranges in value from about 0.2 to 4, depending on the proximity to the critical state. To account for compressibility, the ideal gas equation is modified to *Equation 2.14*.

PV = znRT *Equation 2.14*

Calculations with non-ideal gases involve the use of the *reduced temperature* ($T_r = T/T_c$) and *reduced pressure* ($P_r = P/P_c$) in relatively complex non-linear equations of state that are beyond the scope of this text (see *Refs. 4–5*). As a rough guide, if $T_r \geq 2$ and $P_r < 10$ the compressibility factor ranges from about 0.9 to 1.2, so the gas can be modelled as an ideal gas with accuracy better than +/–20%.

From *Table 2.06* you can see that around normal atmospheric conditions (i.e. 293 K, 101 kPa(abs)), hydrogen, oxygen, nitrogen, air and methane behave effectively as ideal gases. Butane and water are liquids at 293 K, 101 kPa(abs), but in multi-component gas systems above 273 K their vapours approximate the ideal gas law if the partial pressure is below about 10% of the critical pressure (i.e. $P_r < 0.1$).

Table 2.06. Critical temperature and pressure of common substances.

Substance	T_c	P_c
	K	kPa(abs)
Hydrogen (H_2)	33.3	1297
Oxygen (O_2)	154.4	5035
Nitrogen (N_2)	126.2	3394
Methane (CH_4)	190.7	4653
Carbon dioxide (CO_2)	304.2	7385
n-Butane (C_4H_{10})	425.2	3796
Water (H_2O)	647.3	22110

[8] Specifically, a volumetric equation of state. Equations of state for single-phase, single-component systems may relate any three intensive variables.

[9] The *critical temperature* (T_c) is the temperature above which the gas cannot be liquefied by increasing the pressure.
The *critical pressure* (P_c) is the minimum pressure required to liquefy the gas at its critical temperature.

EXAMPLE 2.09 Using the ideal gas law to calculate a gas volume.

Problem: Dry air[10] contains 21 vol% O_2 (M = 32.00 kg.kmol^{-1}) + 79 vol% N_2 (M = 28.01 kg.kmol^{-1}). Calculate the volume occupied by 5.00 kg of dry air at 400 K, 500 kPa(gauge). Assume ideal gas behaviour.

Solution: By the ideal gas law:

PV = nRT (see *Equation 2.13*)

Molar mass of dry air (see *Equation 2.18*):

$M_m = \Sigma(M(j).y(j)) = (32.00)(0.21) + (28.01)(0.79)$ $= 28.85$ kg.kmol^{-1}

Amount of dry air:

$n = m / M_{av} = 5$ kg / 28.85 kg.kmol^{-1} $= 0.173$ kmol

Absolute pressure (see *Equation 2.02*):

P = Atmospheric pressure + gauge pressure = 101.3 kPa + 500 kPa(gauge) = 601.3 kPa(abs)

Solve *Equation 2.13* for V and substitute values.

V = nRT/P = (0.173 kmol)(8.314 kJ.kmol^{-1}.K^{-1})(400 K)/(601.3 kPa(abs)) = **0.959** m^3

NOTE: The ideal gas law does not apply to liquids or solids.

LIQUIDS AND SOLIDS

For most purposes the pressure–temperature–volume relation for liquids and solids can be approximated by *Equation 2.15*.

$$\mathbf{v \cong v_o(1 + \theta(T - T_o) - \sigma(P - P_o))} \qquad \textit{Equation 2.15}$$

where: v = specific volume m^3/kg

v_o = specific volume at $P = P_o, T = T_o$ m^3/kg

θ = coefficient of thermal expansion of the material $= 1/v_o(dv/dT)$ K^{-1}

σ = coefficient of compressibility of the material $= -1/v_o(dv/dP)$ kPa^{-1}

Both θ and σ are slightly affected by temperature and pressure, and this fact should be accounted for in accurate calculations, especially when T and/or P change over a wide range.

DENSITY

Density is the ratio of mass to volume.

$$\boldsymbol{\rho = m/V} \qquad \textit{Equation 2.16}$$

where: ρ = density m = mass V = volume

The density of a material depends on its phase, the temperature and the pressure. *Table 2.07* shows values of density for some earthly materials, which ranges from about 0.09 kg.m^{-3} for hydrogen gas at room conditions to 22,000 kg.m^{-3} for solid osmium.

---

[10] More accurately the composition of dry air is: N_2 = 78.08, O_2 = 20.95, Ar = 0.93, CO_2 = 0.04 vol%.

Table 2.07. Density and specific volume of some materials.

Material	Phase	M	Pressure	Temperature	Density	Specific Volume
		$kg.kmol^{-1}$	kPa(abs)	K	$kg.m^{-3}$	$m^3.kg^{-1}$
Hydrogen	gas	2	101	273	0.089	11.2
Air	gas	28.8	101	273	1.28	0.781
Chlorine	gas	71	101	273	3.16	0.316
Pentane	liquid	72	101	291	630	0.00159
Water	liquid	18	101	277	1000	0.00100
Bromine	liquid	253	101	293	3119	0.000321
Mercury	liquid	201	101	293	13550	0.0000738
Octadecane	solid	254	101	298	775	0.00129
Water	solid	18	101	273	920	0.00109
Iron	solid	56	101	293	7860	0.000127
Osmium	solid	190	101	293	22500	0.0000444

Variations in temperature and pressure have a small effect on the density of solids and liquids, as measured respectively by the coefficient of thermal expansion and the coefficient of compressibility. For example, for liquid water at 298 K:

Coefficient of thermal expansion $= (1/v)(dv/dT) = d(\ln(v))/dT = 257E\text{-}6\ K^{-1}$

Coefficient of compressibility $= -(1/v)(dv/dP) = -d(\ln(v))/dP = 45E\text{-}8\ kPa^{-1}$

The density of gases is strongly affected by variation in temperature and pressure. For ideal gases the density can be calculated from *Equation 2.17*.

$$\rho = MP/RT \qquad \textit{Equation 2.17}$$

where:

ρ	=	density of ideal gas or gas mixture	$kg.m^{-3}$
M	=	molar mass of gas or gas mixture	$kg.kmol^{-1}$
P	=	pressure	kPa(abs)
R	=	gas constant = 8.314	$kJ.kmol^{-1}.K^{-1}$
T	=	temperature	K

For a gas mixture the molar mass is the mean value defined by *Equation 2.18*

$$M_m = \Sigma(M(j)y(j)) \qquad \textit{Equation 2.18}$$

where:

M_m	= mean molar mass ("molecular weight") of gas mixture	$kg.kmol^{-1}$
M(j)	= molar mass of component "j"	$kg.kmol^{-1}$
y(j)	= mole fraction of component "j"	–

The density of pure solids, liquids and non-ideal gases are relatively difficult to calculate (see *Ref. 5*) and are best obtained by measurement or from tabulations such as those in *Refs. 1, 2* and *3*. For *ideal mixtures*[11] of

[11] In an *ideal mixture* the partial volumes of the components can be added to give the total volume of the mixture. This is not the case for a *non-ideal* mixture.

substances (solids, liquids or gases) the density of the mixture can be calculated from the individual component densities by *Equation 2.19*.

$$\rho_m = 1 / \Sigma (w(j) / \rho (j))$$ *Equation 2.19*

where: ρ_m = density of ideal mixture — $kg.m^{-3}$
$w(j)$ = mass fraction of component "j" — –
$\rho(j)$ = density of component "j" — $kg.m^{-3}$

Table 2.08. Density of sulphuric acid — water mixtures at 293 K, 101 kPa(abs).

H_2SO_4 conc.	Actual density	Density from Eqn 2.19
wt%	$kg.m^{-3}$	$kg.m^{-3}$
0	998	998
10	1066	1048
50	1395	1249
90	1814	1691
100	1831	1831

For *non-ideal mixtures* of substances the density of the mixture is only approximated by *Equation 2.19* and is best obtained by measurement or from tabulations in the literature. For example, *Table 2.08* shows a comparison of the actual density of sulphuric acid — water liquid mixtures with values calculated by the ideal mixture *Equation 2.19*.

SPECIFIC GRAVITY

Specific gravity is the ratio of the density of a material to the density of a reference substance.

$$SG = \rho / \rho_{ref}$$ *Equation 2.20*

where: SG = specific gravity of material — –
ρ = density of material — $kg.m^{-3}$
ρ_{ref} = density of reference substance — $kg.m^{-3}$

For solids and liquids the reference substance is conventionally liquid water at 277 K, 101 kPa(abs), with a density of 1000 $kg.m^{-3}$. For gases the reference substance is usually air at 293 K, 101 kPa(abs), with density 1.20 $kg.m^{-3}$. Due to this potential ambiguity you should check the reference substance when you use values of specific gravity.

SPECIFIC VOLUME

Specific volume is the ratio of volume to mass, i.e. the inverse of density.

$$v = V/m = 1/\rho$$ *Equation 2.21*

where: m = mass — kg
v = specific volume — $m^3.kg^{-1}$
V = volume — m^3

Molar volume

Molar volume is the volume of one mole of a substance and equal to the product of the specific volume and molar mass.

$$\mathbf{v_m = M\,v = M/\rho}$$ *Equation 2.22*

where: v_m = molar volume — $m^3.kmol^{-1}$
M = molar mass — $kg.kmol^{-1}$
ρ = density — $kg.m^{-3}$

Concentration

Concentration is the measure of the relative amount of a component in a mixture (i.e. the mixture composition). *Table 2.09* summarises the common ways of expressing concentration. With so many different ways to express concentration there is much potential for ambiguity. For a given component in a given mixture, each of the measures in *Table 2.09* will usually give a different value, so it is important to specify the measure when giving values of concentration.

Table 2.09. Common measures of concentration.

Name	Definition	Symbol	Units
Mass fraction	Mass of component / total mass of mixture	w	dimensionless
Mole fraction	Moles of component / total moles of mixture	x, y	dimensionless
Volume fraction	Partial volume of component / total volume of mixture	$\tilde{v}$	dimensionless
Molarity	Moles of component / volume of mixture	M	$mol.l^{-1}$ ($kmol.m^{-3}$)
Molality	Moles of component / mass of solvent	*m*	$mol.kg^{-1}$
Parts per million	1E6 w(j) 1E6 $\tilde{v}$(j)	ppm(wt) ppm(vol)	dimensionless
Parts per billion	1E9 w(j) 1E9 $\tilde{v}$(j)	ppb(wt) ppb(vol)	dimensionless
Mass per volume	Mass of component / volume of mixture	–	$kg.m^{-3}$
Mass per mass	Mass of component / mass of solvent	–	dimensionless
Activity	Function of chemical potential (see *Refs. 5–6*)	*a*	dimensionless

The mass fraction and mole fraction are well defined measures of concentration commonly used in process engineering.

$$\mathbf{w(j) = m(j) / \Sigma\, m(j)}$$ *Equation 2.23*

$$\mathbf{x(j) = n(j) / \Sigma\, n(j) = (m(j) / M(j)) / \Sigma\, (m(j) / M(j))}$$ *Equation 2.24*

where: m(j) = mass of component "j" — kg
M(j) = molar mass of component "j" — $kg.kmol^{-1}$
w(j) = mass fraction (i.e. weight fraction) of component "j" — –
x(j) = mole fraction of component "j" — –

Note that:[12] $\Sigma (w(j)) = 1$ *Equation 2.25*

$\Sigma (x(j)) = 1$ and $\Sigma (y(j)) = 1$ *Equation 2.26*

Inter-conversion of mass fraction with mole fraction in a given mixture can be done by *Equations 2.27* and *2.28*, as shown in *Example 2.10 A* and *B*.

To convert mass fraction to mole fraction:

$$x(j) = (w(j) / M(j)) / \Sigma (w(j) / M(j))$$ *Equation 2.27*

To convert mole fraction to mass fraction:

$$w(j) = (x(j) M(j)) / \Sigma (x(j) M(j))$$ *Equation 2.28*

Fractional concentrations are often expressed as percentages, where:

Mass % = weight % = wt% = 100 w(j) *Equation 2.29*

Mole % = 100 x(j) *Equation 2.30*

In the case of ideal gas mixtures the mole fraction and volume fraction have the same value, i.e.

$$\tilde{v}(j) = y(j) = p(j) / P \quad \text{(ideal gas mixtures)}$$ *Equation 2.31*

where: $\tilde{v}(j)$ = partial volume fraction of component j in gas mixture –

y(j) = mole fraction of component j in gas mixture –

p(j) = partial pressure of component j kPa(abs)

P = total pressure kPa(abs)

Equation 2.31 applies approximately to non-ideal gas mixtures, but does not apply to solid or liquid mixtures.

EXAMPLE 2.10 Converting between mass fraction and mole fraction compositions.

Problem A: Gunpowder is a mixture of sulphur (29 wt%), carbon (14 wt%) and potassium nitrate. Find the mole fraction of potassium nitrate (KNO_3) in the gunpowder mixture.

Solution: Define the components.

Component	S	C	KNO_3
Number (j)	1	2	3
w(j)	0.29	0.14	w(3)
M(j)	32	12	101

From *Equation 2.25* w(3) = 1 – (0.29 + 0.14) = 0.57

From *Equation 2.27* x(3) = w(3)/M(3)/Σ(w(j)/M(j)) = 0.57/101 / (0.29/32 + 0.14/12 + 0.57/101) = **0.20**

Problem B: Air (gas) is a mixture of oxygen (21 vol%) and nitrogen. Find the mass fraction of nitrogen in air.

Solution: Define the components.

Component	O_2	N_2
Number	1	2
y(j)	0.21	y(2)
M(j)	32	28

[12] Mole fraction is usually given the symbol "x" in solid and liquid mixtures, "y" in gas mixtures.

From *Equation 2.26* y(2) = 1 – 0.21 = 0.79
From *Equation 2.28* w(2) = (y(j) M(j)) / Σ (y(j) M(j)) = (0.79)(28) / (0.21*32 + 0.79*28) = **0.77**

Other measures of concentration are used in various situations. For example, low concentrations are conventionally given as *ppm(wt)* in liquids and solids and as *ppm(vol)* in gases. Concentrations of toxic dusts in air are given as micrograms per cubic metre and solubilities of solids in liquids are commonly listed as grams of solute per 100 grams of solvent. The concentrations of solutions used in the chemical laboratory are usually given as molarity (M) or *normality*[13] (N), whereas thermodynamic calculations involving solutions use the *molality*, or more generally the dimensionless *activity*.

MULTI-SPECIES, MULTI-PHASE EQUILIBRIA

When two (or more) phases are in contact they tend to an equilibrium in which the concentration of each species in one phase is related to its concentration in the other phase(s). The equilibrium relations for the distribution of species between phases are used in M&E balances to calculate the composition of process streams associated with multi-phase systems.

The equilibrium distribution of species between the phases of multi-phase systems is strongly temperature dependent, and the temperature dependence is usually non-linear.

Some of the relations commonly used to model equilibrium in multi-phase systems are as follows:

GAS–LIQUID SYSTEMS

Vapour pressure

The vapour pressure of a substance is its partial pressure in the gas phase in equilibrium with a condensed phase (solid or liquid) of the pure substance. The vapour pressure of a substance is determined by the nature of the substance and the temperature. Whenever a pure (single component) condensed substance is in equilibrium with a gas the partial pressure of that substance in the gas phase will be its vapour pressure at the specified temperature. [Note that this condition is modified (cf. the Kelvin equation) as the radius of curvature of the condensed phase becomes very small (e.g. < 1 micron), due to the effect of surface energy in such systems.] Vapour pressures of pure substances are tabulated and correlated with temperature in sources such as *Refs. 1 to 5* of this chapter.

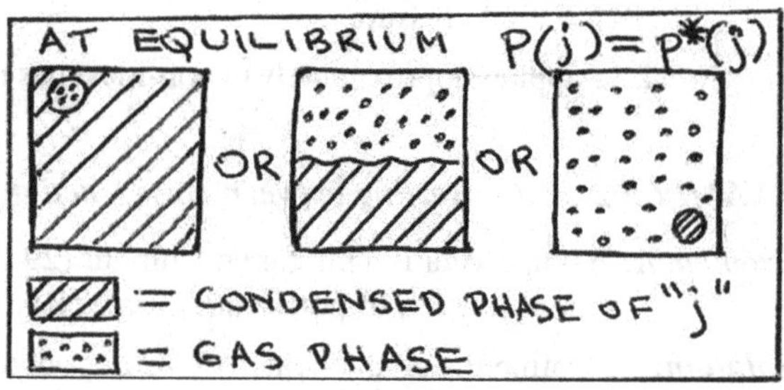

Vapour pressures of many substances can be estimated by empirical *equations*[14] such as the Antoine equation:

$$p^*(j) = \exp[A(j) - B(j) / (T + C(j))] \quad \text{(Antoine equation)} \qquad \textit{Equation 2.32}$$

[13] The normality (N) is the gram equivalents of solute per litre of solution, where the solute equivalent weight is defined relative to the type of reaction in which the solute is used, for example, an acid/base neutralization or an oxidation/reduction reaction.

[14] Vapour pressure can also be estimated (less accurately) by the Clausius-Clapeyron equation: $\ln(p^*_1/p^*_2) = (h_v/R)(1/T_2 - 1/T_1)$.

where: $p^*(j)$ = vapour pressure of pure substance j kPa(abs)
T = temperature K
A(j), B(j), C(j) = Antoine constants for substance j –, K, K

The Antoine equation defines a line that is the locus of the unique set of points where a pure liquid and its vapour can co-exist at equilibrium. *Table 2.10* shows values for the Antoine constants and the vapour pressure of some common substances.

The vapour pressures of solids are much lower than those of liquids, typically less than 0.1 kPa(abs) at temperatures below the melting point. However some solids such as camphor, iodine and naphthalene do exert substantial vapour pressure (e.g. up to 10 kPa(abs)) at temperatures below the melting point.

The phase relations for mixtures of substances (i.e. multi-species systems) range from simple to complex. In mixtures of mutually insoluble compounds (e.g. L/L dispersions or emulsions) each liquid exerts its own vapour pressure independent of others.

Table 2.10. Vapour pressure of some common substances.

Condensed phase	T_m	Antoine constants#			Temperature K			
		A	B	C	250	300	500	3000
	K	–	K	K	Vapour pressure kPa(abs)			
Acetone (liq)	178	14.7171	2975.95	-34.5228	2.5	33.4	4121	–
Ethanol (liq)	159	16.1952	3423.53	-55.7152	0.24	8.9	4863	–
n-Hexane (liq)	178	14.0568	2825.42	-42.7089	1.53	21.7	2639	–
Methanol (liq)	175	16.4948	3593.39	-35.2249	0.79	18.6	6395	–
n-Octane (liq)	216	14.2368	3304.16	-55.2278	0.07	2.1	905	–
Toluene (liq)	178	14.2515	3242.38	-47.1806	0.18	4.2	1201	–
Water (liq)	273	16.5362	3985.44	-38.9974	–	3.5	2673	–
Mercury (liq)	234	–	–	–	1.8E-6	3E-4	5.2	–
Water (solid)	273	–	–	–	0.08	–	–	–
Iodine (solid)	387	–	–	–	–	0.07	–	–
Tungsten (solid)	3655	–	–	–	–	–	–	9E-6
T_m = melting point temperature. # p^* = kPa(abs), T = K								

When liquids are mutually soluble they form homogeneous single-phase liquid systems, classified as either *"ideal"* or *"non-ideal"* mixtures. An *ideal* mixture is a mixture in which the molecules of component species do not interact with molecules of other component species. A *non-ideal* mixture involves interactions between component species that affect their chemical activity.

The vapour/liquid equilibrium for homogeneous *ideal mixtures* in the *condensed phase*[15] can be described by linear relations such as Raoult's law:

$$p(j) = x(j)\, p^*(j) \quad \text{(Raoult's law)} \qquad \textit{Equation 2.33}$$

[15] The *condensed phase* is the liquid or solid phase in contact with the gas.

where: p(j) = partial pressure of pure substance j in the gas phase kPa(abs)
p*(j) = vapour pressure of pure substance j at the specified temperature kPa(abs)
x(j) = mole fraction of substance j in ideal liquid phase mixture at equilibrium with the gas phase –

Many multi-component systems of practical interest involve homogeneous non-ideal mixtures. Over small concentration ranges (e.g. x(j) < 0.05) the phase equilibria of non-ideal mixtures can be approximated by linear relations, such as Henry's law for liquid/gas systems:[16]

$$\mathbf{p(j) = K_H x(j) = f(T)} \quad \text{(Henry's law)} \qquad \textit{Equation 2.34}$$

where:
K_H(j) = Henry's constant for solute substance j (in a specified solvent at a specified temperature) kPa
x (j) = mole fraction of substance j in (non-ideal) liquid phase mixture at equilibrium with the gas phase –

Values of Henry's constant are tabulated in various sources (see *Refs. 1,5*) and some of these values are given in *Table 2.11.*

EXAMPLE 2.11 Using Raoult's and Henry's laws to calculate composition in a gas/liquid systems.

A closed vessel contains hydrogen gas plus an ideal liquid mixture of 40 mol% n-hexane + 60 mol% n-octane in equilibrium with the gas phase at 400 K, with a total pressure P = 500 kPa(abs). Henry's constant for H_2 dissolution in the liquid mixture = 1E6 kPa.

Table 2.11. Henry's constant for some liquid–gas systems.

Solvent	Solute	Temperature K		
		298	333	363
		Henry's constant kPa		
Water(l)	O_2	4.4E6	6.5E6	7.1E6
Water(l)	N_2	8.7E6	12.1E6	12.7E6
Water(l)	H_2	7.1E6	7.7E6	7.5E6
Benzene(l)	H_2	0.37E6	–	–
Benzene(l)	CO_2	0.01E6	–	–

Problem:

Calculate the composition of the gas phase in the vessel and the concentration of H_2 in the liquid phase. [mol%.]

Solution:

Identify the components as: hydrogen: j = A, n-hexane: j = B, n-octane: j = C
Take Antoine constants from *Table 2.10*:
Vapour pressure n-hexane at 400 K = p*(B) = exp(14.0568 – 2825.42/(400 – 42.7089)) = 468 kPa(abs)
Vapour pressure n-octane at 400 K = p*(C) = exp(14.2368 – 3304.16/(400 – 55.2278)) = 105 kPa(abs)
Assume negligible effect of the hydrogen content of the liquid on the partial pressures of n-hexane and n-octane in the gas. [The low concentration of H_2 in the liquid (0.025 mol%) justifies the assumption.]

Then by Raoult's law (*Equation 2.33*):

p (B) = x(B)p*(B) = (0.4)(468 kPa) = 187 kPa(abs)
p (C) = x(C)p*(C) = (0.6)(105 kPa) = 63 kPa(abs)

[16] Henry's law is paricularly useful to calculate the solubility of *non-condensable gases* in liquids.

By Dalton's law (see *Equation 2.09*): p (A) = P – (p(B) + p(C)) = 500 – (187 + 63) = 250 kPa(abs)
By Henry's law (see *Equation 2.34*): p (A) = K_Hx(A)
x (A) = p(A)/K_H = 250 kPa/1E6 kPa = 2.5E-4

Component	Liquid phase	mol%	Gas phase	mol%
H_2	–	**0.03**	250 kPa/500 kPa = 0.50	**50**
n-hexane	–	**39.99**	187 kPa/500 kPa = 0.37	**37**
n-octane	–	**59.98**	63 kPa/500 kPa = 0.13	**13**

Non-ideal mixtures in general show complex non-linear behaviour that is dealt with by sophisticated thermodynamic calculations beyond the scope of this text (see *Refs. 5, 11*). Empirical data are tabulated for common non-ideal mixtures (see *Ref. 1*) and will be used in some examples in this text.

Bubble-point and dew-point

The *bubble-point* is the condition of pressure and temperature at which a liquid mixture just begins to "boil" (i.e. first bubbles of vapour form). If the pressure is decreased at a fixed temperature then the pressure at which the first bubbles of vapour form is the *bubble-point pressure* at that temperature. If the temperature is increased at a fixed pressure then the temperature at which the first bubbles of vapour form is the *bubble-point temperature,* at that pressure.
For an ideal liquid mixture [17] the bubble point condition can be calculated from *Equation 2.35:*

$$1 = \Sigma[x(j)p^*(j)] / P \qquad \textit{Equation 2.35}$$

where: P = total pressure kPa(abs)
$p^*(j)$ = f (T_B) = vapour pressure of component "j" kPa(abs)
T_B = bubble-point temperature K
x(j) = mole fraction of component "j" in the liquid mixture –

The *dew-point* is the condition of pressure and temperature at which a vapour mixture forms the first drops of liquid (i.e. dew drops). If the pressure is raised at a fixed temperature then the pressure at which liquid drops first form is the *dew-point pressure*. If the temperature is lowered at a fixed pressure then the temperature at which liquid drops first form is the *dew-point temperature*. When the condensate is an ideal liquid mixture [17] the dew-point condition can be calculated from *Equation 2.36*:

$$1 = P(\Sigma[y(j)/p^*(j)]) \qquad \textit{Equation 2.36}$$

where: P = total pressure kPa(abs)
y(j) = mole fraction of component "j" in the vapour mixture –
$p^*(j)$ = f (T_D) = vapour pressure of component "j" kPa(abs)
T_D = dew-point temperature K

[17] *Equations 2.35* and *2.36* come from Raoult's law, with respectively: Σ [y(j)] = 1 and Σ [x(j)] = 1.

In *Equations 2.35* and *2.36*, if the total pressure is fixed, the Antoine equation (*Equation 2.32*) can be inserted as the vapour pressure function $f(T_B)$ or $f(T_D)$ and the resulting non-linear equations solved for T_B or T_D. When some components of the liquid mixture have very low vapour pressures (relative to the others) the bubble-point calculation is simplified by setting their vapour pressures as zero in *Equation 2.35*. For example, the bubble point of a solution of sugar (sucrose) in water is determined only by the mole fraction of water in the solution. Similarly, when some components of a gas are non-condensable[18] or have a very high vapour pressure (relative to the others) then the dew-point calculation is simplified by setting their vapour pressures at infinity in *Equation 2.36*. For example, in the case of humid air the vapour pressures of N_2 and O_2 are set at infinity and *Equation 2.36* becomes:

$$1 = P(y_w / p_w^*) \qquad \textit{Equation 2.37}$$

where:			
	P	= total pressure	kPa(abs)
	y_w	= mole fraction of water vapour in humid air	–
	p_w^*	= $f(T_D)$ = vapour pressure of water	kPa(abs)

Equation 2.37 is also useful to find the dew-point for combustion product gas mixtures of N_2, O_2, CO_2 and H_2O.

NOTE: For water mixed with "non-condensable" gases, *Equation 2.37* is easily "solved" by looking up the vapour pressure of water in a *steam table* (*Table 2.20*).

[18] Non-condensable gases are those above their critical temperature.

EXAMPLE 2.12 Calculating the bubble-point and dew-point conditions of ideal mixtures.

Problem: Calculate

A. The bubble-point temperature of a mixture of 20 mol% n-hexane + 80 mol% n-octane under a total pressure = 300 kPa(abs).
B. The dew-point pressure of a combustion exhaust gas with 10 vol% H_2O, 15 vol% CO_2, 73 vol% N_2, 2 vol% O_2 at 400 K.

Solution:

A. Substitute values to the bubble-point *Equation 2.35*: $1 = \Sigma [x(j)p^*(j)] / P$
$300 = (0.20) \exp(14.0568 - 2825.42 / (T_B - 42.7089)) + (0.8) \exp(14.2368 - 3304.16 / (T_B - 55.2278))$
Solve this non-linear equation for T_B (e.g. by Excel "Solver") to give: T_B = **435 K**

B. Substitute values into the dew-point *Equation 2.36*: $1 = P(\Sigma [y(j) / p^*(j)])$
Note that for each component CO_2, N_2 and O_2 $p^*(j) >> p^*(H_2O)$
[In fact CO_2, N_2 and O_2 are *non-condensable* at 400 K]
Then: $1 \approx P_D[0.1/\exp(16.5362 - 3985.44 / (400 - 38.9974)) + 0.15/\infty + 0.73/\infty + 0.02/\infty]$
Solve this equation to give: P_D = **2440 kPa(abs)**

Note that the same value of P_D is obtained by examination of the saturated liquid/vapour data in the steam table (*Table 2.20*). A similar approach can be used to find the dew-point temperature under a given pressure.

GAS–SOLID SYSTEMS

Surface concentration

Gases are adsorbed on solid surfaces to form mono- or multi-molecular layer adsorbate films. The equilibrium between a species adsorbed on a solid surface and its gas is represented by an "adsorption isotherm" such as:

$$\Gamma = aK_A p(j) / (1 + K_A p(j))$$ (Langmuir adsorption isotherm (fixed temperature)) *Equation 2.38*

where: a = empirical constant (relates surface coverage to surface concentration) $kmol.m^{-2}$
Γ = equilibrium surface concentration of the adsorbed species (i.e. the adsorbate) $kmol.m^{-2}$
K_A = empirical adsorption constant kPa^{-1}
$p(j)$ = equilibrium partial pressure of the adsorbate in the gas phase kPa(abs)

Adsorption isotherms are used, for example, in calculation of the distribution of adsorbed species between gases or liquids and microporous solids (e.g. activated carbon, silica gel, molecular sieves) that have specific surface areas of the order 10E3 to 1000E3 m^2/kg. These adsorption equilibria are important in the design of processes such as solid catalysed gas phase reactions (e.g. synthesis of ammonia) and gas separations (e.g. pressure swing adsorption to remove nitrogen from air).

Liquid–liquid systems

Distribution coefficient

Liquid–liquid (L–L) systems typically consist of two mutually insoluble liquids (like oil and water) in contact with each other. The distribution of a solute species between the two liquids at equilibrium is determined by the value of the distribution coefficient.

The distribution coefficient (also called the "partition coefficient") is defined as:

$$\mathbf{D(j) = x(j)_A / x(j)_B = f(T)} \qquad \textit{Equation 2.39}$$

where: D(j) = liquid–liquid distribution coefficient for species j –
$x(j)_A$ = mole fraction of solute species j in liquid A at equilibrium –
$x(j)_B$ = mole fraction of solute species j in liquid B at equilibrium –

Values of the distribution coefficient for many systems are tabulated in several sources (see *Ref. 1*) and *Table 2.12* gives values of the distribution coefficients for some common L–L systems.

Table 2.12. Distribution coefficients for some common systems.

Solute	Liquid A	Liquid B	Temperature K					
			273	298	303	313	333	343
			Distribution coefficient (dimensionless)					
Acetic acid	benzene	water	–	0.033	0.098	0.102	0.064	–
Ethanol	ethyl acetate	water	0.026	0.50	–	–	–	0.455
Methanol	n-butanol	water	0.600	–	0.510	–	0.682	–
Toluene	aniline	n-heptane	0.577	–	–	0.425	–	–

Liquid–liquid (L–L) distribution calculations are used in the design of liquid–liquid separation (a.k.a. liquid extraction or solvent extraction) processes, such as the recovery of penicillin from fermentation liquor (*Fig. 3.01*), the separation of aromatics (e.g. benzene and toluene) from petroleum and the reprocessing of spent fuel from nuclear reactors to recover unused uranium, plutonium and other isotopes.

Liquid–solid systems

Solubility

The solubility of a material (the solute) is the amount of that material that is dissolved in a saturated solution with a specified solvent. Solubility values vary over a wide range and are expressed in several ways. One common way of expressing solubility is given in *Equation 2.40.*

$$\mathbf{S'(j) = m(j) / m_s = f(T)} \qquad \textit{Equation 2.40}$$

where: $S'(j)$ = solubility of solute species j in solvent — kg solute (kg.solvent)$^{-1}$
$m(j)$ = mass of solute in the saturated solution — kg
m_s = mass of solvent in the saturated solution — kg

Solubility data are tabulated in sources such as *Refs. 1–2*. *Table 2.13* gives values of solubility for some common materials in water. The solubility of gases in liquids (at fixed pressure) generally decreases with increasing temperature (cf. *Table 2.11*). The solubility of most solids increases with temperature, although a few solids (e.g. $Ca(OH)_2$) show a negative temperature coefficient of solubility. Note that these tabulated solubility values are for *saturated* solutions under equilibrium conditions at the specified temperature. By suppressing crystal nucleation it is possible to obtain supersaturated solutions in which the concentration of solute exceeds the tabulated solubility value and such solutions are fairly common in practical processes.

Table 2.13. Solubility of some common materials in water.

Solute	Solvent	Temperature K					
		273	293	313	333	353	373
		Solubility kg solute. (kg solvent)$^{-1}$					
O_2 (1atm)	Water	70E-6	44E-6	33E-6	28E-6	26E-6	25E-6
CO_2 (1atm)	Water	3.35E-3	1.69E-3	0.97E-3	0.58E-3	–	0.00
NH_3 (1atm)	Water	0.88	0.53	0.32	0.18	–	–
$CaCO_3$	Water	–	12E-6	–	–	–	20E-6
$Ca(OH)_2$	Water	1.85E-3	1.65E-3	1.41E-3	1.16E-3	0.94E-3	0.77E-3
$CaCl_2$	Water	0.595	0.745	–	1.37	1.47	1.59
NaCl	Water	0.357	0.360	0.366	0.373	0.384	0.398

CHEMICAL EQUATIONS AND STOICHIOMETRY

Chemical equations represent the transformation of species in a chemical reaction. They are conventionally written with the reactants (initial species) on the left hand side (LHS) and the products (final species) on the right hand side (RHS):

$$\alpha A + \beta B \rightarrow \chi C + \delta D \qquad \textit{Reaction 2.01}$$

In *Reaction 2.01* the reactants are species A and B, the products are species C and D. The numbers α, β, χ and δ are the stoichiometric coefficients of the reaction. *Reaction 2.01* shows that:

α molecules of species A reacts with β molecules of species B to generate χ molecules of species C plus δ molecules of species D.

In terms of mole[19] quantities *Reaction 2.01* means that:

α moles A + β moles B react to generate χ moles + δ moles D

---

[19] A *mole* is the amount of a substance containing the same number of "*elementary particles*" as there are atoms in 0.012 kg of carbon 12. (i.e. Avogadro's number = 6.028E23 particles.) The number of moles in a given quantity of a molecular species (or element) is its mass divided by its molar mass, i.e. n = m / M. For molecular and ionic compounds in chemical processes (e.g. H_2O, H_2, O_2, CH_4, NaCl) the "*elementary particles*" are molecules. For unassociated elements and for ions the "elementary particles" are atoms (e.g. C, Na^+, Cl^-) or charged groups of atoms (e.g. NH_4^+, ClO^-). Refer to a basic chemistry text for more comprehensive information of atoms, ions, molecules and chemical bonding.

In terms of mass quantities *Reaction 2.01* means that:

α M(A) kg of A + β M(B) kg of B react to generate χ M(C) kg of C + δ M(D) kg of D

where M(j) = molar mass (molecular weight) of species j   kg $kmol^{-1}$

Chemical reactions transform molecules, but the atoms that constitute the molecules are conserved. A properly written chemical equation must thus be *balanced* with respect to its atoms, i.e. for each element:

Number of atoms on LHS = Number of atoms on RHS

Also, the conservation of mass requires that for a balanced chemical reaction:

$$\alpha\ \mathbf{M(A)} + \beta\,\mathbf{M(B)} = \chi\,\mathbf{M(C)} + \delta\,\mathbf{M(D)} \qquad \textit{Equation 2.41}$$

or more generally:

$$\Sigma\,(\nu\,(j)M(j)) = 0 \qquad \textit{Equation 2.42}$$

where: ν (j) = stoichiometric coefficient of species j, with a (–) sign for reactants and a (+) sign for products.

EXAMPLE 2.13 Calculating reaction stoichiometry for the oxidation of ammonia.

Problem: Calculate the mass of NH_3 consumed, the mass of NO produced and the mass of water produced when 320 kg of O_2 is consumed by the reaction:

$$4NH_3 + 5O_2 \rightarrow 4NO + 6H_2O \qquad \textit{Reaction 1}$$

Solution:

NH_3 consumed:
= (320 kg O_2 / 32 kg O_2/kmol O_2)(4 kmol NH_3/5 kmol O_2)(17 kg NH_3 / kmol NH_3) = **<u>136 kg NH_3</u>**

NO produced:
= (320 kg O_2 / 32 kg O_2/kmol O_2)(4 kmol NO/5 kmol O_2)(30 kg NO / kmol NO) = **<u>240 kg NO</u>**

H_2O produced:
= (320 kg O_2 / 32 kg O_2/kmol O_2)(6 kmol H_2O/5 kmol O_2)(18 kg H_2O / kmol H_2O) = **<u>216 kg H_2O</u>**

Check the overall mass balance:

Mass INITIAL = 320 kg O_2 + 136 kg NH_3 = 456 kg
Mass FINAL = 240 kg NO + 216 kg H_2O = 456 kg   OK

NOTE: In this example the molar mass values are rounded for simplicity. The more accurate values are:

O_2 = 32.000 NH_3 = 17.030   NO = 30.006 H_2O = 18.016 kg/kmol

LIMITING REACTANT

When reactants are mixed in ratios corresponding to their stoichiometric coefficients they are said to be in *stoichiometric proportions*. If reactants are not mixed in *stoichiometric proportions* then the reactant present in the smallest relative amount will determine the maximum *extent of reaction*. That reactant is called the *limiting reactant* and the remaining reactants are called *excess reactants*. *Example 2.14* shows how to find the limiting reactant.

EXAMPLE 2.14 Calculating the limiting reactant in the combustion of ethane.

A batch of 6 kmol of C_2H_6 is burned with 14 kmol of O_2 and the limiting reactant is completely consumed by the reaction:

$2C_2H_6(g) + 7O_2(g) \rightarrow 4CO_2(g) + 6H_2O(g)$ *Reaction 1*

Problem: A. What is the limiting reactant?
B. Calculate the composition of the reaction product mixture. [mole%.]

Solution:

A. Stoichiometric O_2 required for 6 kmol $C_2H_6(g)$ = (6 kmol $C_2H_6(g)$)(7 kmol O_2/2 kmol $C_2H_6(g)$) = 21 kmol O_2
O_2 supplied = 14 kmol O_2 < 21 kmol O_2 required i.e. **limiting reactant = O_2**

B. By mole balances on the closed system: ACC = IN – OUT + GEN – CON IN = OUT = 0

			mol%
Balance on O_2	ACC = Final – Initial = Final – 14 = GEN – CON = 0 – 14	Final O_2 = 0 kmol	**0.0**
Balance on C_2H_6	ACC = Final – Initial = Final – 6 = GEN – CON = 0 – (14)(2/7)	Final C_2H_6 = 2 kmol	**9.1**
Balance on CO_2	ACC = Final – Initial = Final – 0 = GEN – CON = (14)(4/7) – 0	Final CO_2 = 8 kmol	**36.4**
Balance on H_2O	ACC = Final – Initial = Final – 0 = GEN – CON = (14)(6/7) – 0	Final H_2O = 12 kmol	**54.5**
		Total = 22 kmol	**100.0**

Check the overall mass balance:
Mass INITIAL = 6 kmol C_2H_6 (30 kg/kmol) + 14 kmol O_2 (32 kg/kmol) = **628 kg**
Mass FINAL = 2 kmol C_2H_6 (30 kg/kmol) + 0 kmol O_2 (32 kg/kmol)
+ 8 kmol CO_2 (44 kg/kmol) + 12 kg H_2O (18 kg/kmol) = **628 kg**

CONVERSION, EXTENT OF REACTION, SELECTIVITY AND YIELD

In real chemical processes the progress of a reaction depends on the chemical equilibria and reaction rates. Reactions may not go to completion and reactants may engage in secondary reactions to give undesired products. The terms "conversion", "extent of reaction", "selectivity" and "yield" are used to specify the progress of the reaction. You will see various definitions of some of these terms in different places, so be clear on the definitions when you use these terms.

Conversion **X(j)** = $\dfrac{\textbf{amount of reactant species j converted to all products}}{\textbf{amount of reactant species j introduced to the reaction}}$ *Equation 2.43*

Extent of reaction $\varepsilon(\ell)$ = $\dfrac{\textbf{amount of species j converted in a specified reaction } \ell}{\textbf{stoichiometric coefficient of species j in the specified reaction } \ell}$ *Equation 2.44*

Selectivity **S(j, q)** = $\dfrac{\textbf{amount of reactant species j converted to product species q}}{\textbf{amount of reactant species j converted to all products}}$ *Equation 2.45*

Yield **Y(j, q)** = $\dfrac{\textbf{amount of reactant species j converted to product species q}}{\textbf{amount of reactant species j introduced to the reaction}}$ *Equation 2.46*

where: X(j) = conversion of reactant species j — dimensionless
$\varepsilon(\ell)$ = extent of reaction for the specified reaction ℓ (independent of species) — kmol
S(j, q) = selectivity for product species q from reactant species j — dimensionless
Y(j, q) = yield of product species q from reactant species j — dimensionless

Note that by the above definitions the conversion must be defined relative to a specific reactant and the extent of reaction refers to a specific reaction, while the selectivity and yield must all be defined relative to a specific reactant and a specific product. Then $X(j) \leq 1$, $S(j, q) \leq 1$, $Y(j, q) \leq 1$ and

$$\mathbf{Y(j, q) = X(j)\,S(j, q)} \qquad \textit{Equation 2.47}$$

EXAMPLE 2.15 Conversion, extent of reaction, selectivity and yield from simultaneous reactions.

Hydrogen (H_2) can react with oxygen (O_2) by *Reaction 1* to produce hydrogen peroxide (H_2O_2) or by *Reaction 2* to produce water (H_2O).

$$H_2 + O_2 \rightarrow H_2O_2 \qquad \textit{Reaction 1}$$

$$2H_2 + O_2 \rightarrow 2H_2O \qquad \textit{Reaction 2}$$

Reactions 1 and 2 are competitive reactions that occur in parallel when hydrogen reacts with oxygen on a special catalyst. There are no other reactions.

The initial reaction mixture (before reaction) contains: 5 kmol H_2 + 3 kmol O_2
The final reaction mixture (after reaction) contains: 1 kmol H_2O_2 + 2 kmol H_2O + unreacted H_2 and O_2

Problem: Calculate the conversion of H_2, the extent of *Reaction 2*, the selectivity for H_2O_2 from H_2 and the yield of H_2O_2 from H_2.

Solution: Note that the initial reaction mixture does not have stoichiometric proportions of H_2 and O_2 for either *Reaction 1* or *Reaction 2.*

H_2 converted in *Reaction 1* = (1 kmol H_2O_2)(1 kmol H_2/kmol H_2O_2) = 1 kmol
H_2 converted in *Reaction 2* = (2 kmol H_2O)(1 kmol H_2/kmol H_2O) = 2 kmol
Total H_2 converted = **3 kmol**

O_2 converted in *Reaction 1* = (1 kmol H_2O_2)(1 kmol O_2/kmol H_2O_2) = 1 kmol
O_2 converted in *Reaction 2* = (2 kmol H_2O)(0.5 kmol O_2/kmol H_2O) = 1 kmol
Total O_2 converted = **2 kmol**

Conversion of H_2 = H_2 converted / H_2 initial = 3 kmol / 5 kmol = 0.60 ≡ **60%**

Extent of *Reaction 2* = moles H_2 converted in *Reaction 2* / stoich. coeff. of H_2 in *Reaction 2*
= 2 kmol / 2 = **1 kmol**

Selectivity for H_2O_2 from H_2 = H_2 converted to H_2O_2 / Total H_2 converted = 1 kmol / 3 kmol = 0.33 ≡ **33%**

Yield of H_2O_2 from H_2 = H_2 converted to H_2O_2 / H_2 initial = 1 kmol / 5 kmol = 0.20 ≡ **20%**

or

Yield of H_2O_2 from H_2 = (Conversion of H_2)(Selectivity for H_2O_2 from H_2) = (0.60)(0.33) = 0.20 ≡ **20%**

EQUILIBRIUM CONVERSION AND ACTUAL CONVERSION

Many chemical reactions are *reversible*[20] and *tend to* an equilibrium state defined by the temperature dependent equilibrium constant.

$$\alpha A + \beta B \leftrightarrow \chi C + \delta D \qquad \textit{Reaction 2.01}$$

$$K_{eq} = (a_C^{\chi} a_D^{\delta})/(a_A^{\alpha}/a_B^{\beta}) = f(T) \quad \text{(reaction equilibrium constant)} \qquad \textit{Equation 2.48}$$

where:

K_{eq}	=	reaction equilibrium constant = $\exp(-\Delta G^o_T/RT)$	– (dimensionless)
a(j)	=	activity of species j in the reaction mixture *at equilibrium*	– (dimensionless)
ΔG^o_T	=	f (T) = standard free energy change for the reaction at temperature T	$kJ.kmol^{-1}$
R	=	gas constant	$kJ.(kmol.K)^{-1}$
T	=	reaction temperature	K

For most calculations the activities in *Equation 2.48* can be approximated as follows:

Reactions in the solid phase: $a(j) \approx x(j)$ *Equation 2.49*

Reactions in the liquid phase: $a(j) \approx [J] / (1\ kmol.m^{-3})$ *Equation 2.50*

[20] A *reversible* reaction is one that can go in both directions, shown by the two-way arrow ↔. Strictly all chemical reactions are reversible, but for many the equilibrium conversion is near 100% (i.e. >99%) so these are usually described as irreversible reactions and indicated by a one-way arrow →.

Reactions in the gas phase: $\boldsymbol{a(j) \approx p(j) / (101.3\ kPa)}$ *Equation 2.51*

where: [J] = concentration of species j kmol.m^{-3}
p(j) = partial pressure of species j kPa(abs)
x(j) = mole fraction of species j –

Note that:

A. The denominators in *Equations 2.50* and *2.51* refer to the species' *standard state*[21] and have the effect of making the activity a dimensionless number. Consequently, the equilibrium constant is also a dimensionless number.
B. The equilibrium constant is a strong function of temperature but is nearly independent of pressure.
C. For endothermic reactions ΔG^o_T decreases and K_{eq} increases with rising temperature.
For exothermic reactions ΔG^o_T increases and K_{eq} decreases with rising temperature.

The *equilibrium conversion* is the hypothetical conversion that would be obtained if the reaction reached equilibrium at a specified temperature. However all reactions proceed at finite net rates that tend to zero as the mixture approaches equilibrium, and thus can never reach true equilibrium. The *actual conversion* is the conversion obtained in reality, as determined by the reaction rate and residence time in the reactor. The actual conversion is always less than the equilibrium conversion, i.e.:

$$\mathbf{X_{act} / X_{eq} < 1} \qquad \textit{Equation 2.52}$$

where: X_{act} = actual conversion
X_{eq} = equilibrium conversion

In practice, chemical reactions can be roughly divided into four categories:

1. **High equilibrium constant + high**[22] **rate** e.g. acid-base neutralisation at 293 K, actual conversion of limiting reactant exceeds 99% within milliseconds.

2. **High equilibrium constant + low rate** e.g. fermentation of glucose at 303 K, actual conversion of limiting reactant reaches 99% in several days.

3. **Low equilibrium constant + high rate** e.g. ammonia synthesis at 700 K, $X_{eq} \approx 0.2$, actual conversion of limiting reactant exceeds 99% of X_{eq} within seconds.

4. **Low equilibrium constant + low rate** e.g. esterification at 293 K, $X_{eq} \approx 0.5$, actual conversion of limiting reactant reaches 99% of X_{eq} in several minutes.

The rate of reaction nearly always increases with temperature (there are very few exceptions) and as a rule of thumb the reaction rate roughly doubles for each 10 K rise in temperature.

The calculation of actual conversion in a chemical reaction requires a knowledge of reaction kinetics plus heat and mass transfer dynamics that is generally beyond the scope of this text (see *Refs. 1, 9–10*). A simple case showing the comparison between equilibrium conversion and actual conversion in a batch reactor is given here in *Example 2.16*. *Example 7.03* takes a more detailed look at the time dependent conversion in a batch reactor.

---

[21] The *standard state* of a species is its normal state at 101.3 kPa(abs) and (usually) 298.15 K [some texts define the standard pressure as 1 bar = 100.0 kPa(abs), but this difference has negligible effect in practical calculations].

[22] The terms "fast rate" and "slow rate", often seen in technical writing, are grammatically incorrect. You should not use them.

EXAMPLE 2.16 Comparing the equilibrium conversion with the actual conversion in a chemical reaction.

The reversible liquid phase *Reaction 2.07* is carried out at 400 K in an isothermal batch reactor with constant volume.

A(l) ↔ B(l) The reactor initially contains a solution of A with zero B *Reaction 2.07*

The equilibrium constant for *Reaction 2.07* at 400 K is $K_{eq} = 5.0$ and the reaction rates[23] at 400 K are:

Forward rate = $d[A]/dt = -k_1[A]$

Reverse rate = $d[B]/dt = k_2[B]$

where: k_1 = forward reaction rate constant = 0.03 s^{-1}

k_2 = reverse reaction rate constant = k_1/K_{eq} = 0.006 s^{-1}

$[A], [A]_o, [A]_{eq}$ = concentration of A at time t, at time zero, at equilibrium $kmol.m^{-3}$

$[B], [B]_o, [B]_{eq}$ = concentration of B at time t, at time zero, at equilibrium $kmol.m^{-3}$

t = time s

Note that *at equilibrium* (t = ∞): Forward rate = Reverse rate. Thus: $K_{eq} \approx [B]_{eq}/[A]_{eq} = k_1/k_2$

Problem: Calculate:

A. X_{eq} = the equilibrium conversion of A [%]

B. X_{act} = the actual conversion of A at reaction time = 10 seconds [%]

Solution:

A. The *equilibrium conversion* at 400 K depends only on the value of K_{eq} at 400 K.

Define the system = contents of batch reactor (a closed system)

Specify the quantity = (i) total mass (ii) moles of A

(i) Integral balance on total mass:

ACC = Final – Initial = IN – OUT + GEN – CON

$V_R(M_A[A] + M_B[B]) - V_R M_A[A]_o = 0 - 0 + 0 - 0$ [mass is conserved]

Since $M_A = M_B$ $[A] + [B] - [A]_o = 0$ [1]

M_A, M_B = molar mass of A, B $kg.kmol^{-1}$ V_R = reaction volume m^3 (specified as constant)

(ii) Integral balance on moles of A:

$V_R([A]_{eq} - [A]_o) = 0 - 0 + 0 - X_{eq} V_R [A]_o$ $X_{eq} = ([A]_o - [A]_{eq}) / [A]_o$ [2]

From equation [1]: $[B]_{eq} = [A]_o - [A]_{eq} = X_{eq}[A]_o$

Equilibrium constant: $K_{eq} = [B]_{eq}/[A]_{eq} = X_{eq}[A]_o / (1 - X_{eq})[A]_o = X_{eq} / (1 - X_{eq})$ [3]

(*Equations 2.48* and *2.50*) Substitute $K_{eq} = 5.0$ Solve equation [3] for: $X_{eq} = 0.83 \equiv$ **83%**

---

[23] Reaction rate is the rate of consumption of a reactant or generation of a product with respect to time. This rate depends on the instantaneous concentration(s) of reactive species, the temperature and the effect of any catalyst that may be present.

B. The *actual conversion* at 400 K depends on the reaction rate and the residence time:
Define the system = contents of batch reactor (a closed system)
Specify the quantity = amount (moles) of A
Differential balance on moles of A:
Rate ACC = Rate IN – Rate OUT + Rate GEN – Rate CON
$V_R d[A]/dt = 0 - 0 + V_R k_2[B] - V_R k_1[A]$
Substitute equation [1]:
$d[A]/dt = -[A]_o d(X_{act})/dt = k_2[B] - k_1[A] = k_2[A]_o X_{act} - k_1[A]_o(1 - X_{act})$
where: $X_{act} = ([A]_o - [A])/[A]_o$
$dX_{act}/dt = k_1 - (k_1 + k_2)X_{act}$ [4]
Substitute values: $dX_{act}/dt = 0.03 - (0.036)X_{act}$ Limits: $X_{act} = 0$ at $t = 0$
Integrate and solve for X_{act}
$X_{act} = 0.83[1 - \exp(-0.036\, t)]$ [5]
When t = 10 s $X_{act} = 0.25 \equiv$ **25%**

Note that $X_{act} \to X_{eq}$ only when $t \to \infty$ (infinity) i.e. complete equilibrium cannot be reached in a finite time!

THERMOCHEMISTRY

Each molecule of any species contains chemical energy. When a chemical reaction occurs this chemical energy is redistributed among the reactant and product species and partially converted to heat. The study of such thermal effects in chemical systems is called "thermochemistry". Thermochemistry provides several measures that are useful in the calculation of energy balances.[24]

Table 2.14. Thermochemical values*.

Substance	Phase	$h_f^o{}_{,298K}$	$h_c^o{}_{,298K}$	$C_p^o{}_{298K}$	$h_{m,Tm}$	$h_{v,Tb}$	$h_{v,298K}$	T_c
		kJ.kmol^{-1}	kJ.kmol^{-1}	kJ.kmol^{-1}.K^{-1}	kJ.kmol^{-1}	kJ.kmol^{-1}	kJ.kmol^{-1}	K
			[Gross]					
H_2	G	0.0	-285.8E3	28.8	117	903	–	33.3
O_2	G	0.0	0.0	29.4	443	6.8E3	–	154.4
H_2O	G	-241.8E3	0.0	33.6	–	–	–	647.4
H_2O	L	-285.8E3	0.0	75.3	–	40.6E3	44.0E3	647.4
H_2O	S	-291.8E3	0.0	37 (273K)	6.00E3	–	–	647.4
CO_2	G	-393.5E3	0.0	37.1	8.36E3	25.2E3	–	304.2
CH_4	G	-74.8E3	-889.5E3	35.3	936	8.1E3	–	190.7
CH_3OH	L	-238.7E3	-726.6E3	81.6	3.16E3	34.8E3	37.8E3	513.2
C_6H_6	L	+49E3	-3264E3	136	9.91E3	30.7E3	33.8E3	562.6

* *The superscript "o" means at the standard state pressure of 101.3 kPa(abs).*
T_m = normal melting point, T_b = normal boiling point, T_c = critical temperature.

[24] Many sources quote these values in units of kJ.mol^{-1}. To preserve dimensional consistency this text uses units of kJ.kmol^{-1} (see *Table 2.01*).

Heat (i.e. enthalpy) of formation

The heat of formation of a compound (h_f) is the heat absorbed by a reaction mixture (i.e. the heat input to the mixture) when that compound is formed from its elements, at a specified condition of temperature, pressure and phase. Heat of formation values for many compounds are tabulated in sources such as *Refs. 1–4*. By convention these tables give the heat of formation of each compound in the *standard state* at 298.15 K, 101.3 kPa(abs). The effect of pressure on the heat of formation is small, but the effect of temperature is large — so the temperature is specified in the subscript, e.g. $h_{f,\ 298K}$. Values of the heat of formation of some common substances are given in *Table 2.14*.

Some sources designate the heats (i.e. enthalpies) of combustion, formation, reaction, etc. with a "Δ", e.g. $\Delta h_{f,298K}^{o}$, to indicate that the quantity is a change in enthalpy relative to a reference state. That symbolism is not used in this text.

Heat (i.e. enthalpy) of reaction

The heat of reaction (h_{rxn}) is the heat absorbed by a reaction mixture (i.e. the heat input to the mixture) when the reaction occurs, with complete conversion of the reactants in stoichiometric proportions, at a specified condition of temperature, pressure and phase.

$$h_{rxn} = \Sigma\,(\nu(j)h_f(j))\ \text{products} - \Sigma\,(\nu(j)h_f(j))\ \text{reactants} \qquad \textit{Equation 2.53}$$

where: h_{rxn} = heat of reaction — $kJ.kmol^{-1}$

$h_f(j)$ = heat of formation of species j at the specified conditions — $kJ.kmol^{-1}$

$\nu(j)$ = stoichiometric coefficient of species j (in the balanced reaction) — -

$\nu(j)$ is taken as positive for reactants and for products in *Equation 2.53*

Note that the units of h_{rxn} can cause confusion. When the units are given as kJ per kmol the question is: "Per kmol of what?" One way to resolve this ambiguity is to state h_{rxn} per kmol of a specific reactant or product, as shown in *Example 2.17*.

EXAMPLE 2.17 Calculating the standard heat of reaction in the oxidation of hydrogen by oxygen.

Problem: Calculate the standard heat of reaction for *Reaction 2.01*.

Note that the phase of each species is shown in brackets.

$$2H_2(g) + O_2(g) \rightarrow 2H_2O(g) \qquad \textit{Reaction 2.01}$$

Solution: By *Equation 2.53*, $h_{rxn} = \Sigma\,(\nu(j)h_f(j))\ \text{products} - \Sigma\,(\nu(j)h_f(j))\ \text{reactants}$ *Equation 2.53*

For the reaction as written, with data from *Table 2.14*:

$$h_{rxn,\ 298K} = (2)(-242E3\ kJ.kmol^{-1}) - [(2)(0\ kJ.kmol^{-1}) + (1)(0\ kJ.kmol^{-1})] = \underline{-484E3\ kJ.(kmol\ rxn)^{-1}}$$

For a specific reactant (e.g. H_2):

$$h_{rxn,\ 298K} = -484E3\ kJ.(kmol\ rxn)^{-1} / (2\ kmol\ H_2 / kmol\ rxn) = \underline{-242E3\ kJ.(kmol\ H_2)^{-1}}$$

The algebraic sign of h_{rxn} shows whether the reaction is *endothermic* or *exothermic.*
An *endothermic* reaction (h_{rxn} = positive) requires heat input to the reaction mixture to keep the temperature at 298 K as the reaction occurs.
An *exothermic* reaction (h_{rxn} = negative) requires heat output from the reaction mixture to keep the temperature at 298 K as the reaction occurs.

HEAT (I.E. ENTHALPY) OF COMBUSTION

The heat of combustion of a fuel (h_c) is the heat of reaction for the *complete combustion* of the fuel in oxygen (or air), at specified conditions of temperature, pressure and phase. For hydrocarbon fuels *complete combustion* means conversion of all carbon to CO_2 and all hydrogen to H_2O, in a reaction such as *Example 1.03.* When fuels contain other elements such as sulphur (S) and nitrogen (N), their combustion products should be specified, for example, as SO_2 and N_2.

The *standard heat of combustion* is the heat of combustion at standard state 298 K, 101.3 kPa(abs) which is tabulated in sources such as *Refs. 1– 2, 4.* These tables usually give the *gross heat of combustion* as kJ (or kCal, where 1 kCal = 4.18 kJ) per mole of fuel.

The *gross heat of combustion* (-ve higher heating value) is the heat of reaction obtained when the product water is a liquid. The *net heat of combustion* (-ve lower heating value) is the heat of reaction obtained when the product water is a gas. The difference between the gross and net heats of combustion corresponds to the heat of vaporisation of the product water.

$$h_{c,net} = h_{c,gross} + n_w h_v \qquad \textit{Equation 2.54}$$

where:
$h_{c,gross}$ = gross heat of combustion kJ.kmol^{-1}
$h_{c,net}$ = net heat of combustion kJ.kmol^{-1}
n_w = amount of water produced in the reaction kmol(kmol fuel)$^{-1}$
h_v = heat of vaporisation of water kJ.kmol^{-1}

NOTE: The combustion of fuels is always an exothermic process, but beware that some sources (e.g. the *Chemical Engineers' Handbook*) reverse the sign of h_c and tabulate it with a positive value.

HEATS (ENTHALPIES) OF PHASE CHANGES

The energy inputs required to melt and to vaporise a substance are called respectively the *(latent) heat of fusion* and the *(latent) heat of vaporisation.*

Heat of fusion (melting) (h_m) is the heat input required to change the phase of a substance from solid to liquid at specified conditions of temperature and pressure.

Heat of vaporisation (h_v) is the heat input required to change the phase of a substance from a condensed phase (solid or liquid) to gas in equilibrium with the condensed phase at specified conditions of temperature and pressure. Unless otherwise stated the condensed phase is usually a liquid.

Values for the heat of fusion and heat of vaporisation are tabulated in sources such as *Refs 1, 2* and *4.* These values are usually given, respectively, at the *normal melting point* [25] and *normal boiling point* of the substance.

[25] The *normal* melting and boiling points are the melting and boiling temperatures under a pressure of 1 standard atmosphere, i.e. 101.3kPa(abs).

Heat of fusion is affected little by changes in temperature, but heat of vaporisation decreases with increasing temperature according to *Equation 2.55*, and becomes zero at the critical temperature.

$$h_{v,T} = h_{v,298K} + \int_{298\,K}^{T} (C_{p,g} - C_{p,l})dT \qquad \textit{Equation 2.55}$$

where:

$h_{v,T}$	= heat (enthalpy) of vaporisation at T	$kJ.kmol^{-1}$
$h_{v,298K}$	= heat (enthalpy) of vaporisation at 298	$kJ.kmol^{-1}$
$C_{p,g}$	= heat capacity of gas	$kJ.kmol^{-1}.K^{-1}$
$C_{p,l}$	= heat capacity of liquid	$kJ.kmol^{-1}.K^{-1}$

Note that as a rule: $C_{p,g} < C_{p,l}$ and h_v decreases as T increases, as exemplified for the case of water in *Figure 2.03*.

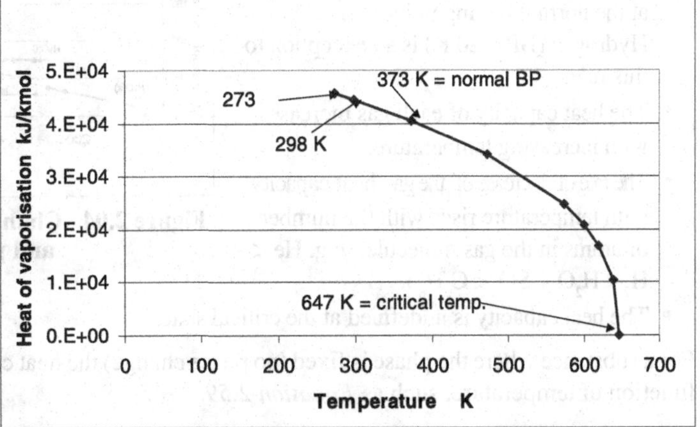

Figure 2.03. Effect of temperature on the heat of vaporisation of water.

HEAT CAPACITY [26]

The heat capacity of a substance is effectively the heat input required to raise the temperature of a specified amount of that substance by 1 degree, *without a change of composition or of phase*.

Heat capacity is defined more precisely by a pair of *partial differential equations*.[27]

$$C_v = [\partial u / \partial T]_V \quad \text{(no phase change, no reaction)} \qquad \textit{Equation 2.56}$$

$$C_p = [\partial h / \partial T]_P \quad \text{(no phase change, no reaction)} \qquad \textit{Equation 2.57}$$

where:

C_v	= heat capacity at constant volume	$kJ.kmol^{-1}.K^{-1}$
C_p	= heat capacity at constant pressure	$kJ.kmol^{-1}.K^{-1}$
u	= specific *internal energy*	$kJ.kmol^{-1}$
h	= specific *enthalpy*	$kJ.kmol^{-1}$

For a given substance in the solid or liquid phase, C_v and C_p have nearly the same value (i.e. $C_p \cong C_v$), but in the gas phase C_v differs from C_p, according to *Equation 2.58*.

$$C_p = C_v + R \quad \text{(Ideal gas)} \qquad \textit{Equation 2.58}$$

where: R = universal gas constant = 8.314 $kJ.kmol^{-1}.K^{-1}$

[26] Heat capacity cannot be defined over a phase change because a phase change involves the transfer of heat to a substance without a change in its temperature.

[27] A *partial differential equation* represents the rate of change of a variable (x) with respect to another variable (y) while other variables that affect (x) are held constant.

In energy balance calculations involving gases, C_v is used to calculate energy in closed systems and for the accumulation in open systems, whereas C_p is used for the flowing streams in open (continuous) systems.

The heat capacity of each substance changes with its temperature, pressure and phase. The effect of pressure is small (except for gases near the critical state) but changes of temperature and phase have a substantial effect on the heat capacity, as shown in *Figure 2.04*. Some points you should note about *Figure 2.04* are as follows:

- Heat capacity is zero at 0 K.
- The discontinuity of heat capacity values at each phase change.
- The liquid heat capacity of a substance is usually higher than its gas heat capacity at the normal boiling point.
  Hydrogen (BP = 20 K) is an exception to this rule.
- The heat capacity of each gas increases with increasing temperature.
- The rate of increase of the gas heat capacity with temperature rises with the number of atoms in the gas molecule, (e.g. He < H_2 < H_2O < SO_3 < C_2H_6).
- The heat capacity is undefined at the critical state.[28]

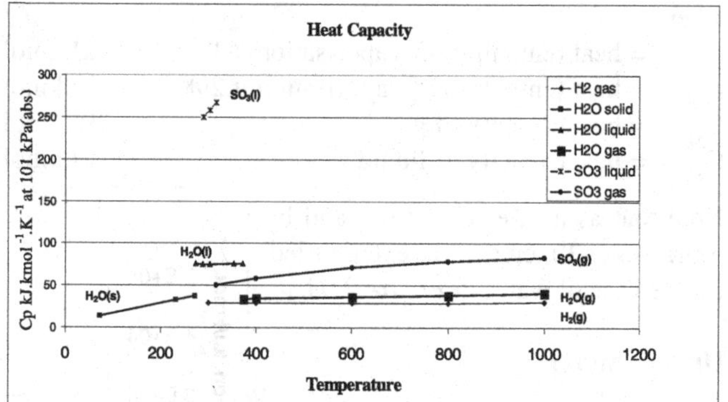

Figure 2.04. Changes of heat capacity with temperature and phase (at 101.3 kPa(abs)).

For a substance where the phase is fixed (no phase change) the heat capacity can be calculated as a polynomial function of temperature, such as *Equation 2.59*.

$$C_p = a + bT + cT^2 + dT^3 \qquad \textit{Equation 2.59}$$

where:

T = temperature K a, b, c, d = empirical constants specific to each substance [appropriate units]

Table 2.15 lists values of a, b, c and d for some common gases.

Table 2.15. Polynomial coefficients for the heat capacity of gases at 101.3 kPa(abs).

Gas	a	b	c	d	Range	C_p at 298 K
					K	kJ kmol^{-1} K^{-1}
H_2	28.79	-0.092E-3	0.879E-6	0.544E-9	298 – 1500	28.9
N_2	29.58	-5.52E-3	13.85E-6	-5.27E-9	298 – 1500	29.0
O_2	26.02	11.3E-3	-1.55E-6	0.921E-9	298 – 1500	29.3
H_2O	33.89	-3.01E-3	15.19E-6	-4.85E-9	298 – 1500	34.2
CO_2	21.51	64.4E-3	-41.6E-6	10.1E-9	298 – 1500	37.3
CH_4	21.09	39.0E-3	37.1E-6	-22.5E-9	298 – 1500	35.7

$C_p = a + bT + cT^2 + dT^3$ kJ.kmol^{-1} K^{-1} T = K

[28] The critical temperatures and critical pressures of H_2, SO_2 and H_2O are respectively 33 K, 1.3E3 kPa(abs), 431 K, 7.88E3 kPa(abs) and 647 K, 22.1E3 kPa(abs). These substances do not approach the critical state in *Figure 2.04*.

For energy balance calculations it is often satisfactory to use the *mean heat capacity* defined by *Equation 2.60.*

$$C_{p,m} = \int_{T_{ref}}^{T} C_p \, dT / (T - T_{ref}) \quad \text{(no phase change, no reaction)} \qquad \textit{Equation 2.60}$$

where: $C_{p,m}$ = mean heat capacity over the temperature range T_{ref} to T kJ.kmol^{-1}.K^{-1}

Some values of mean heat capacity are listed in *Table 2.16.*
Be careful when you use the heat capacity that the value applies to the correct phase, and it is not taken across a phase change.

Table 2.16. Mean heat capacities of gases at 101.3 kPa(abs) relative to 298 K.

Temperature	H_2(g)	N_2(g)	O_2(g)	H_2O(g)	CO_2(g)	CH_4(g)	C_2H_6(g)	SO_2(g)
K	$C_{p,m}$ kJ.kmol^{-1}.K^{-1}							
298	28.8	29.1	29.4	33.8	37.2	35.8	52.8	39.9
573	29.2	29.4	30.5	34.4	42.3	43.1	70.0	44.4
873	29.3	30.2	31.9	36.3	46.2	51.3	86.1	47.9
1173	29.6	31.1	32.9	38.1	49.1	58.7	99.1	50.2
1473	30.2	31.9	33.8	39.8	51.3	64.8	109.4	51.8
1773	30.7	32.6	34.3	41.4	53.1	–	–	–
2070	31.2	33.2	34.9	42.8	54.1	–	–	–
2373	31.7	33.6	35.4	44.0	55.1	–	–	–

INTERNAL ENERGY

Internal energy (U) is a "catch-all" term that includes all the energy in a substance from the intra-atomic, inter-atomic and inter-molecular forces that give it coherence.[29] Internal energy does not include *potential energy* due to the position or *kinetic energy* due to the motion of a mass of the substance.

The internal energy of a substance is a conceptual *state function*[30] whose absolute value cannot be measured or calculated. When the internal energy is used in energy balance calculations, its value is a relative value defined with respect to a *reference state* whose internal energy is arbitrarily set at zero.

ENTHALPY

The enthalpy of a substance is a state function defined by *Equation 2.61*.

$$H = U + PV \qquad \textit{Equation 2.61}$$

where: H = enthalpy kJ P = pressure kPa
U = internal energy kJ V = volume m^3

[29] The concept of internal energy originated with J.W. Gibbs (1839–1903) before the study of nuclear energy.

[30] A *state function* is a quantity whose value depends only on the state of a substance and is independent of the route taken to reach that state.

In energy balance calculations, enthalpy must be defined relative to a specified *reference state*. The enthalpy of pure substances is given in *thermodynamic tables* that are available in sources such as *Refs. 1* and 2, but when you use such tables you should be aware of the *reference state*, as discussed below.

A convenient *reference state* for energy balance calculations on chemical processes is the chemical elements in their standard states at 101.3 kPa(abs), 298.15 K. With this basis the specific enthalpy of a pure substance can be *estimated* by *Equation 2.62*.

You can see from *Equation 2.62* why enthalpy is sometimes called the "heat content" of a substance.

$$h^* \approx \int_{T_{ref}}^{T} C_p \cdot dT + h^{\circ}_{f,T_{ref}} = C_{p,m}(T - T_{ref}) + h^{\circ}_{f,T_{ref}}$$ [Respect the phase] *Equation 2.62*

where:

h^*	= specific enthalpy of the substance at temperature T in phase Π w.r.t. *elements* at the standard state	$kJ.kmol^{-1}$
C_p	= heat capacity at constant pressure of the substance in phase Π	$kJ.kmol^{-1}.K^{-1}$
$h^{\circ}_{f,Tref}$	= enthalpy of formation of the substance at T_{ref} in phase Π	$kJ.kmol^{-1}$
T	= temperature	K
T_{ref}	= reference temperature (usually 298 K)	K

Note that *Equations 2.62* and *2.65* are approximate because they do not account for the (usually small) effect of pressure on enthalpy.

The value of $h^{\circ}_{f,Tref}$ in *Equation 2.62* includes the latent heat of any phase changes required to convert the substance from its standard state at T_{ref} to the phase Π at T_{ref}. For example, if the substance is a solid at its standard state but the phase Π is a gas, then:

$$\mathbf{h^{\circ}_{f,Tref}\ (gas) = h^{\circ}_{f,Tref}\ (solid) + h_{m,Tref} + h_{v,Tref}}$$ *Equation 2.63*

Or if the substance is a liquid in its standard state, but phase Π is a gas, then:

$$\mathbf{h^{\circ}_{f,Tref}\ (gas) = h^{\circ}_{f,Tref}\ (liquid) + h_{v,Tref}}$$ *Equation 2.64*

where:

$h^{\circ}_{f,Tref}$ (gas)	= heat of formation of substance in gas phase at T_{ref}	$kJ.kmol^{-1}$
$h^{\circ}_{f,Tref}$ (liquid)	= heat of formation of substance in liquid phase at T_{ref}	$kJ.kmol^{-1}$
$h^{\circ}_{f,Tref}$ (solid)	= heat of formation of substance in solid phase at T_{ref}	$kJ.kmol^{-1}$
$h_{m,Tref}$	= heat of fusion of substance at T_{ref}	$kJ.kmol^{-1}$
$h_{v,Tref}$	= heat of vaporisation of substance at T_{ref}	$kJ.kmol^{-1}$

In general the enthalpy defined by *Equation 2.62* can be considered in three parts:

- Sensible heat = $\int C_p.dT = C_{p,m}(T - T_{ref})$ required to change temperature at constant phase
- Latent heat = h_m and/or h_v required to change phase at constant temperature
- Heat of formation = h_f required to form a compound (at specified phase) from its elements

The specific enthalpy (h*) defined by *Equation 2.62* is called the "total specific enthalpy" to differentiate it from the "specific enthalpy" (h) found in many sources of *thermodynamic data.* The specific enthalpy reported in thermodynamic tables and charts, such as the steam table (*Table 2.20*), enthalpy-pressure diagrams and psychometric charts (*Figure 2.07*) *does not include heat of formation.* This specific enthalpy "h" is measured relative to *compounds* at a reference state and can be *estimated*[31] by *Equation 2.65.*

$$h \approx \int_{T_{ref}}^{T} C_p.dT + h_{p,T_{ref}} = C_{p,m}(T - T_{ref}) + h_{p,T_{ref}}$$ [Respect the phase] *Equation 2.65*

where:

h	= specific enthalpy of the substance at temperature T in phase Π w.r.t. *compounds* at the reference state	$kJ.kmol^{-1}$
C_p	= heat capacity at constant pressure of the substance in phase Π	$kJ.kmol^{-1}.K^{-1}$
$h_{p,Tref}$	= latent heat of any phase changes to convert the substance from its reference state at T_{ref} to the phase Π at T_{ref}.	$kJ.kmol^{-1}$
T	= temperature	K
T_{ref}	= reference temperature	K

Figure 2.05 shows the relation between the total specific enthalpy and temperature for a pure substance as it passes through phase changes at constant pressure. You should examine *Figure 2.05* to see how the total specific enthalpy (h*) can be calculated in several ways, by summing the enthalpy changes along different paths,[32] to give the same final result as that of *Equation 2.62*. In *Figure 2.05* you can also see that shifting the reference state (baseline) from the elements to the compound would give the specific enthalpy (h) of *Equation 2.65*.

EXAMPLE 2.18 Finding the enthalpy of water by calculation and from the steam table.

Problem: Calculate the total specific enthalpy of pure water at:

(A) 200 kPa(abs), 400 K (B) 3000 kPa(abs), 600 K

(i) By *Equation 2.62*. (ii) From steam tables (*Table 2.20*, see page 72)

Data: For H_2O (l) $h^o_{f,298K}$ = -285.8E3 kJ/kmol $h_{v,298K}$ = 44E3 $kJ.kmol^{-1}$ (at 101.3 kPa(abs))

For H_2O (g) $C_{p,m}$ ref 298 K = 35.2 $kJ.kmol^{-1}.K^{-1}$ 400 K 36.3 $kJ.kmol^{-1}.K^{-1}$ 600 K (at 101.3 kPa(abs))

---

[31] The values of "h*" and "h" from *Equations 2.62* and *2.65* are not exact because they do not account for the effect of pressure on C_p and h_p.

[32] Since enthalpy is a state function a change in enthalpy depends only on the initial and final states and is independent of the path between those states (this condition is the basis for Hess's Law).

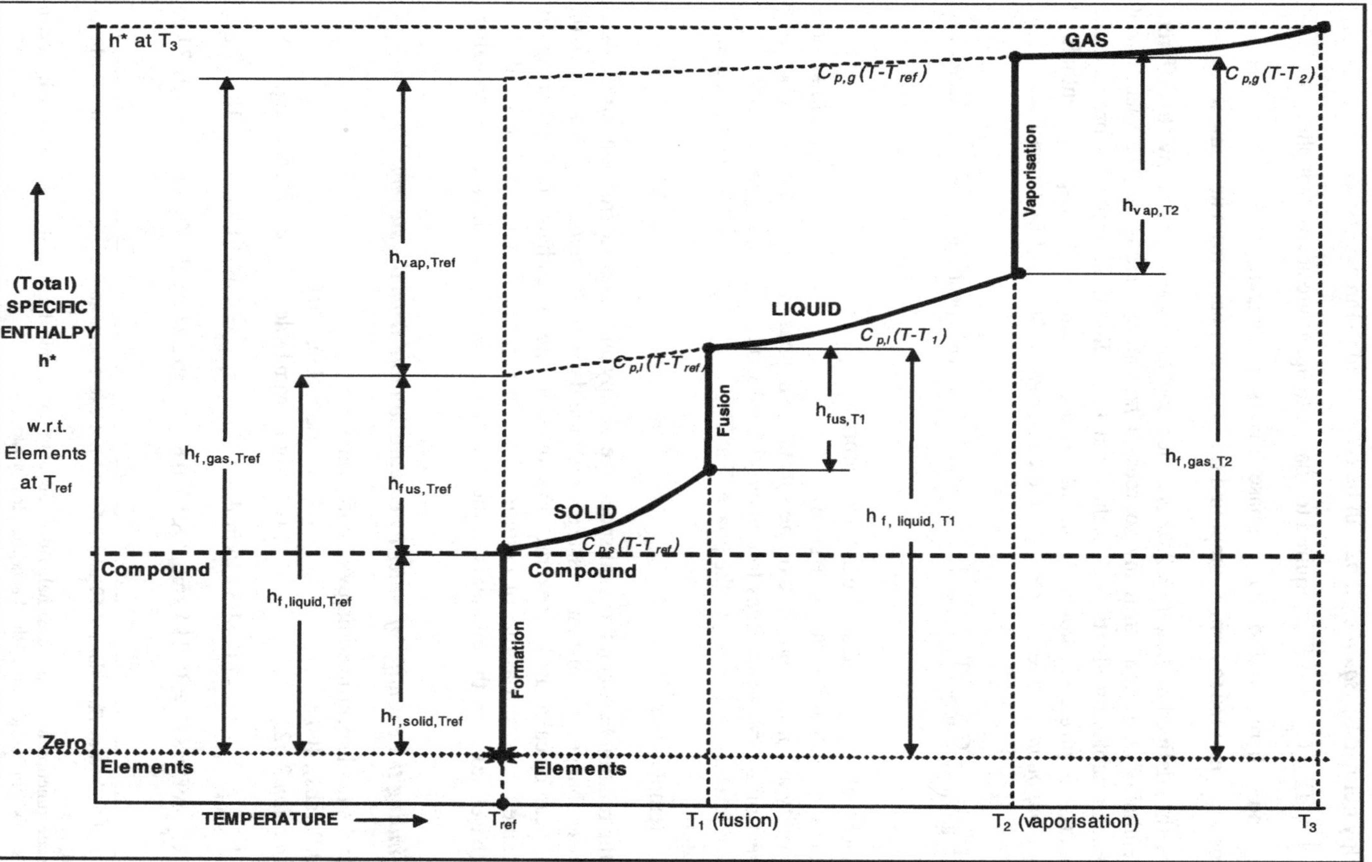

Figure 2.05. Thermodynamic values (h, h_f, h_v, h_m, C_p, etc.) at constant pressure.

Graphical interpretation of the specific enthalpy (h^*) for a single pure substance at a constant pressure.

Reference condition = elements in standard state at reference temperature T_{ref}

NOTE: For simplicity the enthalpy of formation is shown here as positive, although in many compounds the heat of formation is negative.

Solution:

A. Define the reference state for total specific enthalpy (h*)
= elements (H_2 and O_2) at 101.3 kPa(abs), 298.15 K, i.e. standard state.

(i) The vapour pressure of water at 400 K is 246 kPa(abs), so at 200 kPa(abs), 400 K water is a gas.

Equation 2.64: $h_{f,Tref}$ (gas) = $h_{f,Tref}$ (liquid) + $h_{v,Tref}$ = -285.8E3 + 44E3 = -241.8E3 kJ.kmol^{-1}

Equation 2.62: h* = $C_{p,m}(T - T_{ref}) + h_{f,Tref}$ = (35.2 kJ.kmol^{-1}.K^{-1})(400 − 298)K + (-241.8E3 kJ.kmol^{-1})
= -238.2E3 kJ.kmol^{-1}

The steam table gives more accurate values of the sensible and latent heats than do *Equations* 2.62 and 2.64, but these enthalpy values are for a reference condition of liquid water at 0.61 kPa(abs), 273.16 K and do not include the heat of formation of water.

(ii) To find the enthalpy w.r.t. the elements at 101.3 kPa(abs), 298.15 K consider a process path as follows:

1. H_2O(l) is formed from H_2 and O_2 at 101.3 kPa(abs), 298.15 K.
Δh = -285.8E3 kJ.kmol^{-1} (*Table 2.14*)

2. H_2O(l) at 101.3 kPa(abs), 298.15 K is cooled to 273.16K at 0.6 kPa(abs).
Δh = -1.86E3 kJ.kmol^{-1} (*Table 2.20*)

3. H_2O(l) at 0.6 kPa(abs), 273.16 K is vaporised and heated to 400 K, 200 kPa(abs)
Δh = +48.96E3 kJ.kmol^{-1} (*Table 2.20*)

Then
Total specific enthalpy of water at 200 kPa(abs), 400 K = h* = Sum **= -238.7E3 kJ.kmol^{-1}**

B. Define the reference state for total enthalpy = elements (H_2 and O_2) at 101.3 kPa(abs), 298.15 K, i.e. standard state.

(i) The vapour pressure of water at 600 K is 12330 kPa(abs), so at 3000 kPa(abs), 600 K water is a gas.

Equation 2.64: $h_{f,Tref}$ (gas) = $h_{f,Tref}$ (liquid) + $h_{v,Tref}$ = -285.8E3 + 44E3 = -241.8E3 kJ.kmol^{-1}

Equation 2.62: h* = $C_{p,m}(T - T_{ref}) + h_{f,Tref}$ = (36.3 kJ.kmol^{-1}.K^{-1})(600 − 298)K + (-241.8E3 kJ.kmol^{-1})
= -230.8E3 kJ.kmol^{-1}

(ii) To find the enthalpy w.r.t. the elements at 101.3 kPa(abs), 298.15 K consider a process path as follows:

1. H_2O(l) is formed from H_2 and O_2 at 101.3 kPa(abs), 298.15 K
Δh = -285.8E3 kJ.kmol^{-1} (*Table 2.13*)

2. H_2O(l) at 101.3 kPa(abs), 298.15 K is cooled to 273.16 K at 0.6 kPa(abs).
Δh = -1.86E3 kJ.kmol^{-1} (*Table 2.20*)

3. H_2O(l) at 0.6 kPa(abs), 273.16 K is vaporised and heated to 3000 kPa(abs), 600 K
Δh = +55.07E3 kJ.kmol^{-1} (*Table 2.20*)

Then
Total specific enthalpy of water at 3000 kPa(abs), 600 K = h* = Sum **= -232.6E3 kJ/kmol**

The difference between the values in (i) and (ii) is mostly due to the variation of C_p of the gas with pressure, which is not accounted for in (i).

Potential energy and kinetic energy

Potential energy is the energy contained in a mass of material due to its position in a force field, such as an electric, gravitational or magnetic field. The most common of these is the gravitational potential energy, calculated by *Equation 2.66.*

$$\mathbf{E_p = (1E\text{-}3)mgL'} \qquad \textit{Equation 2.66}$$

where: E_p = potential energy — kJ
m = mass of material — kg
g = gravitational constant (= 9.81 $m.s^{-2}$ on Earth) — $m.s^{-2}$
L′ = elevation of the mass above a reference level — m

Kinetic energy is the energy contained in a mass due to its motion and is calculated by *Equation 2.67.*

$$\mathbf{E_k = (1E\text{-}3)0.5\,m\tilde{u}^2} \qquad \textit{Equation 2.67}$$

where: E_k = kinetic energy — kJ
m = mass of material — kg
$\tilde{u}$ = velocity of the mass — $m.s^{-1}$

Surface energy

Due to the existence of unbalanced cohesive forces at phase boundaries, material within about one nanometer (1E-9 m) of an interface has a slightly higher specific internal energy than the same material in a bulk phase. The corresponding excess surface energy of a multi-phase system is calculated by *Equation 2.68.*

$$\mathbf{E_i = \gamma A_i} \qquad \textit{Equation 2.68}$$

where: E_i = excess surface energy — kJ
γ = interfacial tension — $kN\,m^{-1}$
A_i = interfacial area (i.e. area of phase boundaries) — m^2

For multi-phase systems in which the disperse phase exists as mono-disperse spherical bubbles, drops or particles the interfacial area is given by *Equation 2.69.*

$$\mathbf{A_i = (6e/d)V} \qquad \textit{Equation 2.69}$$

where: A_i = interfacial area (i.e. area of phase boundaries) — m^2
e = volume fraction of the disperse phase — –
d = diameter of dispersed bubble, drop or particle — m
V = total volume of system — m^3

Surface energy is an "exotic" energy term that is negligible in most chemical processes (e.g. for water/air at 298 K, $\gamma = 72E\text{-}6\ kN.m^{-1}$) but can be significant in systems with high specific surface area, such as in *colloidal dispersions*[33] (e.g. emulsions, foams, mists).

MIXING

When two (or more) pure substances are mixed together several things can happen. If the substances are, and remain in, different phases the mixture is called a *dispersion.* Dispersions are classified into various types, such as:

- Aerosol = solid or liquid in gas dispersion (e.g. smog, smoke, mist)
- Conglomerate or Composite = solid in solid dispersion (e.g. nut chocolate, fibreglass)
- Emulsion = liquid in liquid dispersion (e.g. mayonnaise, milk, oily water)
- Foam = gas in liquid or solid dispersion (e.g. soap lather, foamed rubber)
- Slurry or Suspension = solid in liquid dispersion (e.g. muddy water, yoghurt)

If the substances dissolve to form a single phase the mixture becomes a *solution.* Solutions are classified into three phases:

- Solid solution (e.g. metal alloys)
- Liquid solution (e.g. sea water, gasoline, whiskey)
- Gas solution — usually called a *gas mixture* — because all gases are mutually soluble (e.g. air).

The mixed substances may engage in chemical reaction(s) that transform their molecules. Chemical reactions are classified as:

- Homogeneous reaction = a reaction that occurs in a single phase.
- Heterogeneous reaction = a reaction that occurs across a phase boundary.

In reality nearly all substances have a degree of mutual solubility that varies from very low (e.g. parts per million (wt) for mineral oil in water) to infinity (e.g. ethanol in water) — so the separate phases of a dispersion nearly always contain species from the other phases in solution. Often, the degree of mutual solubility is so low that it can be ignored, but sometimes even a very low solubility is important, for example, when dealing with catalyst poisons in chemical reactors or toxic materials in the environment.

You must be careful because the word "mixture" is ambiguous. In scientific and engineering jargon, unless a reaction is specified, a *mixture* implies that no chemical reactions occur between the components. The terms *ideal mixture* and *non-ideal mixture* generally mean multi-species, single phase solutions, whereas, for example, a *mixture* of oil and water usually means a two-phase oil-water *dispersion* and a *mixture* of sand and water is a two-phase *slurry* of sand in water.

---

[33] A colloidal dispersion is a multi-phase mixture in which the dispersed phase has bubbles, drops or particles with diameter in the range (approximately) 1 to 5000 nm. Most ions and molecules (except large polymers) have diameters below about 1 nm.

HEAT OF MIXING

When two (or more) pure substances are mixed to form a solution (see above), the resulting mixture can be classified as ideal or non-ideal.

- **Ideal mixture** No interactions between molecules (or atoms) of the individual components.
- **Non-ideal mixture** Interactions between molecules (or atoms) of the individual components.

The formation of a non-ideal mixture at constant pressure and temperature involves an input of an amount of heat called the *heat of mixing*. When the mixture is formed by dissolving a gas or a solid in a liquid the heat of mixing is called the *heat of solution*.

For substances that are soluble in water the heat of mixing is added to the heat of formation of the solute to give the heat of formation of the solute at *infinite dilution*,[34] which is shown in thermodynamic tables as h_f (aq). *Table 2.17* lists values of the heat of mixing with corresponding heat of formation for sulphuric acid in water and *Table 2.18* gives some values of heat of solution in water. Heats of mixing are tabulated in sources such as *Refs. 1* and *2*.

Table 2.17. Heat of mixing sulphuric acid with water at 298 K.

Amount of water	Heat of mixing	$h^o_{f,298K}$
moles H_2O/mole H_2SO_4	kJ.kmol^{-1} H_2SO_4	kJ.kmol^{-1} H_2SO_4
0	0.00	-811E3
1	-28E3	-839E3
2	-42E3	-853E3
10	-67E3	-878E3
Infinity	-96E3	-907E3*

* *Heat of formation of H_2SO_4 at infinite dilution*

NOTE: *The heat of formation of H_2SO_4(aq) does not include the heat of formation of water.*

Table 2.18. Heat of solution of some solids in water at 291 K.

Substance		NaOH	$NaClO_3$	KNO_3	Citric Acid	Sucrose	Quinone
Heat of solution at infinite dilution	kJ.kmol^{-1}	-43E3	+22E3	+36E3	+23E3	+5.5E3	+17E3

Values of the heat of mixing can be positive (+) or negative (–). A positive value means that the mixing process is endothermic, so if no heat is supplied the system temperature will drop as the mixture is formed. This occurs, for example, when KNO_3 dissolves in water. If the mixing process is exothermic (H_{mix} is negative) and no heat is removed the system temperature will rise as the mixture is formed. This occurs when NaOH is dissolved or H_2SO_4 is mixed into water. The addition of water to cold sulphuric acid can release enough heat to vaporise the water and cause an explosion.

In some cases (e.g. sulphuric acid-water) the heat of mixing is substantial and should be included in the enthalpy values used for energy balance calculations. Enthalpy-concentration diagrams, which are described below, contain data for heat of mixing.

---

[34] *Infinite dilution* means the presence of a large excess (hypothetically infinite) of solvent.

THERMODYNAMIC TABLES AND REFERENCE STATES

Many sources, [e.g. *Refs. 1– 5*] record the thermodynamic properties of pure substances. The information may be in the form of diagrams, tables or equations that correlate the properties with temperature and/or pressure. *Table 2.14* lists some general thermodynamic data for a few substances and *Table 2.20* presents a condensed form of a thermodynamic table for water commonly called the "*steam table*". Other sources of thermodynamic data are enthalpy-concentration diagrams, enthalpy-pressure diagrams, enthalpy-entropy diagrams (called Mollier diagrams) and psychrometric charts.

The thermodynamic literature uses several different reference states that create ambiguity and confusion among students. *Table 2.19* lists common reference conditions with their usual context. When you use thermodynamic data for energy balance calculations you should always be aware of the reference conditions and ensure that they cancel from the balance equation (see Chapter 5).

Table 2.19. Conventional reference conditions for thermodynamic data.

Name	Condition	Application
Standard temperature and pressure [STP]	101.3 kPa(abs), 273.15 K	Gas law
Normal temperature and pressure [NTP]	101.3 kPa(abs), 293 K	Gas metering (used in industry)
Triple-point of water	Liquid water 0.611 kPa(abs), 273.16 K	Steam tables
Standard state (elements)	Element at 101.3* kPa(abs), 298.15 K	Heat of formation, reactive energy balances
Standard state (compounds)	Compound at 101.3* kPa(abs), 298.15 K	Heat of mixing, non-reactive energy balances
Absolute zero temperature	Compounds at 0 K	Mollier diagrams
Miscellaneous conditions	Compounds at 233 K	Refrigerants
Ice point	Compounds at 101.3 kPa(abs) 273.15K	Enthalpy-concentration diagrams.

* *Some sources use 1 bar = 100 kPa(abs) as the standard state pressure.*

The reference condition for both enthalpy and internal energy values in the steam table is the liquid compound water at its triple-point. The enthalpy and internal energy values in the steam table ***do not include the heat of formation of water***.

Table 2.20. Condensed version of the steam table. Properties of pure water.
Reference condition for "h" and "u" is liquid water at 0.611 kPa(abs), 273.16 K.

T	P	v_l	v_g	h_l	h_g	h_v	u_l	u_g
K	kPa(abs)	$m^3.kg^{-1}$	$m^3.kg^{-1}$	$kJ.kg^{-1}$	$kJ.kg^{-1}$	$kJ.kg^{-1}$	$kJ.kg^{-1}$	$kJ.kg^{-1}$
Subcooled liquid water								
273.15	5000	0.000998	–	5.1	–	–	0.07	–
273.15	50000	0.000977	–	49.4	–	–	0.52	–
400	500	0.001067	–	532.8	–	–	532.3	–
400	50000	0.001040	–	567.3	–	–	515.3	–
500	3000	0.001202	–	975.7	–	–	972.1	–
500	50000	0.001150	–	991.1	–	–	933.6	–
600	14000	0.001524	–	1499.4	–	–	1478.1	–
600	50000	0.001359	–	1454.0	–	–	1386.1	–
Saturated liquid-vapour water								
^{T}273.16	0.611*	0.001000	206.1	0.0	2500.9	2500.9	0.0	2375.7
300	3.536*	0.001004	39.10	111.7	2550.1	2438.4	112.5	2412.5
350	41.66*	0.001027	3.844	322.5	2637.9	2315.4	321.7	2478.4
400	245.6*	0.001067	0.7308	532.7	2715.6	2182.9	532.6	2536.2
450	931.5*	0.001123	0.2080	749.0	2774.9	2025.9	748.2	2579.8
500	2637*	0.001202	0.07585	975.6	2803.1	1827.5	972.3	2601.7
550	6122*	0.001322	0.03179	1219.9	2782.6	1562.7	1212.3	2589.3
600	12330*	0.001540	0.01375	1504.6	2677.1	1172.5	1486.9	2511.6
^{C}647.3	21830*	0.003155	0.003155	2098.8	2098.8	0.0	2037.3	2037.3
Superheated gas water								
600	1.0	–	276.9	–	3130.1	–	–	2853.2
1100	1.0	–	507.7	–	4221.7	–	–	3714.1
600	101.3	–	2.727	–	3128.0	–	–	2851.6
1100	101.3	–	5.009	–	4221.3	–	–	3713.8
600	3000	–	0.08626	–	3059.6	–	–	2891.4
1100	3000	–	0.1684	–	4209.4	–	–	3704.7
600	10000	–	0.02008	–	2818.3	–	–	2617.5
1100	10000	–	0.04992	–	4180.3	–	–	3681.2

T = Triple-point of water (solid/liquid/vapour) C = Critical point of water

* *Vapour pressure over pure liquid water (with zero surface curvature.)*[35]

Values between points in the full version of the steam table are usually obtained by linear interpolation, i.e. calculated assuming the points are connected by a straight line.

e.g. For $T_1 < T < T_2$, $h_T = h_{T_1} + (h_{T_2} - h_{T_1})((T - T_1)/(T_2 - T_1))$

[35] The effect of curvature on vapour pressure becomes significant when the radius of curvature of the condensed phase is small (e.g. <1 micron cf. the Kelvin equation).

The columns of *Table 2.20* are interpreted as follows:

T = temperature	K	h_g = specific enthalpy of gas water	kJ.kg^{-1}
P = pressure	kPa(abs)	$h_v = h_g - h_l$ = heat of vaporisation of liquid water	kJ.kg^{-1}
v_l = specific volume of liquid water	m^3.kg^{-1}	u_l = specific internal energy of liquid water	kJ.kg^{-1}
v_g = specific volume of gas water	m^3.kg^{-1}	u_g = specific internal energy of gas water	kJ.kg^{-1}
h_l = specific enthalpy of liquid water	kJ.kg^{-1}		

The line obtained by plotting P versus T from the liquid–vapour section of *Table 2.20* is the liquid/vapour (L/V) equilibrium line for water, shown in *Figure 2.02*. When the pressure and temperature of pure water are on the L/V line then water at this condition can exist as two phases (liquid and vapour) in equilibrium and is said to be *"saturated"*. The mass fraction of vapour in the saturated liquid/vapour mixture is called the "*quality*" of the mixture. The relation between quality and specific enthalpy (h), internal energy (u), volume (v) or entropy (s) of the two-phase mixture is:

$$\mathbf{Z_m = w_q Z_g + (1 - w_q) Z_l} \qquad \textit{Equation 2.70}$$

where: Z_m = value of h, u, v or s for the mixture (mass basis)
Z_l = value of h, u, v or s for the liquid phase alone
Z_g = value of h, u, v or s for the gas (vapour) phase alone
w_q = quality = mass fraction vapour in mixture

If the temperature is below the saturation temperature at a given pressure, then pure water can exist <u>at equilibrium</u> only as a liquid and is said to be *"subcooled"*. If the temperature exceeds the saturation temperature at a given pressure then pure water can exist <u>at equilibrium</u> only as a gas and is said to be *"superheated"*. As an exercises you should consider the L/V equilibrium of water described above in terms of the *phase rule*.

The thermodynamic properties of subcooled liquid water depend mainly on the temperature and are only slightly affected by pressure, whereas the properties of superheated gas water depend on both pressure and temperature. For most purposes the properties of subcooled liquid water can be taken as those of the liquid in the saturated liquid–vapour table, at the specified temperature. Thermodynamic data for superheated gas water are tabulated in a separate part of the steam table and should be used for calculations involving superheated conditions (see *Table 2.20*).

Note that enthalpy in the steam table is based on a reference condition of liquid water at its triple-point and does not include the heat of formation of water. This value of enthalpy is fine for energy balance calculations involving only physical changes to water (e.g. compression, expansion, evaporation or heating) but is not sufficient when water is involved in chemical reactions. If you use the steam table as a source of enthalpy data in energy balances where water is a reactant or a reaction product then you must add the heat of formation and consider the implications of the different reference conditions for h_l, h_g and h_f; or otherwise account for the heat of reaction, which is usually a major term in the energy balance.[36]

---

[36] See Chapter 5 for more on energy balance calculations.

EXAMPLE 2.19 Using the steam table to find the phase, enthalpy and volume of water.

Problem A: 1 kg of pure water at 1000 kPa(abs), 400K is heated to 101.3 kPa (abs), 1100 K. Specify the phase at each condition then calculate the change of volume and enthalpy for the process.

Solution:

At 1000 kPa(abs), 400 K pure water is below its saturation temperature, i.e. subcooled liquid.
PHASE = liquid $v_{in} = 0.001061\ m^3/kg$, $h_{in} = 533.1$ kJ/kg [From the full steam table]
(NOTE: the similarity to saturated liquid at 400 K)

At 101.3 kPa(abs), 1100 K pure water is above its saturation temperature, i.e. superheated gas.
PHASE = gas, $v_{fin} = 5.009\ m^3/kg$, $h_{fin} = 4221.3$ kJ/kg
Change of volume = final volume – initial volume $= v_{fin} - v_{in} = 5.009 - 0.001061 =$ **5.008 m³**
Change in enthalpy = final enthalpy – initial enthalpy $= h_{fin} - h_{in} = 4221.3 - 533.1 =$ **3687.2 kJ**

Problem B: 1 kg of pure water at 500 K has a volume of 0.06 m³ at equilibrium. Specify the phase and the pressure then calculate the quality and the specific enthalpy.

Solution:

From *Table 2.20* at 500 K: $v_l \leq 0.001202\ m^3/kg$, and $v_g \geq 0.07585\ m^3/kg$
Since 500 K > 273.16 K (triple-point), then water with v = 0.06 m³/kg must be a mixture of liquid and gas.
PHASE = saturated (liquid + gas), Pressure = saturation pressure = 2637 kPa(abs)

Let w_q = quality

Then by *Equation 2.70* $v = (1 - w_q)v_l + w_q v_g$ $0.06 = 0.001202(1 - w_q) + 0.07585 w_q$ $w_q =$ **0.79**
Specific enthalpy $= h = (1 - w_q)h_l + w_q h_g = (1 - 0.79)(975.6\ kJ/kg) + (0.79)(2803.1\ kJ/kg)$ h = **2419.3 kJ.kg⁻¹**

Problem C: A gas mixture of 2 kg H_2 + 16 kg O_2 at 101.3 kPa(abs), 298.15 K is reacted to give to 1 kmol of water at 10000 kPa(abs),1100 K.
Calculate the change of enthalpy using *Equation 2.62* with *Tables 2.14* and *2.20*.

Solution:

Water is involved in a chemical reaction, so the steam table alone is not sufficient to calculate the enthalpy change. From *Table 2.20*, at 10000 kPa(abs), 1100 K water is a superheated gas.
Define the reference condition = elements (H_2 and O_2) in standard state at 101.3 kPa(abs), 298.15 K.

Initial specific enthalpy by *Equation 2.62*: $h^*_{in} = C_{p,m}(T - T_{ref}) + h^o_{f,Tref}$

For H_2: $h^*_{in,H_2} = (1\ kmol)(28.8\ kJ/(kmol.K))(298 - 298)\ K + 0\ kJ.kmol^{-1}))$ $= 0$ kJ
For O_2: $h^*_{in,O_2} = (0.5\ kmol)((29.4\ kJ/(kmol.K))(298 - 298)\ K + 0\ kJ.kmol^{-1}))$ $= 0$ kJ
Total $= h^*_{in}$ $= h^*_{in,H_2} + h^*_{in,O_2}$ $= 0 + 0$ $= 0$ kJ

Final specific enthalpy $= h^*_{fin} = [h_g$ at 10000 kPa, 1100K $- h_l$ at 101.3 kPa, 298K$] + h_{f,}$ liquid water at 101.3 kPa, 298K

Values of h_g and h_l are from the steam table (*Table 2.20*), whose enthalpy values include the sensible heat and latent heat effects but *exclude the heat of formation.* Value of h_f comes from the thermodynamic property *Table 2.14.*

$$h^*_{fin} = 18\ \text{kg/kmol}\ [4180.3\ \text{kJ/kg} - 105\ \text{kJ/kg}] + (-285.8\text{E}3\ \text{kJ/kmol}) = \underline{\mathbf{-212.4E3\ kJ/kmol\ H_2O}}$$

THERMODYNAMIC DIAGRAMS

Thermodynamic diagrams present complex thermodynamic data for materials over wide ranges of conditions. Such charts are used to simplify difficult thermodynamic calculations. The list of thermodynamic diagrams includes:

- Enthalpy-concentration diagrams.
- Psychrometric diagrams (properties of gas/vapour mixtures)
- Enthalpy-pressure diagrams
- Entropy-temperature diagrams
- Enthalpy-entropy diagrams (called Mollier diagrams)

As a rule the reference conditions for thermodynamic charts are the pure compounds at some specified pressure and temperature. Thus the enthalpies found from these charts *do not include heat of formation.*

ENTHALPY-CONCENTRATION DIAGRAMS

Enthalpy-concentration diagrams show thermodynamic data for multi-component non-ideal systems. Such systems are fairly common in practical processes and cause difficulties in calculations because the properties of the mixture are not a simple combination of the properties of the pure components. Enthalpy-concentration diagrams give a simple method to account for the *"heat of mixing"* that results when two pure components are combined to form a non-ideal mixture.

Figure 2.06 shows the enthalpy-concentration diagram for sulphuric acid-water mixtures. The reference condition for "enthalpy" in enthalpy-concentration and similar mixing diagrams is the pure

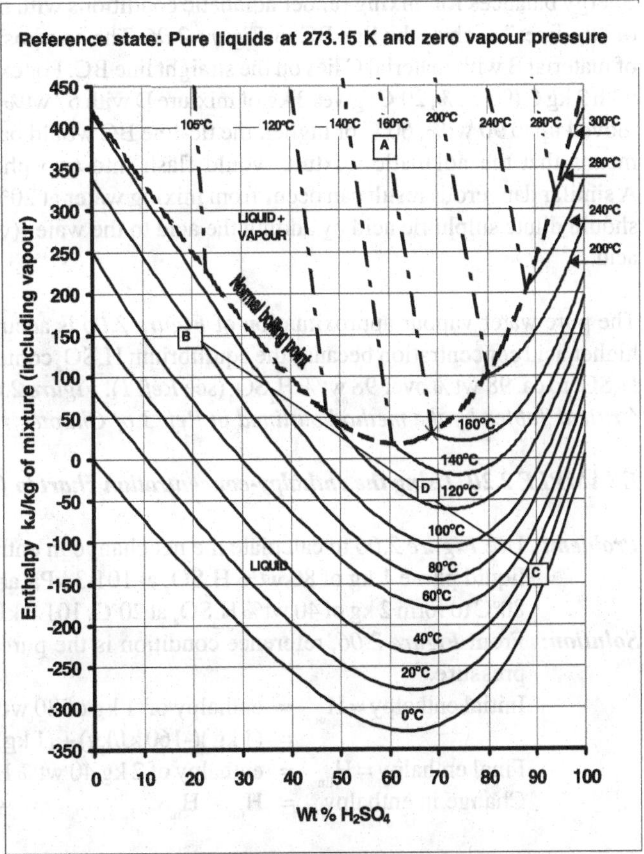

Figure 2.06. Enthalpy-concentration diagram for H_2SO_4-H_2O mixtures at 101.3 kPa(abs)

component compounds at a specified pressure and temperature. This mixture enthalpy *does not include the heat of formation for any of the components.* If you use a mixture enthalpy (from an enthalpy-concentration diagram) in an energy balance where the components are involved in chemical reactions you should add appropriate heats of formation, and beware of different reference conditions. *Example 2.20* demonstrates a basic mixing calculation using *Figure 2.06.*

NOTE: The gas phase of the [liquid + vapour] region in *Figure 2.06* is taken as pure superheated water vapour. For example, a mixture with 60 wt% H_2SO_4 at 101.3 kPa(abs), 160°C (point A) has a specific enthalpy of ca. 375 kJ/kg and consists of a superheated water vapour in equilibrium with a liquid solution of ca. 68 wt % H_2SO_4. The phase split is calculated from a mass balance, which in this case gives the original mixture as 88 wt% liquid + 12 wt% gas.

Energy balances for mixing (under adiabatic conditions with no chemical reactions) are simplified by the use of "tie-lines" such as the line BC in *Figure 2.06.* The composition and enthalpy of an adiabatic combination of material B with material C lies on the straight line BC. For example, a combination of 1 kg B (20 wt%, 80°C) with 2 kg C (90 wt%, 20°C) gives 3 kg of mixture D with 67 wt% H_2SO_4 at ca. 125 °C. Observe that if point C were moved up to 90 wt%, 60°C or higher, the tie-line BC would pass through the [liquid + vapour] region, which means that the adiabatic mixture would flash into two phases with possible dangerous consequences. A similar dangerous result can occur from mixing water at 20°C with 98 wt% H_2SO_4 at 20°C. That is why you should dilute sulphuric acid by adding the acid to the water (while mixing), NOT by adding the water to the acid.

The pure water vapour approximation of *Figure 2.06* is accurate below 90 wt% H_2SO_4, but less reliable at higher acid concentration because the equilibrium H_2SO_4 content of the gas rises from ca. 5 wt% over 90 wt% H_2SO_4 to ca. 98 wt% over 98 wt% H_2SO_4 (see *Ref. 1*). *Figure 2.06 was prepared from data in the International Critical Tables by the method outlined in Ref. 3 of Chapter 1.*

EXAMPLE 2.20 Using the enthalpy-concentration chart to find heat load to process sulphuric acid-water.

Problem: Use *Figure 2.06* to calculate the net change in enthalpy that occurs by (mixing + cooling) in the liquid phase 1 kg of 80 wt% H_2SO_4 at 101.3 kPa(abs), 60°C with 1 kg of water at 101.3 kPa(abs), 80°C to form 2 kg of 40 wt% H_2SO_4 at 20°C, 101.3 kPa(abs).

Solution: From *Figure 2.06*, reference condition is the pure liquid components at 273.15 K, zero vapour pressure.

Initial enthalpy = H_{in} = enthalpy of 1 kg of 80 wt% H_2SO_4 at 60 °C + 1 kg of water at 80 °C
= (1 kg)(-160 kJ/kg) + (1 kg)(334 kJ/kg) = + 174 kJ

Final enthalpy = H_{fin} = enthalpy of 2 kg 40 wt% H_2SO_4 at 20 °C = (2 kg) (-200 kJ/kg) = - 400 kJ

Change in enthalpy = $H_{fin} - H_{in}$ = -400 – (174) = **<u>-574 kJ</u>**

Note that the tie-line for the initial materials shows that under adiabatic conditions the mixture would reach about 110 °C (i.e. nearly boiling).

Psychrometric Charts

Psychrometric charts contain thermodynamic data for gas/vapour mixtures at a fixed pressure. The most common pyschrometric chart is that for water vapour in air (i.e. humid air), which is used, for example, in the design of air conditioning systems and water-cooling towers. *Figure 2.07* shows a psychrometric chart for water–air mixtures.

Figure 2.07 shows the following data for water–vapour–air mixtures at the standard pressure of 101.3 kPa(abs):

- dry-bulb temperature °C
- wet-bulb temperature °C
- relative humidity %
- moisture content of humid air kg/kg dry air
- specific volume of humid air m^3/kg dry air
- enthalpy of humid air kJ/kg dry air, ref. liquid water and gaseous air at 101.3 kPa(abs), 273.15 K

Dry-bulb temperature (T_{DB}) is the ordinary temperature as measured by a dry thermometer.

Wet-bulb temperature (T_{WB}) is the temperature measured with a thermometer whose bulb is covered by a water-wet wick and moved rapidly through the air (>5 m/s). This procedure establishes a balance between heat transfer from air to bulb and mass transfer of water from bulb to air (see *Refs. 2,4*). Evaporation of water into the air lowers the wet-bulb below the dry-bulb temperature to a degree that depends on the relative humidity of the air. If the relative humidity is 100% then the wet-bulb temperature equals the dry-bulb temperature.

Relative humidity is defined as:

$$\mathbf{rh = p_w/p_w^*}$$ *Equation 2.71*

where: rh = relative humidity –
p_w = partial pressure of water vapour in the water–vapour–air mixture kPa(abs)
p_w^* = saturated vapour pressure of water at the dry-bulb temperature kPa(abs)

Note that the enthalpy values in *Figure 2.07* are per kg of dry air, plus the *enthalpy* is defined relative to liquid water and gaseous air at 101.3 kPa(abs), 0°C (273.15 K), and *does not include the heat of formation* of water.

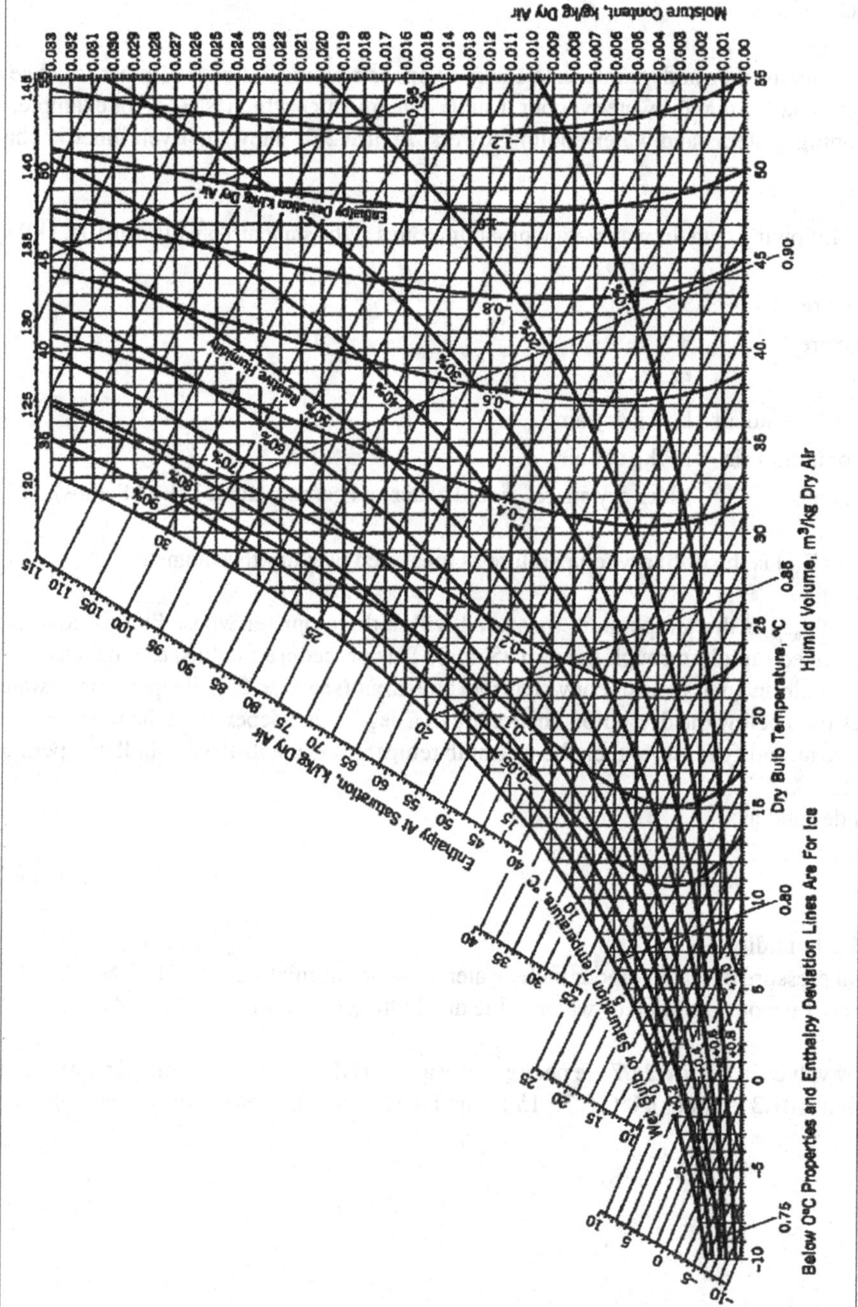

Figure 2.07. Psychrometric chart for H_2O vapour–Air.
Reference states: H_2O(l), Dry Air (g) at 101.3 kPa(abs), 273.15 K.
[Reprinted with permission of Carrier Corporation]

EXAMPLE 2.21 Using the psychrometric chart to find properties of humid air.

5 kg of humid air with dry-bulb temperature = 30°C and wet-bulb temperature = 20°C is cooled to 10°C.

Problem: Use the psychrometric chart to find:

A. The relative humidity of the initial humid air at 101.3 kPa(abs).
B. The moisture content of the initial humid air.
C. The enthalpy of the initial humid air.
D. The dew-point temperature of the air at 101.3 kPa(abs).
E. The mass of water condensed from the air when cooled from 30 to 10°C at 101.3 kPa(abs).

Solution:

A. Locate the intersection of the initial dry-bulb (vertical) and wet-bulb (diagonal) temperature lines.
Interpolate the relative humidity (curved) lines to get:
Initial rh $\approx$ **40%**

B. Locate the intersection of the initial dry-bulb (vertical) and wet-bulb (diagonal) temperature lines.
Follow the constant moisture content (horizontal) line across to the moisture content scale to get:
Initial moisture content = **0.0106 kg water/kg dry air**

C. Locate the intersection of the initial dry-bulb (vertical) and wet-bulb (diagonal) temperature lines.
Follow the diagonal line (constant wet-bulb temperature) back to the enthalpy scale to get:
Enthalpy at saturation = 57.6 kJ/kg dry air
Interpolate the enthalpy deviation lines at the dry-bulb/wet-bulb point to get:
Enthalpy deviation = -0.35 kJ/kg dry air
Add the enthalpy deviation to the enthalpy at saturation to get:
Enthalpy = - 0.35 + 57.6 $\approx$ **57.3 kJ/kg dry air**

D. Locate the intersection of the initial dry-bulb (vertical) and wet-bulb (diagonal) temperature lines.
Follow the constant moisture content (horizontal) line back to the saturation curve to get: T_D = **14.5°C**

E. The final air temperature (10°C) is lower than the initial dew point (14.5°C), so at 10°C the air will be saturated. In the final condition: Dry-bulb temperature = wet-bulb temperature = 10°C.
Locate the final condition ($T_{DB} = T_{WB}$ = 10°C) and follow the constant moisture content line to get:
Final moisture content = 0.0076 kg water/kg dry air
The mass of dry air is fixed, while the mass of water it contains is decreased on cooling from 30 to 10°C.
Mass of dry air = (5 kg humid air)(1 / (1 + 0.0106)) kg dry air/kg humid air = 4.948 kg dry air
Initial mass of water in air = (4.948 kg dry air)(0.0106 kg water/kg dry air) = 0.0524 kg water
Final mass of water in air = (4.948 kg dry air)(0.0076 kg water/kg dry air) = 0.0376 kg water
Mass of water condensed = Initial mass – final mass = **0.0148 kg water**

SUMMARY

[1] Keep your calculations true by ensuring equations have consistent units, check that the units "balance" at each step of the calculation and report results with their units, rounded to an appropriate number of significant figures.

[2] Pressure (P) and temperature (T) are usually measured as relative values (e.g. kPa(gauge) and °C in the SI system), whereas most thermodynamic calculations (e.g. gas law) require P and T as absolute values (e.g. kPa(abs) and K in the SI system). The relations between these values (assuming the standard atmospheric pressure) are:

	SI System	Imperial-American System
Pressure:	–	–
Super-atmospheric	kPa(abs) = kPa(g) + 101.3	psia = psig + 14.7
Sub-atmospheric	kPa(abs) = 101.3 – kPa(vac)	psia = 14.7 – psi(vac)
Temperature:	K = °C + 273.1	°R = °F + 459.6 [°R = 1.8K]

[3] The physical condition of any material can be described by a range of properties, including the phase, density, specific volume, molar volume, specific gravity, concentration (by mass, moles or volume), etc. You should know what these measures mean and have an idea of the ranges of their values for common materials.

[4] The ideal gas equation (pV = nRT) is an equation of state that applies with sufficient accuracy for most purposes to any gas when the reduced pressure $P_r = P/P_c \leq 10$ and reduced temperature $T_r = T/T_c \geq 2$.

For gases at conditions near the critical point (P_c,T_c) the ideal gas equation is modified to pV = znRT, where the gas compressibility factor "z" ranges from about 0.2 to 4 depending on the proximity to the critical state, at which: $P_r = T_r = 1$.

Common gases such H_2, N_2, O_2, CH_4, CO and NO have critical temperature below 200 K and can (for most purposes) be treated as ideal gases when the partial pressure is below 1E4 kPa(abs) and temperature above 273 K. Other gases such as H_2O, CO_2, C_4H_{10}, N_2O, NO_2, SO_2 and SO_3 have critical temperatures above 300K and may be condensed near ambient conditions but have critical pressure above 4E3 kPa(abs) and can usually be treated as ideal gases at partial pressure below the vapour pressure and temperature above 273 K.

The ideal gas law does not apply to liquids or solids.

[5] For liquids and solids the equation of state involves relatively small effects of pressure and temperature that are expressed as the coefficients of compressibility ($\delta = -1/v_o\,(dv/dP)$) and of thermal expansion ($\beta = 1/v_o\,(dv/dT)$), whose respective values are typically below 1E-6 kPa^{-1} and 1E-3 K^{-1}.

[6] Heat capacity is specified at constant pressure (C_p) or at constant volume (C_v).

Liquids and solids: $C_p \approx C_v$
both values are nearly independent of pressure and increase with temperature.
Ideal gases: $C_p = C_v + R$
both values are independent of pressure and increase with temperature.
Non-ideal gases: C_p and C_v both depend on pressure and temperature and become infinite at the critical state.

The heat capacity of a material depends strongly on its phase and is not defined over a phase change.

[7] Specific enthalpy and internal energy are state functions that are used to measure the energy content of materials relative to an arbitrarily specified reference condition. Two common reference conditions are:

A. The *elements* of the material in their standards states at 101.3 kPa(abs), 298.15 K.
B. The *compounds* of the material at a specified condition, such as liquid water at its triple point (0.61 kPa(abs), 273.16 K), as used in the steam table.

[8] The enthalpy (h*) defined in 7(A) contains terms for sensible heat, latent heat and heat of formation and is the most useful in energy balance calculations for processes with chemical reactions. The enthalpy (h) defined in 7(B), which does not include the heat of formation, is the value reported in most diagrams and tables of thermodynamic properties and commonly used in energy balance calculations on processes that do not involve chemical reactions.

[9] The phase rule [$F^{\#} = C^{\#} - \Pi^{\#} + 2$] defines a thermodynamic constraint on the number of intensive variables ($F^{\#}$), needed to fully-specify the state of each phase in a system, the number of components in the system ($C^{\#}$) and the number of phases in the system ($\Pi^{\#}$) at equilibrium. For non-reactive systems the number of components is the number of chemical species in the system, but in reactive systems becomes the number of chemical species minus the number of independent reactions.

[10] Mixtures can take the form of multi-phase or single-phase systems. Multiphase mixtures are called dispersions and are exemplified by systems such as emulsions (L/L), foams (G/L) and slurries (S/L) in which the disperse phase consists of drops, bubbles or particles with diameter exceeding about 1 nm. Single-phase mixtures are called solutions and are classified as either ideal or non-ideal mixtures (a.k.a. solutions) depending on the degree of interaction between the constituent molecules.

[11] The formation of non-ideal mixtures (a.k.a. solutions) involves an enthalpy change called the "heat of mixing" that may be positive or negative and must be added to the pure component enthalpies to get the enthalpy of a non-ideal mixture. In many cases the heat of mixing is a relatively small effect that can be ignored in energy balances, but in some cases (such as sulphuric acid/water) the heat of mixing is substantial and should be included in energy balance calculations.

[12] The physical properties of substances depend strongly on phase and in most process calculations it is necessary to know the phases of the substances involved. For single pure substances, the phase can be obtained from thermodynamic diagrams or tables, or by calculations with equilibrium relations such as the Antoine equation for vapour pressure. The phase relations for mixtures of substances (i.e. in multi-species, multi-phase systems) may be simple or complex, depending on the degree of non-ideality of the mixture. Vapour/liquid equilibrium in multi-phase (G/L) systems is given by Raoult's law for ideal mixtures and can be approximated by Henry's law for non-ideal mixtures with low solute concentrations (i.e. < 5 mole%). Henry's law is useful particularly to calculate solubilities on non-condensable gases in liquids.

Equilibrium in other multi-phase systems is characterised by the adsorption isotherm (G/S systems), the distribution coefficient (L/L systems) and the solubility (L/S systems), all of which are strongly temperature dependent and usually a non-linear function of composition.

[13] The calculation of amounts of materials consumed and generated in a chemical reaction begins with a balanced stoichiometric equation:

$$\alpha A + \beta B \rightarrow \chi C + \delta D$$

The numbers α, β, χ and δ are the stoichiometric coefficients of the reaction, in which:

α moles A + β moles B react to generate χ moles C + δ moles D

The conservation of mass dictates that for any balanced chemical reaction:

$$\Sigma[\nu(j)M(j)] = 0$$

where: M(j) = molar mass
ν(j) = stoichiometric coefficient (–ve for reactants, +ve for products)

[14] The progress of chemical reactions is measured by conversion, extent of reaction, selectivity and yield. Conversion and selectivity are dimensionless ratios that measure respectively; the fraction of a reactant consumed and the fraction of a reactant converted to a specified product. The yield of a specific product is then the reactant conversion multiplied by the corresponding selectivity. The extent of reaction is the moles of a species consumed or generated by the reaction, divided by its stoichiometric coefficient. The extent of reaction has dimensions of moles and has the same value for all species in a given reaction.

For a single reaction the conversion and the extent of reaction are essentially measures of the same quantity and either may be used in material balance calculations involving chemical reactions.

[15] Reversible chemical reactions tend to an equilibrium condition defined by the equilibrium conversion (X_{eq}). All reactions occur at finite net rates that decrease to zero at equilibrium, so the equilibrium conversion can never be reached in a real reaction. The actual conversion (X_{act}) depends on the reaction rate and residence time and is always below the equilibrium conversion. The ratio X_{act} / X_{eq} can exceed 0.99 in many practical processes, but in other equally important processes X_{act}/X_{eq} may be as low as 0.05.

[16] The thermodynamic data needed for process calculations comes from equations or from thermodynamic diagrams and tables. The equations used to estimate specific enthalpy are:

$$h^* \approx \int_{T_{ref}}^{T} C_p . dT + h^o_{f,T_{ref}} = C_{p,m}(T - T_{ref}) + h^o_{f,T_{ref}} \qquad \text{(see \textit{Equation 2.62})}$$

[Element reference state] [Respect the phase]

or

$$h \approx \int_{T_{ref}}^{T} C_p . dT + h_{f,T_{ref}} = C_{p,m}(T - T_{ref}) + h_{p,T_{ref}} \qquad \text{(see \textit{Equation 2.65})}$$

[Compound reference state] [Respect the phase]

where:

h^*	= specific enthalpy of the substance at temperature T in phase Π w.r.t. *elements* at the standard state	$kJ.kmol^{-1}$
h	= specific enthalpy of the substance at temperature T in phase Π w.r.t. *compounds* at the reference state	$kJ.kmol^{-1}$
C_p	= heat capacity at constant pressure of the substance in phase Π	$kJ.kmol^{-1}.K^{-1}$
$C_{p,m}$	= mean heat capacity at constant pressure, w.r.t. T_{ref} in phase Π	$kJ.kmol^{-1}.K^{-1}$
$h^o_{f,Tref}$	= heat (enthalpy) of formation of the substance at T_{ref} in phase Π	$kJ.kmol^{-1}$
$h_{p,Tref}$	= latent heat of any phase changes required to convert the substance from its reference state at T_{ref} to the phase Π at T_{ref}	$kJ.kmol^{-1}$
T	= temperature	K
T_{ref}	= reference temperature	K

More accurate values of specific enthalpy (h) are obtained from thermodynamic diagrams and tables. Such values are normally based on the *compound* reference state and thus do not include heats of formation. If the *compound* based enthalpy values are used in energy balances with chemical reactions, it is necessary to add the heats of formation to the enthalpies or to otherwise account for the heat of reaction (see Chapter 5).

FURTHER READING

[1] R. H. Perry *et al.* (Eds.), *Chemical Engineers' Handbook*, 6th Edition, McGraw-Hill, New York, 1985.

[2] R. C. Weast (Ed.), *Handbook of Chemistry and Physics*, 56th Edition, CRC Press, Cleveland, 1975.

[3] C. L. Yaws, *Physical Properties*, McGraw-Hill, New York, 1977.

[4] R. M. Felder and R. W. Rousseau, *Elementary Principles of Chemical Processes*, John Wiley & Sons, New York, 2000.

[5] R. C. Reid, J. M. Praunitz and B. E. Poling, *The Properties of Gases and Liquids*, McGraw-Hill, New York, 1987.

[6] J. N. Noggle, *Physical Chemistry*, HarperCollins, 1989.

[7] H. D. Young, *University Physics*, Addison-Wesley, New York, 1992.

[8] P. M. Doran, *Bioprocess Engineering Principles*, Academic Press, San Diego, 1995.

[9] H. F. Rase, *Chemical Reactor Design for Process Plants*, John Wiley & Sons, New York, 1977.

[10] O. Levenspiel, *Chemical Reaction Engineering*, 2nd Edition, John Wiley & Sons, New York, 1972.

[11] S. I. Sandler, *Chemical and Engineering Thermodynamics*, John Wiley & Sons, New York, 1999.

[12] J. Winnick, *Chemical Engineering Thermodynamics*, John Wiley & Sons, New York. 1997.

CHAPTER THREE

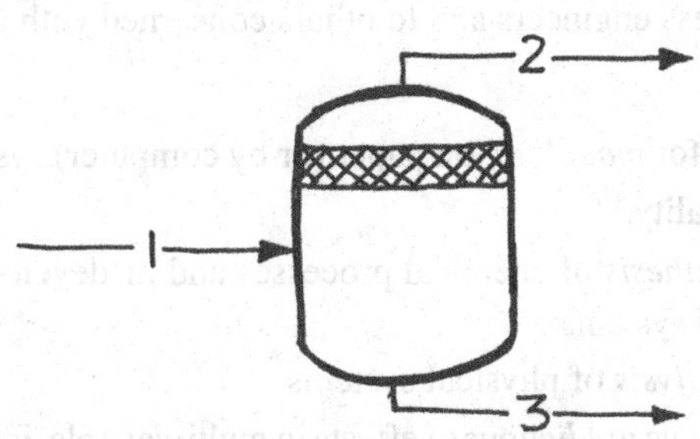

COMPONENT	M	STREAM (kmol/h)		
	kg/kmol	1	2	3
[A] C_5H_{12}	72	80	66.4	13.6
[B] $C_6H_5CH_3$	92	40	18.4	21.6
TOTAL	kmol/h	120	84.8	35.2
TOTAL	kg/h	9440	6475	2965
PHASE	–	L	G	L
PRESSURE	kPa (abs)	6000	800	800
TEMPERATURE	K	510	424	424
VOLUME	m^3/h	15.0	374	4.6
ENTHALPY	kJ/h	-8.70E6	-7.45E6	-1.28E6

MATERIAL AND ENERGY BALANCES IN PROCESS ENGINEERING

This chapter explains the significance of material and energy balances in process engineering and introduces the concepts of flowsheets, process units and stream tables, which are used throughout subsequent chapters.

SIGNIFICANCE OF MATERIAL AND ENERGY BALANCES

The Material and Energy (M&E) balance is a primary tool used in the design of industrial chemical processes. Material and energy balance calculations are based on the general balance equation and have several features that make them useful both to process engineers and to others concerned with understanding the behaviour of physical systems, as follows:

- M&E balances provide a basis for *modelling* on paper (or by computer) systems that would be difficult and/or expensive to study in reality.[1]
- M&E balances assist in the *synthesis* of chemical processes and in developing *conceptual paradigms* for other man-made and natural systems.
- M&E balances guide in the *analysis* of physical systems.
- M&E balances result in *quantitative predictions* of effects in multi-variable, interacting, non-linear systems.
- M&E balances provide a basis for estimating the *economic costs and benefits* of a project and for assessing the physical consequences of system change.

As well, material and energy balances can play an important role in assessing the "triple bottom line", of the industrial economy. The "triple bottom line" is a method of accounting for ecological, economic and social factors in the sustainable development of our civilisation.

FLOWSHEETS

Process engineers usually begin modelling a system by describing it in a flowsheet. Flowsheets may have various levels of complexity, ranging from a simple conceptual flowsheet to a detailed P&ID (process/piping and instrumentation diagram) that is the basis for plant construction. This text is concerned only with conceptual flowsheets.

In a conceptual flowsheet the full system is broken down into a set of singular *process units* interconnected by *streams* that define the paths of material flow between the process units. *Figure 3.01* shows a typical conceptual flowsheet for an industrial process used to make penicillin (see *Refs. 1–4*).

PROCESS UNITS

In process engineering a *process unit* is a piece of hardware that operates on one or more input(s) to produce one or more modified output(s). Actual process units can range from a simple tank to massive and complex pieces of equipment such as fractionation towers, catalytic crackers and multiple effect evaporators (see *Refs. 2, 3*).

---

[1] The process unit models used in this text are "lumped" models, within which the spatial distribution of matter and energy is not resolved. In other words, each process unit is treated as a "black box". More sophisticated M&E balances, with partial differential equations, can be used to develop "distributed" models that map the distribution of matter and energy inside the process unit.

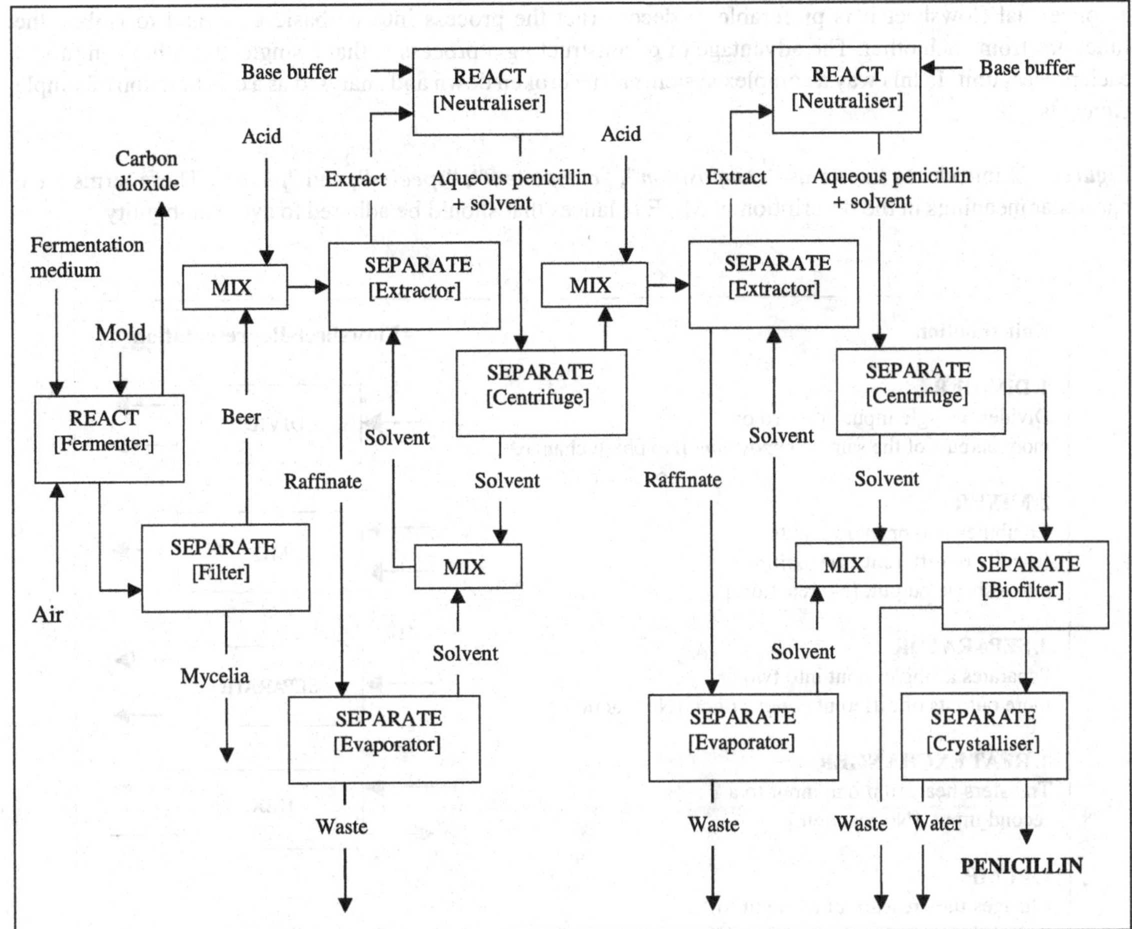

Figure 3.01. Conceptual flowsheet for the industrial production of penicillin.

All process flowsheets can be conceptually simplified to a combination of unit(s), each of which performs one of the six basic functions shown in *Figure 3.02*. The basic unit functions shown in *Figure 3.02* are the building blocks of all conceptual flowsheets and are used throughout this text to model chemical processes. You should study *Figure 3.02* carefully, since your understanding of these six basic unit functions will be critical to your ability to formulate and solve M&E balance problems.

In reality a single piece of equipment may perform several basic functions. For example, a "pump" can be used both to pump and mix fluids at the same time. A "separator" such as an evaporator can include a heat exchanger. A "reactor" can be operated to simultaneously mix, heat, react and separate materials. Such complex multi-function units are modelled in commercial process simulators as "user defined" blocks. Nevertheless, for a conceptual flowsheet it is preferable to deconstruct the process into its basic units and to isolate the functions from each other. The advantage of deconstructing a process is that a single algorithm can model each process unit. In this way a complex system can be broken down and analysed as a combination of simple elements.

Figure 3.02 introduces the terms "*composition*", "*component*", "*species*", and "*phase*". These terms have particular meanings in the description of M&E balances that should be adhered to avoid ambiguity.

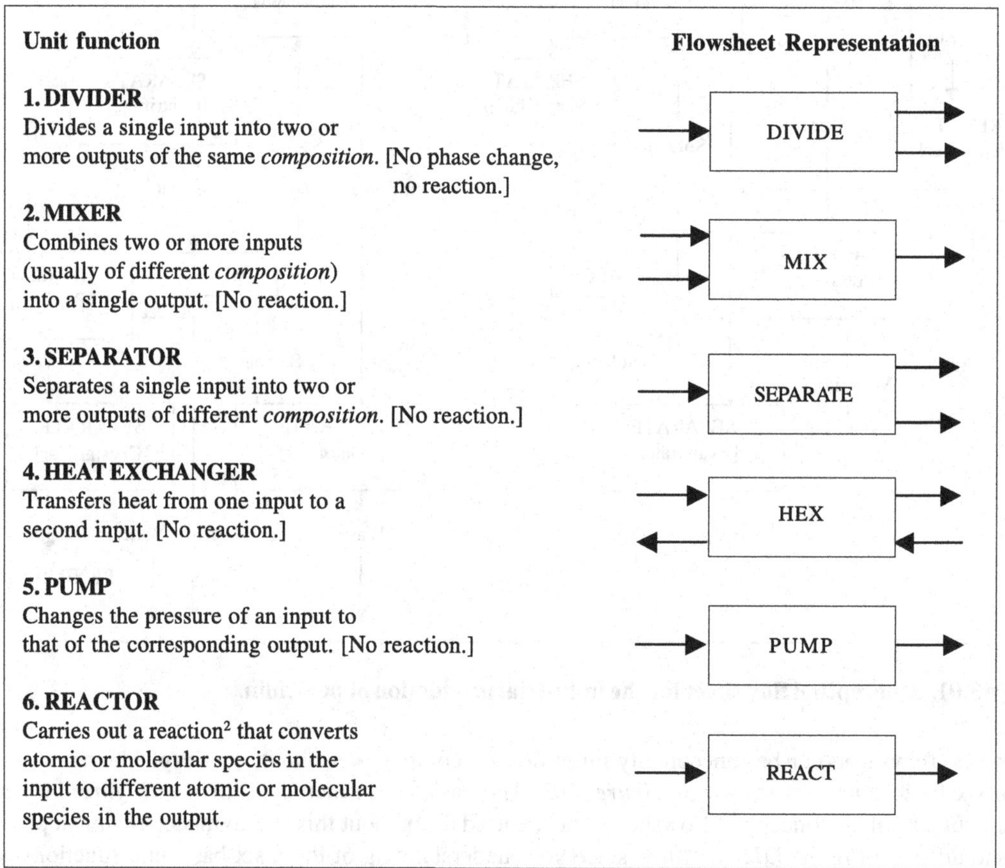

Figure 3.02. The six basic process units.

[2] A reaction can be a nuclear reaction that changes the configuration of atomic nuclei or a chemical reaction that changes the configuration of molecules. Only chemical reactions are considered in this text. Chemical reactions may be biochemical, electrochemical, photochemical or thermochemical reactions.

Composition = the fractional amount of each *component* in a mixture (e.g. mass fraction or mole fraction)
Component[3] = a constituent of a mixture identified as a *phase* or as a *species*
Species = a constituent of a mixture identified by a specific atomic or molecular configuration
Phase = a constituent of a mixture identified as a solid, a liquid or a gas

One of the main difficulties in learning a technical subject such as M&E balances is in the potential for ambiguity incurred by the use of common words with specialised meanings. The specialised meanings are necessary to avoid ambiguity. You should be aware of the specialised meanings of words in your field and take care to "AAA", that is: "Always Avoid Ambiguity".

Dividers (and occasionally separators) are sometimes called "splitters". To prevent confusion of their different functions neither dividers nor separators are referred to as splitters in this text.

In a divider all inputs and outputs have the same composition. For a continuous process a divider works like a tee in a pipe and merely sets the flow rates of the output streams in a desired ratio, keeping all their compositions the same as that of the input stream. A divider does not change phases, separate phases or allow reactions to occur.

A **mixer** adds inputs together to produce a single output. A simple mixer may consist of a tank fitted with a stirrer. A mixer does not allow reactions to occur.

A **separator** performs a more complex function than a divider or a mixer. A separator takes a *multi-component* input and manipulates it to produce a set of outputs of different composition. Some examples of separators are: filters, evaporators and fractionating columns. A separator can change phases and/or separate phases but does not allow reactions to occur.

A **heat exchanger** transfers heat between two inputs that are at different temperatures. Heat is transferred from the input at high temperature to the input at low temperature and there must always be a finite difference in temperature between the two streams, such that $T_{hot} > T_{cold}$. A heat exchanger may or may not cause a change in phase in one or both inputs, but does not allow a reaction. There are many types of heat exchangers, in categories such as direct contact, indirect contact, co-current, counter-current, concentric tube, shell and tube, plate, etc. (*Refs. 2, 3, 8*).

A **pump** makes the pressure of an output fluid higher or lower than that of the corresponding input fluid. Pumps that handle liquids or slurries are usually referred to as pumps while those that handle gases are usually called fans, blowers or compressors. Pumps that reduce gas pressure are called "vacuum pumps". Pumps generally convert mechanical energy into fluid flow energy. Working in reverse, pumps, which convert fluid flow energy into mechanical energy, are called turbines or expansion engines. Again there are many types of pumps, in categories such as centrifugal, diaphragm, piston, gear, etc. (see *Refs. 2, 3, 8*).

3 "Component" has a strict thermodynamic definition, i.e. a species in a mixture whose amount can be independently varied. However the word "component" is used more broadly in some texts to mean either an individual species or a single phase mixture of invariant composition, such as air or sea water.

A **reactor** performs the most complex function of the set of basic process units. Reactors fall into two categories:

- Nuclear reactors, which change the configuration of atomic nuclei; (e.g. U_{235} to U_{236} then to Ba_{144} and Kr_{89} by nuclear fission).
- Chemical reactors, which change the configuration of chemical species[4] (e.g. $C + O_2$ to CO_2).

The chemical reactors treated in this text and are defined broadly to embrace any type of reaction that changes chemical species — including biochemical, electrochemical, photochemical, plasma and thermochemical reactions. Most reactions with which you are probably familiar are thermochemical reactions, such as:

- Neutralisation (acid/base) reaction $HCl + NaOH \rightarrow NaCl + H_2O$ *Reaction 3.01*
- Combustion reaction $CH_4 + 2O_2 \rightarrow CO_2 + 2H_2O$ *Reaction 3.02*
- Organic chlorination reaction $C_2H_4 + Cl_2 \rightarrow C_2H_2Cl_2 + 2HCl$ *Reaction 3.03*

Note that in all chemical reactions the molecules of the reactant species (left hand side) are *transformed* to the products species (right hand side). This transformation of chemical species is the essential feature of a chemical reaction, in which the reactants are *consumed* and the products are *generated*. The properties of the reactants are usually totally different from the properties of the products. For example, in *Reaction 3.01* the reactants are hydrochloric acid (HCl) and sodium hydroxide (NaOH), both corrosive and toxic chemicals, whereas the products are sodium chloride (common salt, NaCl) and water (H_2O).

All of the process units described above in the context of process engineering have analogies in natural systems. For example, in the human body:

- The arteries act as flow *dividers*
- The mouth acts as a *mixer*
- The lungs act as a *separator*
- The skin acts as a *heat exchanger*
- The heart acts as a *pump*
- The stomach acts as a *reactor*

If we consider a system such as the global environment, then:

- A mountain acts as a flow *divider*
- A river confluence acts as a *mixer*
- A lake acts as a *separator*
- Ocean currents act as *heat exchangers*
- The moon's gravity acts as a *pump*
- A forest acts as a *reactor*

[4] "Chemical species" includes atoms, molecules, ions, and radicals.

STREAMS AND STREAM TABLES

Process streams are the flows of material that interconnect process units in a flowsheet. In an industrial chemical process the streams usually represent conduits, conveyors, pipes or mobile containers. The stream characteristics show the nature of the process units and reflect the overall process objective. Some specific characteristics that can define a stream are: density, phase, composition, mass, mole or volume flow, pressure, temperature and enthalpy flow.

Stream characteristics are typically recorded in one of two ways:

A. In a stream table appended to the flowsheet
B. As sets of data written adjacent to each stream in the flowsheet

This text uses *stream tables* as the primary way to record stream characteristics. In a stream table each stream is identified (by number or name) and the corresponding stream characteristics are listed for quick reference in a single table. Any process can thus be completely defined by a *flowsheet* with its associated *stream table*. *Figure 3.03* shows a flowsheet and stream table for an idealised and simplified continuous process used to produce ammonium nitrate from ammonia and nitric acid by *Reaction 3.04*.

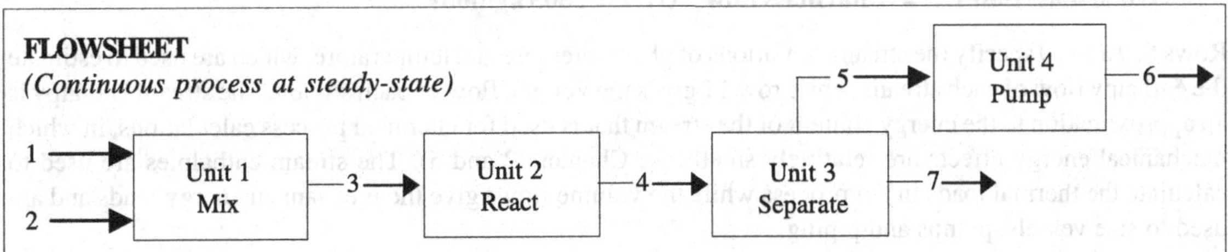

$NH_3 + HNO_3 \rightarrow NH_4NO_3$ *Reaction 3.04*

Row	Component	Molar mass	Stream						
1			*1*	*2*	*3*	*4*	*5*	*6*	*7*
2		kg/kmol				Flow rate kmol/h			
3	[A] NH_3	17.03	50.0	0.00	50.0	0.00	0.00	0.00	0.00
4	[B] HNO_3	63.02	0.00	50.0	50.0	0.00	0.00	0.00	0.00
5	[C] NH_4NO_3	80.05	0.00	0.00	0.00	50.0	0.00	0.00	50.0
6	[D] H_2O	18.02	0.00	100.0	0.00	100.0	100.0	100.0	0.00
7	Total	kg/h	852	4953	5805	5805	1802	1802	4003
8	Phase	–	G	L	L	L	G	G	S
9	Pressure	kPa(abs)	130	130	120	110	10	100	10
10	Temperature	K	298	298	298	298	338	600	338
11	Volume	m^3/h	953	3.59	956.7	4.46	28100	4990	2.32
12	Enthalpy*	kJ/h	-2.31E6	-38.58E6	-40.89E6	-48.05E6	-24.04E6	-23.15E6	-18.28E6
Col >	1	2	3	4	5	6	7	8	9

Figure 3.03. Process flowsheet and stream table for production of ammonium nitrate.

* *Reference condition = elements at standard state, 298K.*

STREAM TABLE NOTES:

1. The molar masses and flows are rounded to keep the stream table uncluttered.
2. In reality NH_3 would react with HNO_3 in the mixer, but in the flowsheet the mixing and reaction function are isolated from each other to clarify and illustrate the calculations. Heats of mixing are assumed negligible.

The stream table in *Figure 3.03* displays the most basic steady-state M&E balance for making ammonium nitrate. Row 1 lists the stream number used to identify each stream on the flowsheet. Column 1 lists the components of the process, which here are the molecular species in each stream, including the solvent water in the feed stream 2. The second column lists the molar mass (molecular weight) of each species, used in the material balance calculations of reaction stoichiometry. Columns 3 to 9 list the composition of each stream in terms of the moles (actually the mole flow rate in kmol/hour) of each species. The "Total" (row 7) shows the total mass flow in each stream and is used to check the integrity of the material balance.

Equation 1.06 is observed over each process unit and the whole system i.e. at steady-state:

For: **Quantity** = **Total mass** **ACC = 0 = IN – OUT + 0 – 0**
Total mass flow IN = **Total mass flow OUT = 5805 kg/hour**

Rows 8, 9 and 10 specify the stream conditions of phase, pressure and temperature, which are used to estimate the enthalpy flow of each stream, while row 11 gives the volume flow of each stream. The stream enthalpy is an approximation to the energy content of the stream that is used for chemical process calculations, in which mechanical energy effects are relatively small (see Chapters 2 and 5). The stream enthalpies are used to calculate the thermal loads in the process while the volume flows give the mechanical energy loads and are used to size vessels, pumps and piping.

PROCESS ECONOMICS

Whether or not the aim is to make a profit, costs and benefits are a major factor in the development of almost all projects. Process engineering always involves a compromise between technical and economic considerations, in which cost and profit are primary measures of the success of a process design. Material and energy balances have a critical role in the estimation of process costs and profitability. For process engineering the *economic viability* of a project is measured by four basic *"figures of merit"*,[5] the *"gross economic potential"*, the *"net economic potential"*, the *"capital cost"* and the *"return on investment"*, which are defined as follows:

- Gross economic potential = **GEP = Value of products – Value of feeds** *Equation 3.01*
- Net economic potential = **NEP = GEP – cost of [*utilities*[6] + *labour* + *maintenance* + *interest*[7]]** *Equation 3.02*
- Capital cost = **Total cost of plant design, construction and installation**
- Return on investment = **ROI = NEP / Capital cost** *Equation 3.03*

---

[5] A *figure of merit* (a.k.a. performance indicator) is a quantity that summarizes value in a specific application.
[6] *Utilities* include items such as electricity, fuel and water needed to operate the plant.
[7] *Interest* is the cost of borrowing money.

The usual units for these quantities are: GEP [$/year] Capital cost [$]
NEP [$/year] ROI [%/year]

The calculation of each of these figures of merit hinges on material and energy balances, for example:

- GEP — Quantities of feeds (inputs) and products (outputs) are found by **material balances**
- NEP — Consumption of *utilities* is found by **energy balances**
- Capital cost. The size (and thus the cost) of each process unit is determined by the amounts of material and energy that it handles. The design and costing of process units involves calculations of *process rates*[8] and equipment size that are beyond the scope of this text (see *Refs. 3, 5, 6, 8, 9*).

Example 3.01 demonstrates how M&E balances are used to estimate the economic viability of a chemical process.

EXAMPLE 3.01 Calculating the economic figures of merit of a chemical process.

Figure 3.03 shows a conceptual flow-sheet and stream table for a process used to make ammonium nitrate from ammonia and nitric acid, with the stream table values based on an assumption of 100% efficiency for all operations in the process. *Table 3.01* gives some cost data for the process.

Table 3.01. Cost data for ammonium nitrate process.

Cost Data					
Component		NH_3	HNO_3	NH_4NO_3	H_2O
Value	$/kg	0.08	0.10	0.25	0.001
Utilities		Electricity	Heating	Cooling	
Value	$/kWh	0.06	0.05	0.001	
Capital cost of plant = 15E6 $US AD 2002					
Operating period = 8000 hours/year Interest rate = 7%/year					

Problem: Estimate: (a) the GEP, (b) the NEP and (c) the ROI of the ammonium nitrate process.

Solution:

A. Value of products = (50 kmol/h) (80.05 kg/kmol) (0.25 $/kg) = 1000 $/h
Value of feeds = (50 kmol/h) (17.03 kg/kmol) (0.08 $/kg) + (50 kmol/h)(63.02 kg/kmol)(0.10 $/kg) = 383 $/h
GEP = Value of products – Value of feeds
= (1000 $/h – 383 $/h) (8000 h/y) **GEP = 4.94E6 $/y**

B. Utility loads are calculated from the energy balance:

	Load kW	Cost	
Unit 2	2000 (cooling)	(2000 kW)(8000 h/y)(0.001 $/kWh)	= 16E3 $/y
Unit 3	1678 (heating)	(1678 kW)(8000 h/y)(0.05 $/kWh)	= 671E3 $/y
Unit 4	246 (pumping)	(246 kW)(8000 h/y)(0.06 $/kWh)	= 118E3 $/y
		Total utilities	= 805E3 $/y

[8] *Process rates* include heat transfer, mass transfer and reaction rates (intrinsic kinetics).

Labour + maintenance	=	5% of capital cost per year = (0.05 1/y)(15E6 \$)	= 750E3 \$/y
Interest	=	(0.07 1/y)(15E6 \$)	= 1050E3 \$/y
NEP	=	GEP – cost of [utilities + labour + maintenance + interest]	
	=	4.94E6 – (805E3 + 750E3 + 1050E3)	**NEP = 2.34E6 \$/y**

C. ROI = Annual NEP / Capital cost = 2.34E6 \$/y / 15E6 \$ = 0.16 **ROI ≈ 16%/y**

Note that the ROI calculated by the method of *Example 3.01* is a very approximate value that would subsequently be refined by more a comprehensive analysis with realistic process efficiencies. The more complex M&E balances required for such work are shown in Chapters 4, 5 and 6 of this text.

SUMMARY

[1] The M&E balance is the primary tool used in the modelling, design and costing of industrial chemical processes.

[2] M&E balances are calculated from the general balance equation and their results displayed using a combination of:

- A *flowsheet*, which is a diagram showing the process units with their interconnecting material flow paths
- A *stream table*, which tabulates the conditions and the quantities of material and energy in each process stream

[3] All process flowsheets can be conceptually simplified to a combination of one or more process units, each of which performs one of the six basic functions. These functions are: DIVIDE, MIX, SEPARATE, HEAT, EXCHANGE, PUMP, REACT. In commercial process simulation, multi-function process units may be modelled by "user" defined blocks that combine two or more of these basic functions.

[4] The information in a process stream table is the basis for calculation of the figures of merit that measure economic viability:

- The *gross economic potential* (GEP) is found from the material balance
- The *net economic potential* (NEP) is found from the GEP plus utility loads calculated by the energy balance
- The *capital cost* is found from design specifications based on both the material and the energy balance
- The *return on investment* (ROI) is estimated as: ROI = NEP/capital cost

FURTHER READING

[1] W. L. Faith, D. H. Keyes and R. L. Clark, *Industrial Chemicals,* John Wiley & Sons, New York, 1966.

[2] J. Kroschwitz *et al.* (Eds.), *Kirk-Othmer Encyclopedia of Chemical Technology,* John Wiley & Sons, New York, 1999.

[3] R. H. T. Perry *et al.* (Eds.), *Chemical Engineers' Handbook,* McGraw-Hill, New York, 1985 and later editions.

[4] T. M. Duncan and J. A. Reimer, *Chemical Engineering Design and Analysis,* Cambridge University Press, 1998.

[5] R. K. Sinnott, *Chemical Engineering Design,* Butterworth-Heinemann, Oxford, 1999.

[6] G. D. Ulrich, *A Guide to Chemical Engineering Process Design and Economics,* John Wiley & Sons, New York, 1984.

[7] J. M. Douglas, *Conceptual Design of Chemical Processes,* McGraw-Hill, New York, 1988.

[8] S. M. Walas, *Chemical Process Equipment — Selection and Design,* Butterworth-Heinemann Boston, 1990.

[9] M. S. Peters and K. D. Timmerhaus, *Plant Design and Economics for Chemical Engineers,* McGraw-Hill, New York, 1980.

[10] L. T. Biegler, I. E. Grossmann and A. W. Westerberg, *Systematic Methods of Chemical Process Design,* Prentice Hall, Upper Saddle Rover, 1997.

[11] W. D. Seider, J. D. Seader and D. R. Lewin, *Process Design Principles,* John Wiley & Sons, New York, 1999.

[12] R. M. Felder and R. W. Rousseau, *Elementary Principles of Chemical Processes,* John Wiley & Sons, New York, 2000.

CHAPTER FOUR

GENERAL MATERIAL BALANCE

Material balance calculations hinge on the *general balance equation,* introduced in Chapter 1 and repeated below as *Equation 4.01.*

For a defined *system* and a specified *quantity:*

Accumulation in system = Input to system − Output from system + Generation in system − Consumption in system *Equation 4.01*

The *system* is a space whose closed boundary is defined to suit the problem at hand. For material balance problems the *quantity* in *Equation 4.01* is an amount of material, usually measured as mass or as moles. *Equation 4.01* then represents the general form of the material balance.

Equation 4.01, in abbreviated form, can be an integral balance or differential balance:

Integral balance: *Equation 4.02*

ACC = IN − OUT + GEN − CON

where:

ACC ≡ (final amount of material − initial amount of material) in the system

IN ≡ amount of material entering the system (input)

OUT ≡ amount of material leaving the system (output)

GEN ≡ amount of material generated in the system (source)

CON ≡ amount of material consumed in the system (sink)

Differential balance: *Equation 4.03*

Rate ACC = Rate IN − Rate OUT + Rate GEN − Rate CON

where:

Rate ACC ≡ rate of accumulation of material in the system, w.r.t. time

Rate IN ≡ rate of material entering the system, w.r.t. time (input)

Rate OUT ≡ rate of material leaving the system, w.r.t. time (output)

Rate GEN ≡ rate of material generated in the system, w.r.t. time (source)

Rate CON ≡ rate of material consumed in the system, w.r.t. time (sink)

In every case the definition of the system boundary is a critical first step in setting up and solving material balance problems. The boundary must be a closed envelope that surrounds the system of interest. All materials crossing the system boundary are considered as input or outputs, and everything occurring inside the system boundary is considered to contribute to the accumulation, generation (source) or consumption (sink) terms in *Equation 4.01.*

Your choice of the system boundary and the quantity balanced can have a big effect on the complexity of the solution of a material balance problem. As a rule you should use the simplest combination of system and quantity that will give the desired result. For some problems the simplest combination is: system = overall process and quantity = mass or moles of specified elements (i.e. an atom balance), whereas more difficult problems may require individual species mole balances on each sub-unit of the process. Sometimes a difficult material balance problem can be easily solved by combining balances on a set of different (carefully chosen) envelopes, each of which contains a part of the system of interest.

In particular cases *Equations 4.02* and *4.03* can be simplified as follows:

	Integral balance	**Differential balance**
Closed system (batch process):	IN = OUT = 0	Rate IN = Rate OUT = 0
Open system (continuous process) at steady-state:	ACC = 0	Rate ACC = 0
Atoms*:	GEN = CON = 0	Rate GEN = Rate CON = 0
Mass or moles of a species with no chemical reaction*:	GEN = CON = 0	Rate GEN = Rate CON = 0
Total mass*:	GEN = CON = 0	Rate GEN = Rate CON = 0
Total moles with no chemical reaction*:	GEN = CON = 0	Rate GEN = Rate CON = 0

** Assumes no nuclear reactions.*

The quantities used in *Equation 4.02* and *4.03* must be clearly specified to avoid ambiguity. For this purpose the nomenclature used in this text is given in *Table 4.01*. With the nomenclature of *Table 4.01* a process *stream table* is described by a two-dimensional matrix, convenient for computer code and spreadsheet calculations.

Table 4.01. Nomenclature for material balances.

Symbol	Meaning	Typical Units
M(j)	molar mass of component j	$kg.kmol^{-1}$
m(j)	mass of component j	kg
n(j)	moles of component j	kmol
V(j)	volume of component j	m^3
$\dot{m}(i, j)$	mass flow rate of component j in stream i	$kg.s^{-1}$
$\dot{n}(i, j)$	mole flow rate of component j in stream i	$kmol.s^{-1}$
$\overline{m}(i)$	total mass flow rate in stream i	$kg.s^{-1}$
$\overline{n}(i)$	total mole flow rate in stream i	$kmol.s^{-1}$
$\dot{V}(i)$	total volume flow rate of stream i	$m^3.s^{-1}$
For example: $\dot{n}(5, A)$ = mole flow rate of component A in stream 5		

CLOSED SYSTEMS (BATCH PROCESSES)

A closed system is defined as one in which, over the time period of interest, there is *zero transfer of material across the system boundary.* Closed systems are also called "controlled mass" systems and correspond to batch processes. Closed systems are common in laboratories and industrial processes but rare in nature. Some examples of closed systems are:

- Batch reactor
- Pressure cooker (before it vents)
- Sealed bottle of beer
- Boiling egg*
- Planet Earth with its atmosphere*

** These natural systems are not strictly closed, since eggs do transpire and some material does enter and leave the perimeter of Earth's atmosphere.*

The integral and differential material balances for a closed system are respectively:

ACC	**= GEN – CON**	(Integral material balance)	*Equation 4.04*
Rate ACC	**= Rate GEN – Rate CON**	(Differential material balance)	*Equation 4.05*

Examples 4.01 and *4.02* illustrate material balance calculations for closed systems.

EXAMPLE 4.01 Material balance on carbon dioxide in a sealed vessel (closed system).

A sealed vessel batch reactor initially contains 3.00 kg carbon + 16.0 kg oxygen + 0.2 m^3 1 M aqueous sodium hydroxide solution. The carbon is ignited and undergoes 100% conversion to carbon dioxide by *Reaction 1*. Carbon dioxide is then absorbed into the solution where it converts 100% of the NaOH to Na_2CO_3 by *Reaction 2*.

$C(s) + O_2(g) \rightarrow CO_2(g)$ *Reaction 1*

$CO_2(g) + 2NaOH(aq) \rightarrow Na_2CO_3(aq) + H_2O(l)$ *Reaction 2*

Problem: Calculate the final mass of carbon dioxide in the vessel.

Solution: Define the system = sealed vessel

Specify the components:	Component	C	O_2	CO_2	NaOH	Na_2CO_3
	Number	1	2	3	4	5
	Molar mass	12	32	44	40	106 (Rounded to whole numbers)
Initial amounts:	kg	3	16	0	8	0
	kmol	0.25	0.5	0	0.2	0

Mole balance on CO_2 [Integral balance], **ACC = IN – OUT + GEN – CON** *Equation 4.02*

IN = OUT	= 0		kmol	(closed system)
GEN = (1) n (1)	= (1) (0.25 kmol)	= 0.25	kmol	(stoichiometry *Reaction 1*)
CON = (0.5) n (4)	= (0.5) (0.2 kmol)	= 0.1	kmol	(stoichiometry *Reaction 2*)
ACC = 0 – 0 + 0.25 – 0.1		= 0.15	kmol	(*Equation 4.02*)
Final mass of CO_2 = (0.15 kmol)(44 kg.kmol^{-1})		= <u>**6.6**</u>	**kg**	

EXAMPLE 4.02 Material balance on water in a space station (closed system).

A space station with a crew of 30 people begins to orbit Earth with a stock of 18.0E3 kg of clean water. During the 500 day period in orbit there is zero import/export of material to/from the space station.

Clean water is generated in the station as follows:

(i) Recovery (by condensation, distillation and sterilisation) then recycle of water from wastes. This source of clean water is equivalent to 90% of the water used in (iii), plus 90% of the water generated through the metabolic bio-oxidation of 0.5 kg/(person/day) of dry food to CO_2 and H_2O.
Food is assumed[1] composed only of carbohydrate, whose reaction is:

$$C_6H_{10}O_5 + 6O_2 \rightarrow 6CO_2 + 5H_2O$$ *Reaction 1*

(ii) Recombination of metabolic CO_2 with all available H_2 from water electrolysis:

$$4H_2 + CO_2 \rightarrow CH_4 + 2H_2O$$ *Reaction 2*

Clean water is consumed in the station as follows:

(iii) Each person uses 8 kg clean water/day for sustenance and hygiene.

(iv) Electrolysis of water to supply 0.6 m^3 STP/(person/day) of oxygen gas:

$$2H_2O \rightarrow 2H_2 + O_2$$ *Reaction 3*

Problem: Calculate the stock of clean water remaining at the end of the 500 day mission. [kg]

Solution: Define the system = space station (a closed system)
Specify the quantity = amount (kmol) of clean water
Integral mole balance on clean water, ACC = IN – OUT + GEN – CON

ACC = ? (Final amount of clean water – Initial amount of clean water) in stock
IN = 0 (closed system)
OUT = 0 (closed system)

[1] For the purpose of this problem. Also assume no change in composition or mass of the crew.

GEN: Recycle = (0.9)(30 persons)(8 kg/day)(500 days) / (18 kg/kmol) = 6.00E3 kmol

Bio-oxidation = (0.9)(30 persons)(500 days)(0.5 kg ($C_6H_{10}O_5$) / (person/day) / (162 kg($C_6H_{10}O_5$) / kmol) x (5 kmol H_2O/kmol ($C_6H_{10}O_5$)) = 208.3 kmol

Recombination = ((30 persons)(0.6 m^3 O_2/day)(500 days) / (22.4 m^3 STP/kmol)) x (2 kmol H_2/kmol O_2) x (0.5 kmol H_2O/kmol H_2) = 401.8 kmol

Total GEN = Recycle + Bio-oxidation + Recombination = 6.00E3 + 208.3 + 401.8 = 6.61E3 kmol

CON: By crew = (30 persons)(8 kg/day)(500 days)/(18 kg/kmol) = 6.67E3 kmol

Electrolysis = ((30 persons)(0.6 m^3/ day)(500 day) / (22.4 m^3/ kmol)) / (0.5 kmol O_2/ kmol H_2O) = 803.6 kmol

Total CON = Crew + Electrolysis = 6.67E3 + 804 = 7.47E3 kmol

ACC = IN – OUT + GEN – CON = 0 – 0 + 6.61E3 – 7.47E3 = - 860 kmol

Final amount = Initial amount + ACC = (18.0E3 kg/18 kg/kmol) – 860 kmol = 140 kmol
= 2.52E3 kg clean water

Check CO_2 balance:

GEN = Metabolism = (6)(0.5)(30)(500)/(162) = 277.7 kmol (*Reaction 1*)
CON = Recombination = (1/4)(803.6) = 200.9 kmol (*Reaction 2*)
ACC = 277.7 – 200.9 = 76.8 kmol

Check O_2 balance:

GEN = Electrolysis = (0.5)(803.6) = 401.8 kmol (*Reaction 3*)
CON = Metabolism = (6)(0.5)(30)(500)/(162) = 277.7 kmol (*Reaction 1*)
ACC = 401.8 – 277.7 = 124.1 kmol

Check H_2 balance:

All H_2 generated by electrolysis (*Reaction 3*) is consumed in reaction (*Reaction 2*).

ACC = 0

These species mole balances predict accumulation of CO_2 and of O_2 in the system, so further processing (and/or venting) may be needed to control levels of CO_2 and O_2 in the space station. Similarly, measures should be taken to deal with the CH_4 generated in *Reaction 2*.

NOTE: Without the water derived from oxidation of food the space station would run out of water in 469 days.

OPEN SYSTEMS (CONTINUOUS PROCESSES)

An open system is defined as one in which, over the time period of interest, *material is transferred across the system boundary*. Open systems are also called "controlled volume" systems and correspond to continuous processes. Open systems are common to our experience and include, for example:

- Automobile engines
- Industrial processes
- Living cells
- Human body and its organs
- Atmosphere of planet Earth

Example 4.03 shows a simple material balance on the global carbon cycle to calculate the change in carbon dioxide concentration in the atmosphere of planet Earth. Similar, though more complex, material balances are used by earth scientists to quantify the other geochemical cycles (e.g. the oxygen, nitrogen and water cycles) that govern life in our biosphere (see *Ref. 1*).

EXAMPLE 4.03 Material balance on carbon dioxide in Earth's atmosphere (open system).

The Earth's atmosphere has a volume of 4.0E18 m^3 STP and in 1850 AD contained 0.028 vol% CO_2. Over the period 1850 to 1990 fossil fuels containing the equivalent of 220E12 kg C were burned to produce CO_2, 440E12 kg CO_2 were vented from volcanoes and by decomposition of biomass, 350E12 kg net CO_2 were consumed by photosynthesis on land and 370E12 kg CO_2 were absorbed by the oceans.

Problem:

If all CO_2 gas is well-mixed into the atmosphere, what is the concentration of CO_2 in the atmosphere in 1990? [ppm vol]. Assume total volume of the atmosphere is constant at 4.0 E18 m^3 STP.

Solution:

Define the system = Atmosphere of Earth (an open system)

Specify the components:

Component	C	CO_2	
Number	1	2	
Molar mass	12	44	[Rounded to whole numbers]

Initial amount of CO_2 = n(2) = PV(2)/RT [Ideal gas law. see *Equation 2.13*]
= (101.3 kPa) (4E18 m^3)(0.028E-2) / ((8.314 kJ.$kmol^{-1}$.K^{-1}) (273K))
= 50.0E12 kmol

Mole balance on CO_2 (Integral balance): **ACC = IN – OUT + GEN – CON** *Equation 4.02*

IN	=	fossil fuel + (volcano + biomass decomposition)	
	=	(220E12 kg C)/(12 kg/kmol) + 440E12 kg CO_2/(44 kg.kmol^{-1})	= 28.3E12 kmol CO_2
OUT	=	photosynthesis + absorption	
	=	350E12 kg CO_2/(44 kg/kmol) + 370E12 kg CO_2/(44 kg.kmol^{-1})	= 16.4E12 kmol CO_2
GEN	=	0	
CON	=	0	
ACC	=	28.3E12 − 16.4E12 + 0 − 0	= 11.9E12 kmol CO_2
Final CO_2	=	Initial CO_2 + ACC CO_2 = 50.0E12 + 11.9E12	= 61.9E12 kmol CO_2

Concentration of CO_2 in atmosphere in 1990 = 0.028% (61.9E12 kmol CO_2 / 50.0E12 kmol CO_2) = 0.0347% (vol)
= **347 ppm (vol)**

As shown in *Example 4.03* and later in *Example 4.16*, the principles used here can be applied to modelling any type of open system. The rest of this chapter focuses on open systems in industrial chemical processes.

Industrial chemical processes normally operate as open systems with continuous or semi-continuous input and output of materials. Large industrial processes (e.g. oil refinery, pulp mill, power plant, integrated circuit plant) operate 24 hours per day for 330+ days per year at conditions close to *steady-state*. Such plants typically run under controls that hold process conditions to *optimise*[2] the production, then shut down for maintenance over a few weeks in an operating cycle that may last from one to five years.

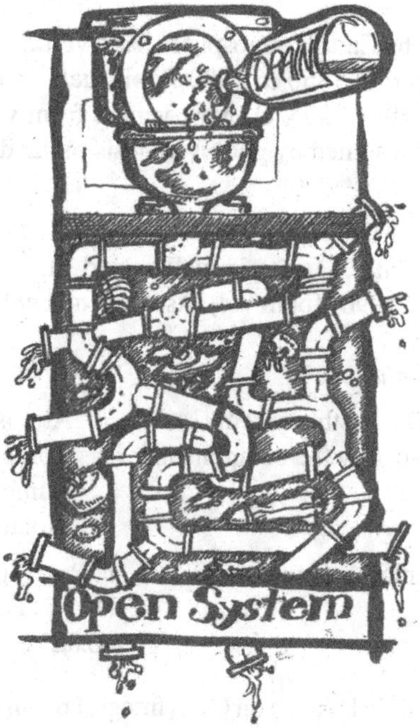

CONTINUOUS PROCESSES AT STEADY-STATE

Material balances for continuous processes at steady-state use the *differential material balance* with Rate ACC = 0.

Rate ACC = 0 = Rate IN − Rate OUT + Rate GEN − Rate CON

Equation 4.03

SINGLE PROCESS UNITS

Chapter 3, *Figure 3.02* describes the six generic process units used to build chemical process flowsheets. Each of these generic units has a material balance specific to its function, as shown in *Figure 4.01* and illustrated in *Example 4.04 A* to *F*.

[2] *Optimise* production does not necessarily mean *maximise* production. Process optimisation is beyond the scope of this text but is treated in *Refs. 9–11*.

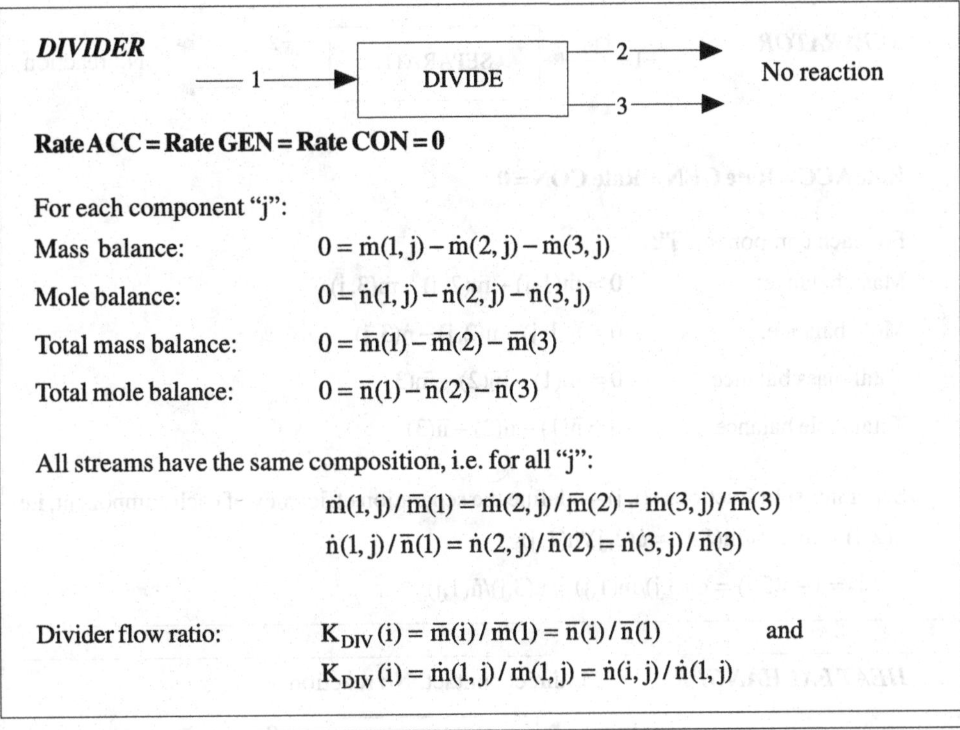

DIVIDER
1
DIVIDE
2
3
No reaction
Rate ACC = Rate GEN = Rate CON = 0
For each component "j":
Mass balance: 0 = ṁ(1, j) − ṁ(2, j) − ṁ(3, j)
Mole balance: 0 = ṅ(1, j) − ṅ(2, j) − ṅ(3, j)
Total mass balance: 0 = m̄(1) − m̄(2) − m̄(3)
Total mole balance: 0 = n̄(1) − n̄(2) − n̄(3)
All streams have the same composition, i.e. for all "j":
ṁ(1, j) / m̄(1) = ṁ(2, j) / m̄(2) = ṁ(3, j) / m̄(3)
ṅ(1, j) / n̄(1) = ṅ(2, j) / n̄(2) = ṅ(3, j) / n̄(3)
Divider flow ratio: K_DIV (i) = m̄(i) / m̄(1) = n̄(i) / n̄(1) and
K_DIV (i) = ṁ(1, j) / ṁ(1, j) = ṅ(i, j) / ṅ(1, j)

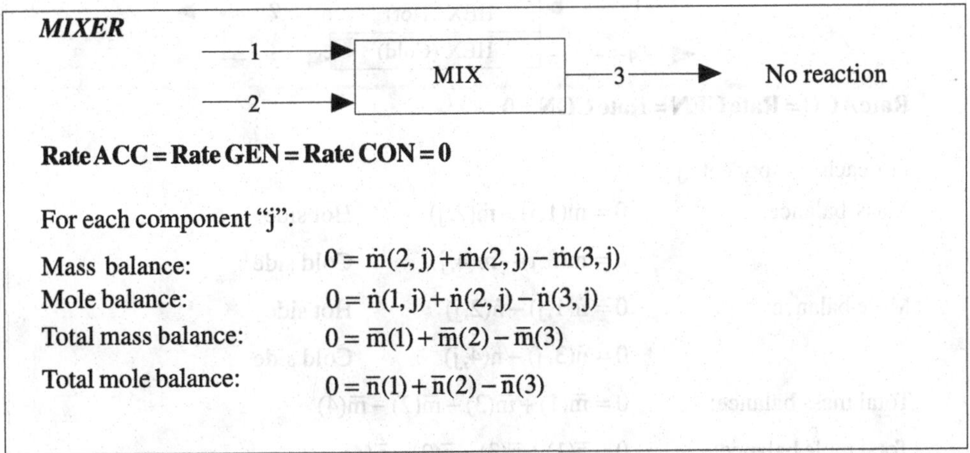

MIXER
1
2
MIX
3
No reaction
Rate ACC = Rate GEN = Rate CON = 0
For each component "j":
Mass balance: 0 = ṁ(2, j) + ṁ(2, j) − ṁ(3, j)
Mole balance: 0 = ṅ(1, j) + ṅ(2, j) − ṅ(3, j)
Total mass balance: 0 = m̄(1) + m̄(2) − m̄(3)
Total mole balance: 0 = n̄(1) + n̄(2) − n̄(3)

Figure 4.01. Material balances on generic process units — continuous processes at steady-state.

SEPARATOR

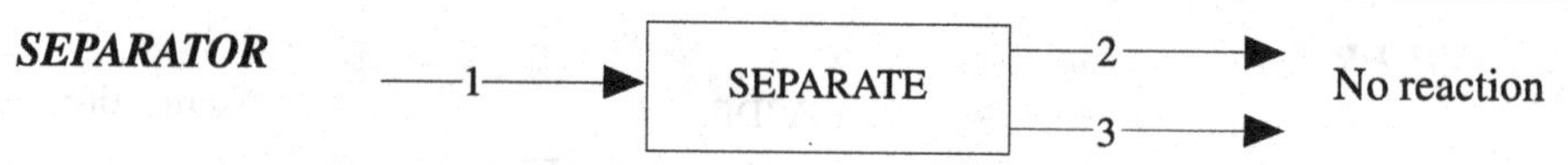

Rate ACC = Rate GEN = Rate CON = 0

For each component "j":

Mass balance:	$0 = \dot{m}(1,j) - \dot{m}(2,j) - \dot{m}(3,j)$
Mole balance:	$0 = \dot{n}(1,j) - \dot{n}(2,j) - \dot{n}(3,j)$
Total mass balance:	$0 = \bar{m}(1) - \bar{m}(2) - \bar{m}(3)$
Total mole balance:	$0 = \bar{n}(1) - \bar{n}(2) - \bar{n}(3)$

Separator split fractions s(i, j) — define the separation efficiency of each component, i.e.

$s(2,j) = \dot{m}(2,j)/\dot{m}(1,j) = \dot{n}(2,j)/\dot{n}(1,j)$

$s(3,j) = 1 - s(2,j) = \dot{m}(3,j)/\dot{m}(1,j) = \dot{n}(3,j)/\dot{n}(1,j)$

HEAT EXCHANGER* Indirect contact. No reaction

1 → HEX (Hot) → 2

4 ← HEX (Cold) ← 3

Rate ACC = Rate GEN = Rate CON = 0

For each component "j":

Mass balance:	$0 = \dot{m}(1,j) - \dot{m}(2,j)$	Hot side
	$0 = \dot{m}(3,j) - \dot{m}(4,j)$	Cold side
Mole balance:	$0 = \dot{n}(1,j) - \dot{n}(2,j)$	Hot side
	$0 = \dot{n}(3,j) - n(4,j)$	Cold side
Total mass balance:	$0 = \bar{m}(1) + \bar{m}(3) - \bar{m}(2) - \bar{m}(4)$	
Total mole balance:	$0 = \bar{n}(1) + \bar{n}(3) - \bar{n}(2) - \bar{n}(4)$	

** The heat exhanger shown here uses countercurrent flow of the hot and cold streams. Values of [T(1) – T(4)] and [T(2) – T(3)] are called respectively the hot and cold "approach".*

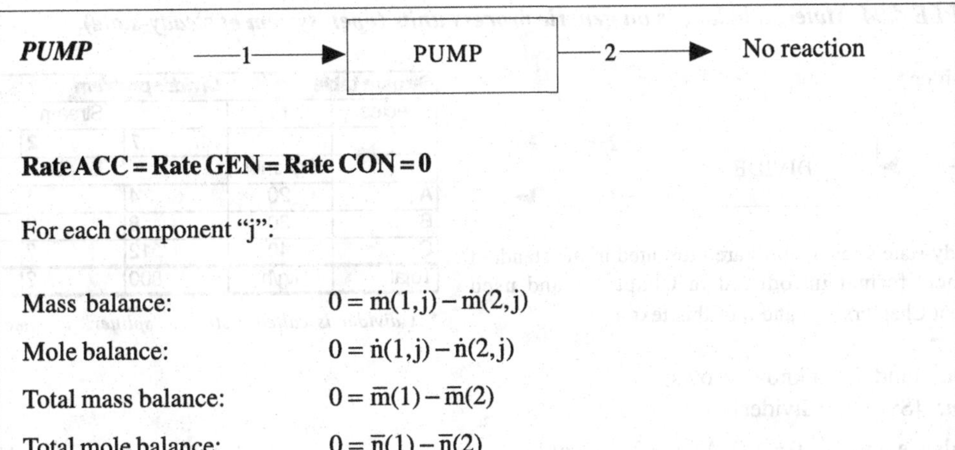

PUMP —1→ PUMP —2→ No reaction

Rate ACC = Rate GEN = Rate CON = 0

For each component "j":

Mass balance: $0 = \dot{m}(1,j) - \dot{m}(2,j)$

Mole balance: $0 = \dot{n}(1,j) - \dot{n}(2,j)$

Total mass balance: $0 = \bar{m}(1) - \bar{m}(2)$

Total mole balance: $0 = \bar{n}(1) - \bar{n}(2)$

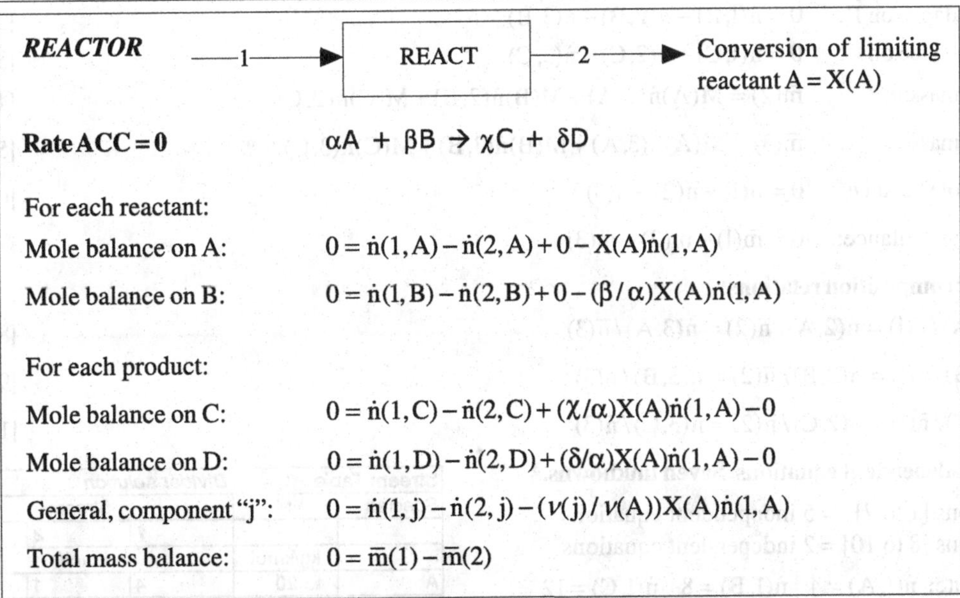

REACTOR —1→ REACT —2→ Conversion of limiting reactant A = X(A)

Rate ACC = 0 $\alpha A + \beta B \rightarrow \chi C + \delta D$

For each reactant:

Mole balance on A: $0 = \dot{n}(1,A) - \dot{n}(2,A) + 0 - X(A)\dot{n}(1,A)$

Mole balance on B: $0 = \dot{n}(1,B) - \dot{n}(2,B) + 0 - (\beta/\alpha)X(A)\dot{n}(1,A)$

For each product:

Mole balance on C: $0 = \dot{n}(1,C) - \dot{n}(2,C) + (\chi/\alpha)X(A)\dot{n}(1,A) - 0$

Mole balance on D: $0 = \dot{n}(1,D) - \dot{n}(2,D) + (\delta/\alpha)X(A)\dot{n}(1,A) - 0$

General, component "j": $0 = \dot{n}(1,j) - \dot{n}(2,j) - (\nu(j)/\nu(A))X(A)\dot{n}(1,A)$

Total mass balance: $0 = \bar{m}(1) - \bar{m}(2)$

NOTE: For all process units:

$$\dot{m}(i,j) = M(j)\dot{n}(i,j)$$
$$\bar{m}(i,j) = \Sigma[\dot{m}(i,j)]$$
$$\bar{n}(i) = \Sigma[\dot{n}(i,j)]$$

In the general reactor balance the value of ν(j) is –ve for reactants and +ve for products. Thus reactants are <u>consumed</u> and products are <u>generated</u> in a reactor.

EXAMPLE 4.04 Material balances on generic process units (open system at steady-state).

A. Divider*

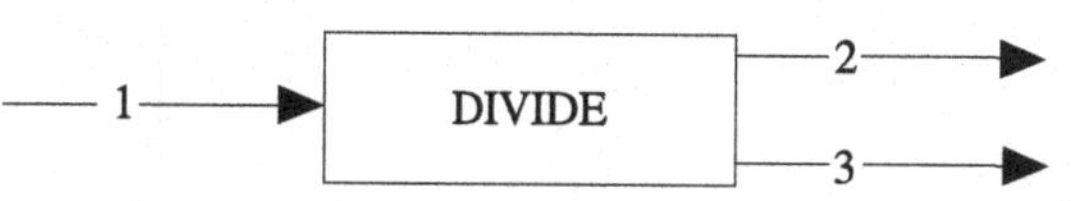

Stream table		Divider problem		
Species	M	Stream		
		1	2	3
	kg/kmol	kmol/h		
A	20	4	1	?
B	30	8	?	?
C	40	12	?	?
Total	kg/h	800	?	?

** A divider is called a stream "splitter" in some texts.*

The steady-state stream tables are presented in the standard spreadsheet format introduced in Chapter 3 and used throughout Chapters 4, 5 and 6 of this text.

Problem: Find the unknown flows.

Solution: [System = divider]

Mole balance on A: $0 = \dot{n}(1,A) - \dot{n}(2,A) - \dot{n}(3,A)$ [1]

Mole balance on B: $0 = \dot{n}(1,B) - \dot{n}(2,B) - \dot{n}(3,B)$ [2]

Mole balance on C: $0 = \dot{n}(1,C) - \dot{n}(2,C) - \dot{n}(3,C)$ [3]

Stream mass: $\overline{m}(2) = M(A)\dot{n}(2,A) + M(B)\dot{n}(2,B) + M(C)\dot{n}(2,C)$ [4]

Stream mass: $\overline{m}(3) = M(A)\dot{n}(3,A) + M(B)\dot{n}(3,B) + M(C)\dot{n}(3,C)$ [5]

Total mole balance: $0 = \overline{n}(1) - \overline{n}(2) - \overline{n}(3)$ [6]

Total mass balance: $0 = \overline{m}(1) - \overline{m}(2) - \overline{m}(3)$ [7]

Divider composition relations:

$\dot{n}(1,A)/\overline{n}(1) = \dot{n}(2,A)/\overline{n}(2) = \dot{n}(3,A)/\overline{n}(3)$ [8]

$\dot{n}(1,B)/\overline{n}(1) = \dot{n}(2,B)/\overline{n}(2) = \dot{n}(3,B)/\overline{n}(3)$ [9]

$\dot{n}(1,C)/\overline{n}(1) = \dot{n}(2,C)/\overline{n}(2) = \dot{n}(3,C)/\overline{n}(3)$ [10]

Seven independent equations. Seven unknowns.[3]

Equations [1 to 7] = 5 independent equations
Equations [8 to 10] = 2 independent equations

Substitute: $\dot{n}(1,A) = 4 \quad \dot{n}(1,B) = 8 \quad \dot{n}(1,C) = 12$

$\dot{n}(2,A) = 1 \quad \overline{m}(1) = 800$

Solve for the unknowns and check closure.

Stream Table		Divider solution		
Species	M	Stream		
		1	2	3
	kg/kmol	kmol/h		
A	20	4	1	3
B	30	8	2	6
C	40	12	3	9
Total	kg/h	800	200	600
Mass IN	= 800 kg/h		Closure	
Mass OUT	= 800 kg/h		800/800 = 100%	

Solution method: Solve [9 and 10] for: $\dot{n}(2,B), \dot{n}(2,C)$ Solve [1, 2 and 3] for: $\dot{n}(3,A), \dot{n}(3,B), \dot{n}(3,C)$
Solve [4 and 5] for: $\overline{m}(2), \overline{m}(3)$

[3] Independent equations are those that cannot be obtained by combining other equations in the (independent) set. The remaining equations are redundant. Redundant equations are shown in these examples to illustrate the concepts of *Figure 4.01* and to indicate that the material balance may be solved using different sets.

B. Mixer

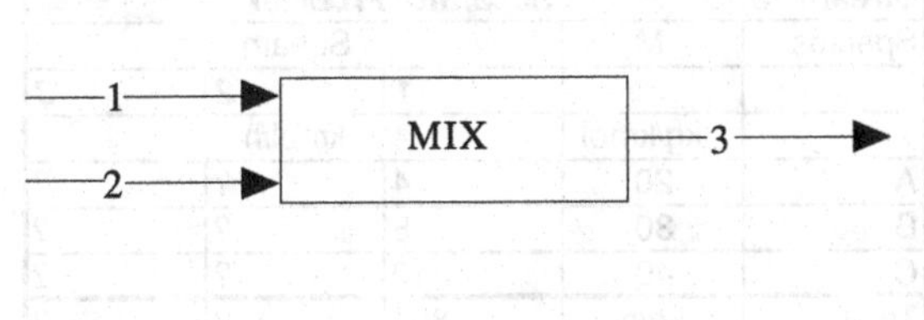

Stream Table	Mixer Problem			
Species	M	Stream		
		1	2	3
	kg/kmol	kmol/h		
A	20	4	3	?
B	30	8	5	?
C	40	12	7	?
Total	kg/h	800	490	?

Problem: Complete the stream table material balances.

Solution: [System = mixer]

Mole balance on A: $0 = \dot{n}(1, A) + \dot{n}(2, A) - \dot{n}(3, A)$ [1]

Mole balance on B: $0 = \dot{n}(1, B) + \dot{n}(2, B) - \dot{n}(3, B)$ [2]

Mole balance on C: $0 = \dot{n}(1, C) + \dot{n}(2, C) - \dot{n}(3, C)$ [3]

Total stream mass: $\overline{m}(3) = M(A)\dot{n}(3, A) + M(B)\dot{n}(3, B) + M(C)\dot{n}(3, C)$ [4]

Total mole balance: $0 = \overline{n}(1) + \overline{n}(2) - \overline{n}(3)$ [5]

Total mass balance: $0 = \overline{m}(1) + \overline{m}(2) - \overline{m}(3)$ [6]

Four independent equations. Four unknowns.

Substitute:

$\dot{n}(1, A) = 4$

$\dot{n}(1, B) = 8$

$\dot{n}(1, C) = 12$

$\overline{m}(1) = 800$

$\dot{n}(2, A) = 3$

$\dot{n}(2, B) = 5$

$\dot{n}(2, C) = 7$

$\overline{m}(2) = 490$

Solve for the unknowns and check closure.

Stream Table	Mixer Solution			
Species	M	Stream		
		1	2	3
	kg/kmol	kmol/h		
A	20	4	3	7
B	30	8	5	13
C	40	12	7	19
Total	kg/h	800	490	1290
Mass IN	=	1290 kg/h	Closure	
Mass OUT	=	1290 kg/h	1290/1290 = 100%	

C. Separator

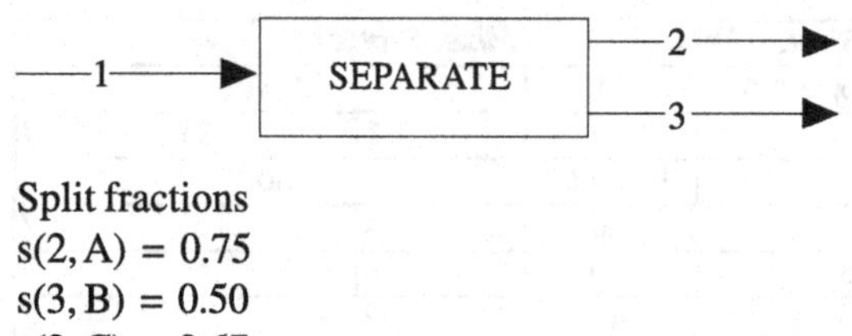

Split fractions
s(2, A) = 0.75
s(3, B) = 0.50
s(3, C) = 0.67

Stream Table		*Separator Problem*		
Species	M	Stream		
		1	*2*	*3*
	kg/kmol	kmol/h		
A	20	4	?	?
B	30	8	?	?
C	40	12	?	?
Total	kg/h	800	?	?

Problem: Complete the stream table material balance.

Solution: [System = separator]

Mole balance on A: $0 = \dot{n}(1, A) - \dot{n}(2, A) - \dot{n}(3, A)$ [1]

Mole balance on B: $0 = \dot{n}(1, B) - \dot{n}(2, B) - \dot{n}(3, B)$ [2]

Mole balance on C: $0 = \dot{n}(1, C) - \dot{n}(2, C) - \dot{n}(3, C)$ [3]

Total mole balance: $0 = \overline{n}(1) - \overline{n}(2) - \overline{n}(3)$ [4]

Total mass balance: $0 = \overline{m}(1) - \overline{m}(2) - \overline{m}(3)$ [5]

Stream mass: $\overline{m}(2) = M(A)\dot{n}(2, A) + M(B)\dot{n}(2, B) + M(C)\dot{n}(2, C)$ [6]

$\overline{m}(3) = M(A)\dot{n}(3, A) + M(B)\dot{n}(3, B) + M(C)\dot{n}(3, C)$ [7]

Split fractions: $s(2, A) = \dot{n}(2, A) / \dot{n}(1, A)$ [8]

$s(3, B) = \dot{n}(3, B) / \dot{n}(1, B)$ [9]

$s(3, C) = \dot{n}(3, C) / \dot{n}(1, C)$ [10]

Eight independent equations. Eight unknowns.

Equations [1 to 7] = 5 independent equations

Equations [8 to 10] = 3 independent equations

Substitute:

$\dot{n}(1,A) = 4$ $\dot{n}(1,B) = 8$

$\dot{n}(1,C) = 12$ $\overline{m}(1) = 800$

$s(2,A) = 0.75$ $s(2,B) = 0.50$

$s(2,C) = 0.67$

Solve for the unknowns and check closure.

Stream Table		*Separator Solution*		
Species	M	Stream		
		1	*2*	*3*
	kg/kmol	kmol/h		
A	20	4	3	1
B	30	8	4	4
C	40	12	4	8
Total	kg/h	800	340	460
Mass IN	800 kg/h			
Mass OUT	800 kg/h	Closure = 800/800 = 100%		

D. Heat exchanger (indirect contact)

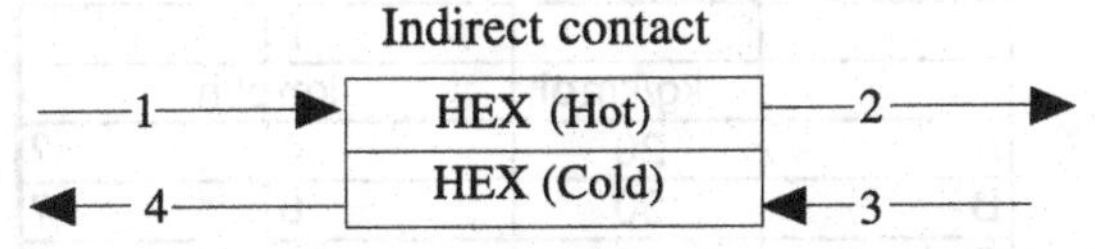

Stream Table		Heat Exchanger Problem			
Species	M	Stream			
		1	2	3	4
	kg/kmol	kmol/h			
A	20	4	?	3	?
B	30	8	?	7	?
C	40	12	?	9	?
Total	kg/h	800	?	630	?

Problem: Complete the stream table material balance.
Solution: [System = HEX hot side]

Mole balance on A: $0 = \dot{n}(1, A) - \dot{n}(2, A)$ [1]

Mole balance on B: $0 = \dot{n}(1, B) - \dot{n}(2, B)$ [2]

Mole balance on C: $0 = \dot{n}(1, C) - \dot{n}(2, C)$ [3]

Total mole balance: $0 = \overline{n}(1) - \overline{n}(2)$ [4]

Total mass balance: $0 = \overline{m}(1) - \overline{m}(2)$ [5]

Stream mass: $\overline{m}(2) = M(A)\dot{n}(2, A) + M(B)\dot{n}(2, B) + M(C)\dot{n}(2, C)$ [6]

[System = HEX cold side]

Mole balance on A: $0 = \dot{n}(3, A) - \dot{n}(4, A)$ [7]

Mole balance on B: $0 = \dot{n}(3, B) - \dot{n}(4, B)$ [8]

Mole balance on C: $0 = \dot{n}(3, C) - \dot{n}(4, C)$ [9]

Total mole balance: $0 = \overline{n}(3) - \overline{n}(4)$ [10]

Total mass balance: $0 = \overline{m}(3) - \overline{m}(4)$ [11]

Stream mass: $\overline{m}(4) = M(A)\dot{n}(4, A) + M(B)\dot{n}(4, B) + M(C)\dot{n}(4, C)$ [12]

Eight independent equations. Eight unknowns.

Equations [1 to 6] = 4 independent equations Equations [7 to 12] = 4 independent equations

Substitute:

$\dot{n}(1, A) = 4$ $\dot{n}(1, B) = 8$

$\dot{n}(1, C) = 12$ $\overline{m}(1) = 800$

$\dot{n}(3, A) = 3$ $\dot{n}(3, B) = 7$

$\dot{n}(3, C) = 9$ $\overline{m}(3) = 630$

Solve for the unknowns and check closure.

Stream Table		Heat Exchanger Solution			
Species	M	Stream			
		1	2	3	4
	kg/kmol	kmol/h			
A	20	4	4	3	3
B	30	8	8	7	7
C	40	12	12	9	9
Total	kg/h	800	800	630	630
		HOT		COLD	
Mass IN	=	800 kg/h	Closure	630 kg/h	Closure
Mass OUT	=	800 kg/h	100%	630 kg/h	100%

E. Pump

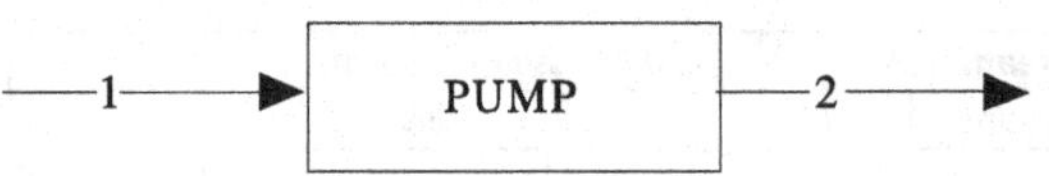

Stream Table		Pump Problem	
Species	M	Stream	
		1	*2*
	kg/kmol	kmol/h	
A	20	4	?
B	30	8	?
C	40	12	?
Total	kg/h	800	?

Problem: Complete the stream table material balance.

Solution: [System = pump]

Mole balance on A: $0 = \dot{n}(1, A) - \dot{n}(2, A)$ [1]

Mole balance on B: $0 = \dot{n}(1, B) - \dot{n}(2, B)$ [2]

Mole balance on C: $0 = \dot{n}(1, C) - \dot{n}(2, C)$ [3]

Total mole balance: $0 = \overline{n}(1) - \overline{n}(2)$ [4]

Total mass balance: $0 = \overline{m}(1) - \overline{m}(2)$ [5]

Stream mass: $\overline{m}(2) = M(A)\dot{n}(2, A) + M(B)\dot{n}(2, B) + M(C)\dot{n}(2, C)$ [6]

Four independent equations. Four unknowns.

Equations [1 to 6] = 4 independent equations
Substitute:

$\dot{n}(1, A) = 4$ $\quad\quad \dot{n}(1, B) = 8$

$\dot{n}(1, C) = 12$ $\quad\quad \overline{m}(1) = 800$

Solve for the unknowns and check closure.

Stream Table		Pump Solution	
Species	M	Stream	
		1	*2*
	kg/kmol	kmol/h	
A	20	4	4
B	30	8	8
C	40	12	12
Total	kg/h	800	800
Mass IN	800 kg/h	Closure	
Mass OUT	800 kg/h	800/800 = 100 %	

F. Reactor

—1→ REACT —2→

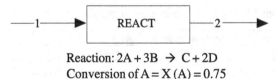

Reaction: $2A + 3B \rightarrow C + 2D$
Conversion of A = X (A) = 0.75

Stream Table		Reactor Problem	
Species	M	Stream	
		1	*2*
	kg/kmol	kmol/h	
A	20	4	?
B	30	8	?
C	40	12	?
D	45	3	?
Total	kg/h	935	?

Problem: Complete the stream table material balance.

Solution: [System = reactor]

Mole balance on A: $0 = \dot{n}(1,A) - \dot{n}(2,A) + 0 - X(A)\dot{n}(1,A)$ [1]

Mole balance on B: $0 = \dot{n}(1,B) - \dot{n}(2,B) + 0 - (3/2)x(A)\dot{n}(1,A)$ [2]

Mole balance on C: $0 = \dot{n}(1,C) - \dot{n}(2,C) + (1/2)x(A)\dot{n}(1,A) - 0$ [3]

Mole balance on D: $0 = \dot{n}(1,D) - \dot{n}(2,D) + \dot{n}(2/2)x(A)\dot{n}(1,A) - 0$ [4]

Total mass balance: $0 = \overline{m}(1) - \overline{m}(2)$ [5]

Stream total: $\overline{m}(2) = M(A)\dot{n}(2,A) + M(B)\dot{n}(2,B) + M(C)\dot{n}(2,C)$ [6]

NOTE: For reactants: Rate GEN = 0
For products: Rate CON = 0

Stream Table		***Reactor Solution***	
Species	M	Stream	
		1	*2*
	kg/kmol	kmol/h	
A	20	4	1
B	30	8	3.5
C	40	12	13.5
D	45	3	6
Total	kg/h	935	935
Mass IN	935 kg/h	Closure	
Mass OUT	935 kg/h	935/935 = 100%	

Five independent equations. Five unknowns.
Equations [1 to 6] = 5 independent equations
Substitute:

$\dot{n}(1,A) = 4$ $\dot{n}(1,B) = 8$

$\dot{n}(1,C) = 12$ $\dot{n}(1,D) = 3$

Solve for the unknowns and check closure.

SUBSIDIARY RELATIONS

The material balances on mixers, indirect heat exchangers and pumps set out in *Figure 4.01* and *Example 4.04* are straightforward and self-explanatory. The material balances on dividers, separators and reactors use subsidiary relations that need explanation.

Dividers

Dividers convert an input stream to multiple output streams, all with the *same composition* as the inlet stream. The fixed composition of the streams gives a set of equations called the "composition restrictions" that can be manipulated in several ways to solve the material balance. For example, the same composition in each stream requires that:

$\dot{m}(1,j)/\overline{m}(1) = \dot{m}(2,j)/\overline{m}(2) = \dot{m}(3,j)/\overline{m}(3)$, etc.

These relations can be re-arranged algebraically to give:

$\dot{m}(1,A)/\dot{m}(1,B) = \dot{m}(2,A)/\dot{m}(2,B) = \dot{m}(3,A)/\dot{m}(3,B)$ or

$\dot{m}(1,A)/\dot{m}(1,C) = \dot{m}(2,A)/\dot{m}(2,C) = \dot{m}(3,A)/\dot{m}(3,C)$, etc. where A, B, C are the components in stream 1.

Analogous relations apply for the mole flows. Also, the divider can be specified by the stream flow ratios $K_{DIV}(i)$, where:

$K_{DIV}(i) = \overline{n}(i)/\overline{n}(1) = \overline{m}(i)/\overline{m}(1) = \dot{m}(i,j)/\dot{m}(1,j) = \dot{n}(i,j)/\dot{n}(1,j)$ for each component "j"

and for the sum over all <u>outlet</u> streams: $\Sigma[K_{DIV}(i)] = 1$ *Equation 4.06*

In some texts the divider is called a stream "splitter". That terminology is not used here because it is ambiguous with respect to the term "split fraction". Split fraction is a term used to specify the performance of separators, which have an entirely different function from that of dividers.

Separators

Separators convert an input stream into multiple outlet streams of *different composition*. They are characterised by the *separation efficiency* for shifting each component into a specified output stream.

$$\text{Separation efficiency for component "j"} = \frac{\text{amount of "j" in specified outlet stream}}{\text{amount of "j" in inlet stream}}$$

Separation efficiency is often indicated in process flowsheets as a *split fraction*, which is defined by *Equation 4.07*:

$$\mathbf{s(i,j) = \dot{n}(i,j)/\dot{n}(1,j) = \dot{m}(i,j)/\dot{m}(1,j)} \qquad \textit{Equation 4.07}$$

where:

s(i, j) = split fraction for component j, from inlet stream 1 to outlet stream i

Separators usually function by concentrating the components into different phases and then separating the phases (e.g. by gravity). Separation efficiency in real separators depends on several factors such as the phase equilibria, mass transfer rates and entrainment rates.[4]

In simplified material balance calculations the separator split fractions are often assumed to be the values set by the equilibrium distribution of components between the phases leaving the separator (i.e. an *equilibrium separator*). This equilibrium distribution can be calculated from the inter-phase equilibrium relations given in Chapter 2, as shown in *Example 4.05*.

[4] Mass transfer is the transport of material across a phase boundary, driven by concentration gradients. Entrainment is the capture of bubbles, drops or particles from one phase into the phase from which it is being separated, for example, micro-drops of liquid from bursting vapour bubbles, that are carried into the gas phase above a boiling liquid.

EXAMPLE 4.05 Calculation of separator split fractions from phase equilibria.

A. Liquid/liquid separator from the L/L distribution coefficient

The solute C partitions between two mutually insoluble solvents A and B by the distribution coefficient:

D(C) = 2.57 (see *Equation 2.39*).

This system is used in the liquid/liquid separator of the flowsheet and stream table below.

Solvent extractor

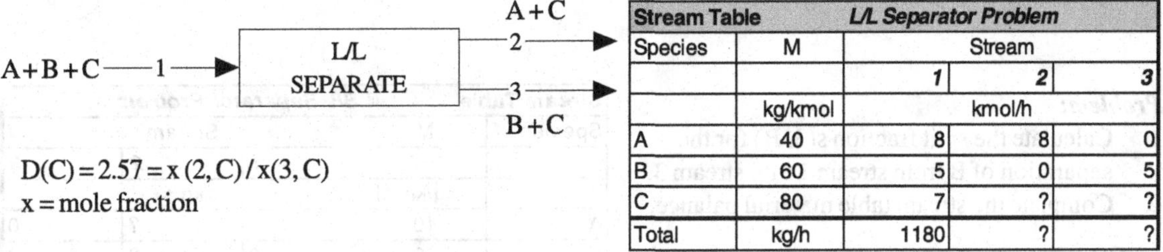

Stream Table		L/L Separator Problem		
Species	M	Stream		
		1	2	3
	kg/kmol	kmol/h		
A	40	8	8	0
B	60	5	0	5
C	80	7	?	?
Total	kg/h	1180	?	?

Problem: 1. Calculate the split fraction s(2,C) for the separation of C from stream 1 into stream 2.
2. Complete the stream table material balance.

Solution: [System = separator]

Mole balance on C: $0 = 7 - \dot{n}(2,C) - \dot{n}(3,C)$ [1]

Equilibrium composition: $D(C) = 2.57 = \dot{n}(2,C)/[8 + \dot{n}(2,C)]/[\dot{n}(3,C)/[5 + \dot{n}(3,C)]]$ [2]

Stream totals: $\bar{m}(3) = (40)(0) + (60)(5) + 80\dot{n}(3,C)$ [3]

$\bar{m}(2) = (40)(8) + (60)(0) + 80\dot{n}(2,C)$ [4]

Four independent equations. Four unknowns.

Simultaneous solution is required for equations [1 and 2].
Combine equations [1 and 2].

Solve the resulting non-linear (quadratic) equation to get:

$\dot{n}(2,C) = 6$ kmol/h

Split fraction: $s(2,C) = \dot{n}(2,C)/\dot{n}(1,C) = 6/7 =$ **0.86**

Solve for the unknowns and check closure.

Stream Table		L/L Separator Solution		
Species	M	Stream		
		1	2	3
	kg/kmol	kmol/h		
A	40	8	8	0
B	60	5	0	5
C	80	7	6	1
Total	kg/h	1180	800	380
Mass IN	1180 kg/h			
Mass OUT	1180 kg/h	Closure = 1180/1180 = 100%		

B. Solid/liquid separator from the S/L solubility

The solute B has a solubility[5] in solvent A of 40 wt% (i.e. 40 kg B/100 kg solution = 66.7 kg B/100 kg A) at 353 K and 20 wt% (i.e. 20 kg B/100 kg solution = 25 kg B/100 kg A) at 293 K.

This system is used in the solid/liquid separator of the flowsheet and stream table below.

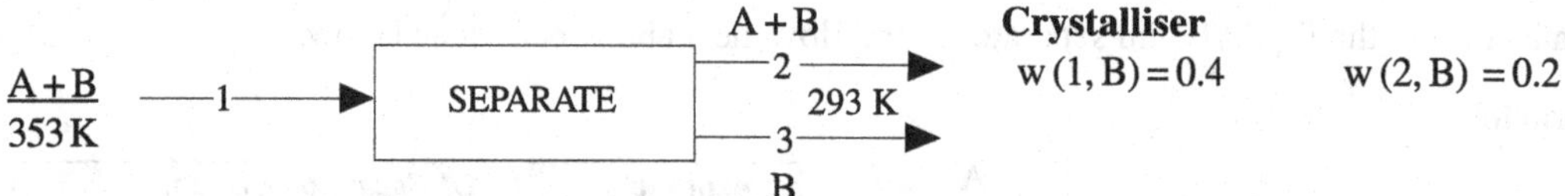

Problem:

1. Calculate the split fraction s(3, B) for the separation of B from stream 1 into stream 3.
2. Complete the stream table material balance.

Stream Table		*S/L Separator Problem*		
Species	M	Stream		
		1	2	3
	kg/kmol	kg/s		
A	40	12	?	0
B	60	8	?	?
Total	kg/s	20	?	?

Solution: [System = separator]

Mass balance on A: $0 = 12 - \dot{m}(2,A) - 0$ [1]

Mass balance on B: $0 = 8 - \dot{m}(2,B) - \dot{m}(3,B)$ [2]

Equilibrium composition at 293 K: $0.2 = \dot{m}(2,B)/[\dot{m}(2,A) + \dot{m}(2,B)]$ [3]

Stream totals: $\dot{m}(2) = \dot{m}(2,A) + \dot{m}(2,B)$ [4]

$\dot{m}(3) = 0 + \dot{m}(3,B)$ [5]

Five independent equations. Five unknowns.

Simultaneous solution is required for equations [1, 2 and 3].

Split fraction:

$s(3, B) = \dot{m}(3,B)/\dot{m}(1,B) = 5/8 = \mathbf{0.63}$

Solve for the unknowns and check closure.
Mass IN = 20 kg/s, Mass OUT = 20 kg/s, Closure = 100%

Stream Table		*S/L Separator Solution*		
Species	M	Stream		
		1	2	3
	kg/kmol	kg/s		
A	40	12	12	0
B	60	8	3	5
Total	kg/s	20	15	5

[5] Solubility = concentration of solute in solution (i.e. in the liquid phase) at equilibrium with excess solid solute, at a specified temperature. Note that "solubility" may be expressed as:
Mass solute/mass solution OR
Mass solute/mass solvent OR
Mass solute/volume solution (or solvent)

A "wt% solution" means 100 [mass solute/mass solution].

Reactors

Reactors differ from other process units because they transform material through chemical reactions, so the mass (or moles) of each *species* involved in the reaction is not conserved. However, there are no nuclear reactions in chemical process reactors so the total mass is conserved, as required for *Equation 4.08.*

Chemical reactor calculations can be based on either the conversion (*Equation 2.43*) or the extent of reaction (*Equation 2.44*), but the dimensionless conversion is used for most of the examples in this text. The reactor material balances shown in *Figure 4.01* and *Example 4.04* hinge on the conversion of a single reactant A. The consumption and generation of every other reactant and product is linked to the conversion of A by the reaction stoichiometry. If reactants are initially in stoichiometric proportions then the conversion will have the same value for all reactants, otherwise each reactant will have a different conversion — so it is important to specify which reactant conversion is considered in the balance equations. To simplify the algebra you should write the individual species material balances for a reactor in terms of mole quantities (not masses).

Conversion in real chemical reactors depends on the reaction rate and residence time[6] of the reaction mixture in the reaction vessel, which drive the reaction to approach equilibrium. In simplified material balance calculations the conversion is often assumed to be the *equilibrium conversion* (X_{eq}) calculated from the reaction equilibrium constant *Equation 2.48* (i.e. an *equilibrium reactor*). For more realistic reactor design the actual conversion (X_{act}) is measured or is calculated from the reaction rate as illustrated in *Example 4.06.* Actual conversion calculations are also shown in *Examples 2.16* and *7.03.*

EXAMPLE 4.06 Conversion in an isothermal continuous stirred tank reactor (CSTR) at steady-state

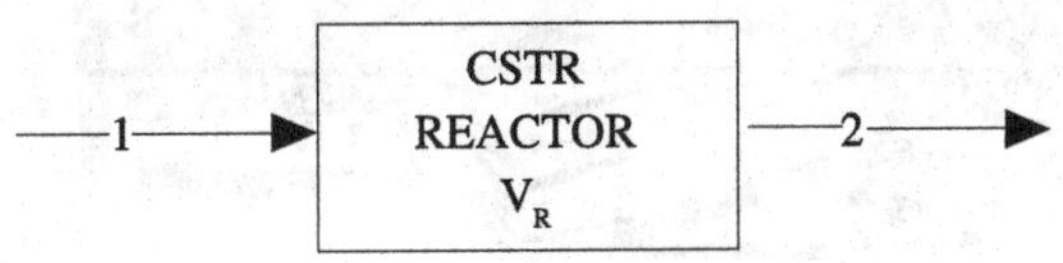

CSTR is "perfectly mixed"
[A] = concentration of A $kmol.m^{-3}$

Reaction: A → Products
Irreversible reaction with first order kinetics.

Reaction rate	= k[A]		$kmol.m^{-3}.s^{-1}$
Rate constant	= k	= 0.20	s^{-1}
Volume flow rate	= $\dot{V}(1) = \dot{V}(2)$	= 0.40	m^3s^{-1}
Reactor volume	= V_R	= 5.0	m^3
Residence time	= $V_R/\dot{V}(2)$	= 5/0.4 = 12.5	s

In an ideal stirred tank the reactants are "perfectly mixed" and the composition of the outlet Stream 2 is the same as that of the fluid in the tank.

[6] Residence time is the time spent in the reactor by the reaction mixture.
In a continuous flow process: Residence time = (reactor volume) / (volume flow rate of reaction mixture).

Problem: Calculate the steady-state conversion of A.
Solution: [System = Reactor]

0 = Rate IN – Rate OUT + Rate GEN – Rate CON *Equation 4.03*

Mole balance on A:	$0 = \dot{n}(1,A) - \dot{n}(2,A) + 0 - k[\dot{n}(2,A)/\dot{V}(2)]V_R$	$kmol.s^{-1}$	[1]
Solve for $\dot{n}(2,A)$:	$\dot{n}(2,A) = \dot{n}(1,A)/[1 + kV_R/\dot{V}(2)]$	$kmol.s^{-1}$	[2]
Conversion[7] of A:	$X(A) = [\dot{n}(1,A) - \dot{n}(2,A)]/\dot{n}(1,A) = 1 - 1/[1 + kV_R/\dot{V}(2)]$	–	[3]
Substitute:	$k = 0.20\ s^{-1}$ $\quad \dot{V}(2) = 0.40\ m^3s^{-1}$ $\quad V_R = 5.0\ m^3$		

$$X(A) = 1 - 1[1 + (0.2/0.4)5] = \underline{\mathbf{0.71}}$$

CLOSURE OF MATERIAL BALANCES

For continuous operation at steady-state the material balance always reduces to *Equation 4.08*.

Equation 4.08 is the overall MASS BALANCE that applies to all chemical processes and individual process units at steady-state.

$$\mathbf{0 = \Sigma[\overline{m}(i)_{in}] - \Sigma[\overline{m}(i)_{out}]}$$ [Steady-state mass balance] *assumes no nuclear reactions* *Equation 4.08*

where:

$[\overline{m}(i)_{in}]$ = sum of all input mass flow rates

$[\overline{m}(i)_{out}]$ = sum of all output mass flow rates

Equation 4.08 gives a simple way to check your material balances.

---

7 Equation [3] in *Example 4.06* is the "design equation" for a CSTR with first order (irreversible) reaction kinetics. This equation is used to find the size of a reactor for a desired conversion and production rate.

For *steady-state* material balances you should always check that:
Total mass flow rate in to the system = Total mass flow rate out of the system
[i.e. the *closure* defined below is within acceptable limits of 100%]

The *system* can be an individual process unit, a collection of process units or a complete plant and must be clearly defined for every case. In practical systems where material flows are based on measurements there is nearly always a discrepancy between mass in and mass out. This discrepancy is measured by the *closure* of the mass balance, defined in *Equation 4.09*:

$$e_M = 100\,\Sigma[\bar{m}(i)_{out}]/\Sigma[\bar{m}(i)_{in}] \qquad \textit{Equation 4.09}$$

where: e_M = mass balance closure %

The acceptable value of "e_M" depends on the type and purpose of the material balance. For exact modelling calculations such as in *Example 4.05* the closure should be 100.0%, but with crude measurements on a real operating system, a closure of 90% to 110% may be acceptable. For chemical process design calculations[8] the closure is typically set above 99.9%. Closure is an important issue in the iterative modelling calculations described later in this chapter.

SIMULTANEOUS EQUATIONS

Steady-state material balances typically generate a set of algebraic equations that must be solved for one or more unknowns. The cases in *Example 4.04* gave simple sets of uncoupled linear algebraic equations that were solved explicitly for each unknown. More complex cases may involve simultaneous linear or non-linear algebraic equations, as shown in *Example 4.07*.

Simultaneous linear equations can be solved by standard procedures in programmable calculators or spreadsheets (see *Ref. 7*). However, in many cases a formidable looking set of material balance linear equations can be reduced to parts that are be easily solved by "hand". The solution of single non-linear equations usually resorts to machine calculations with numerical methods, such as Newton's method or bisection (see *Refs. 3–5*). Numerical methods are used, for example, in the calculator and spreadsheet routines called "Goal Seek" and "Solver". Simultaneous non-linear equations usually require more sophisticated numerical methods of solution, some of which are described in *Refs. 3, 4, 5 and 7*.

---

[8] Detailed process design calculations should have: $99.9\% < e_M < 100.1\%$ or better. Improper closure can mean neglecting small amounts of critical materials that have serious negative effects on the process operation. Preliminary (conceptual) process design may be less rigorous, with say: $90\% < e_M < 110\%$, although a value of e_M much above 100% is usually a sign that something is amiss with the material balance equations.

EXAMPLE 4.07 Material balances with simultaneous equations (open system at steady-state).

A. Separator with simultaneous linear equations (Continuous process at steady-state)

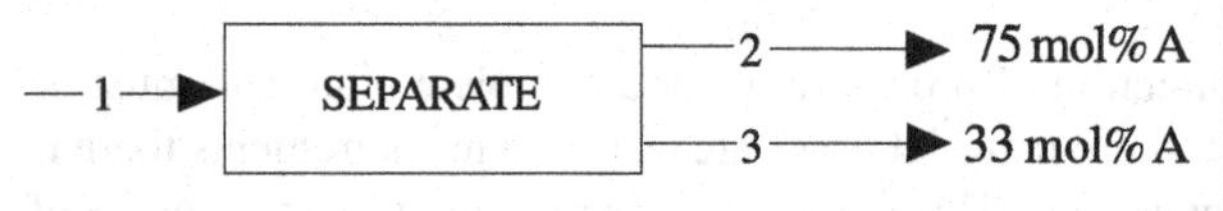

Stream Table	Separator			Problem
Species	M	Stream		
		1	2	3
	kg/kmol	kmol/h		
A	40	40	?	?
B	80	30	?	?
Total	kg/h	4000	?	?

Problem:

Complete the stream table material balances.

Solution: [System = separator]

Mole balance on A: $0 = \dot{n}(1,A) - \dot{n}(2,A) - \dot{n}(3,A) + 0 - 0$ [1]

Mole balance on B: $0 = \dot{n}(1,B) - \dot{n}(2,B) - \dot{n}(3,B) + 0 - 0$ [2]

Overall mole balance: $0 = \bar{n}(1) - \bar{n}(2) - \bar{n}(3) + 0 - 0$ [3]

Overall mass balance: $0 = \overline{m}(1) - \overline{m}(2) - \overline{m}(3) + 0 - 0$ [4]

Stream mass: $\overline{m}(2) = M(A)\dot{n}(2,A) + M(B)\dot{n}(2,B)$ [5]

$\overline{m}(3) = M(A)\dot{n}(3,A) + M(B)\dot{n}(3,B)$ [6]

Stream moles: $\bar{n}(2) = \dot{n}(2,A) + \dot{n}(2,B)$ [7]

$\bar{n}(3) = \dot{n}(3,A) + \dot{n}(3,B)$ [8]

Stream compositions: $x(2,A) = \dot{n}(2,A)/\bar{n}(2)$ [9]

x(i, j) = mole fraction j in i $x(3,A) = \dot{n}(3,A)/\bar{n}(3)$ [10]

$1 = x(2,A) + x(2,B)$ [11]

$1 = x(3,A) + x(3,B)$ [12]

Six independent equations. Six unknowns. Simultaneous solution is required.
[Note that 6 of equations [1 to 12] are redundant]
Combine equations [9–12] with equations [1 and 2]

$$0 = \dot{n}(1,A) - x(2,A)\bar{n}(2) - x(3,A)\bar{n}(3) \quad [13]$$

$$0 = \dot{n}(1,B) - (1 - x(2,A))\bar{n}(2) - (1 - x(3,A))\bar{n}(3) \quad [14]$$

Substitute:

$\dot{n}(1,A) = 40$ $\dot{n}(1,B) = 30$

$x(2,A) = 0.75$ $x(3,A) = 0.33$

$$0 = 40 - 0.75\bar{n}(2) - 0.33\bar{n}(3) \quad [15]$$

$$0 = 30 - 0.25\bar{n}(2) - 0.67\bar{n}(3) \quad [16]$$

Stream Table	Separator			Solution
Species	M	Stream		
		1	2	3
	kg/kmol	kmol/h		
A	40	40	30	10
B	80	30	10	20
Total	kg/h	4000	2000	2000
Mass IN	=	4000 kg/h	Closure %	
Mass OUT	=	4000 kg/h	4000/4000 = 100	

Solve simultaneous linear equations [15 and 16].
Check closure.

B. Separator with simultaneous non-linear equations (isothermal flash split)

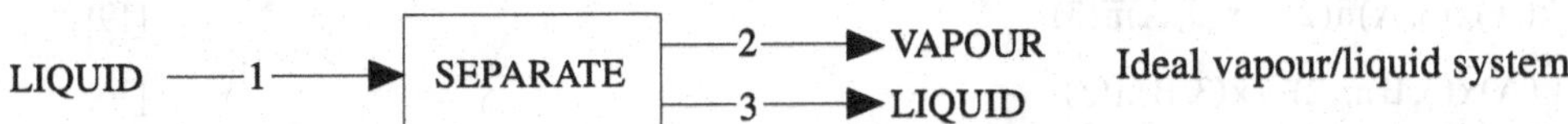

Ideal vapour/liquid system

Fixed pressure and temperature:

k (A) = p*(2, A)/P = 0.1

k (B) = p*(2, B)/P = 10

Stream Table		Separator		Problem
Species	M	Stream		
		1	2	3
	kg/kmol		kmol/h	
A	40	40	?	?
B	80	30	?	?
Total	kg/h	4000	?	?

Solution: [System = Separator]

Mole balance on A: $0 = \dot{n}(1,A) - \dot{n}(2,A) - \dot{n}(3,A) + 0 - 0$ [1]

Mole balance on B: $0 = \dot{n}(1,B) - \dot{n}(2,B) - \dot{n}(3,B) + 0 - 0$ [2]

Overall mole balance: $0 = \overline{n}(1) - \overline{n}(2) - \overline{n}(3) + 0 - 0$ [3]

Overall mass balance: $0 = \overline{m}(1) - \overline{m}(2) - \overline{m}(3) + 0 - 0$ [4]

Stream mass: $\overline{m}(2) = M(A)\dot{n}(2,A) + M(B)\dot{n}(2,B)$ [5]

$\overline{m}(3) = M(A)\dot{n}(3,A) + M(B)\dot{n}(3,B)$ [6]

Stream moles: $\overline{n}(2) = \dot{n}(2,A) + \dot{n}(2,B)$ [7]

$\overline{n}(3) = \dot{n}(3,A) + \dot{n}(3,B)$ [8]

Stream composition: $y(2,A) = \dot{n}(2,A)/\overline{n}(2)$ [9]

$x(3,A) = \dot{n}(3,A)/\overline{n}(3)$ [10]

$y(2,B) = \dot{n}(2,B)/\overline{n}(2)$ [11]

$x(3,B) = \dot{n}(3,B)/\overline{n}(3)$ [12]

V/L equilibrium: $y(2,A) = k(A)x(3,A)$ Raoult's law (see *Equation 2.33*) [13]

$y(2,B) = k(B)x(3,B)$ [14]

$y(i,j)$ = mole fraction j in gas i $\quad 1 = y(2,A) + y(2,B)$ [15]

$x(i,j)$ = mole fraction j in liquid i $\quad 1 = x(3,A) + x(3,B)$ [16]

Six independent equations. Six unknowns. Simultaneous solution is required.

[Note that 10 of equations [1 to 16] are redundant]

Combine equations [1 and 2] with [9 and 10].

$0 = \dot{n}(1,A) - y(2,A)\overline{n}(2) - x(3,A)\overline{n}(3)$ [17]

$0 = \dot{n}(1,B) - y(2,B)\overline{n}(2) - x(3,B)\overline{n}(3)$ [18]

Combine equations [13 and 14] with [17 and 18]

$$0 = \dot{n}(1,A) - k(A)x(3,A)\bar{n}(2) - x(3,A)\bar{n}(3) \quad [19]$$

$$0 = \dot{n}(1,B) - k(A)x(3,B)\bar{n}(2) - x(3,B)\bar{n}(3) \quad [20]$$

Solve equations [19 and 20] for $x(3,A)$ and $x(3,B)$

$$x(3,A) = \dot{n}(1,A)/[k(A)\bar{n}(2) + \bar{n}(3)] \quad [21]$$

$$x(3,B) = \dot{n}(1,B)/[k(B)\bar{n}(2) + \bar{n}(3)] \quad [22]$$

Combine equations [3 and 16] with [21 and 22]

$$1 = \dot{n}(1,A)/[\bar{n}(2)(k(A)-1) + \bar{n}(1)] + \dot{n}(1,B)/[\bar{n}(2)(k(B)-1) + \bar{n}(1)] \quad [23]$$

Substitute:

$\dot{n}(1,A) = 40 \qquad \dot{n}(1,B) = 30 \qquad \bar{n}(1) = 70 \quad k(A) = 0.1 \qquad k(B) = 10$

$$1 = 40/[\bar{n}(2)(0.1-1)+70] + 30/[\bar{n}(2)(10-1)+70] \quad [24]$$

Solve the non-linear equation [24] for $\bar{n}(2)$.

Substitute into prior equations for the full result.

NOTE: Equation [23] is the classic "isothermal flash split", often seen as:

$$1 = \Sigma\{z(j)/[\upsilon(k(j)-1)+1]\}$$

where:

$z(j)$ = mole fraction of j in feed (stream 1)

$k(j)$ = equilibrium constant of j at the conditions in the separator

υ = moles vapour / moles feed $= \bar{n}(2)/\bar{n}(1)$

This method can be extended to a multi-component flash with ($J > 2$), where conditions (P, T) are between the bubble-point and dew-point.

Stream Table		*Separator*		*Solution*
Species	M	Stream		
		1	2	3
	kg/kmol	kmol/h		
A	40	40.0	2.6	37.4
B	80	30.0	26.3	3.7
Total	kmol/h	70	28.9	41.1
Total	kg/h	4000	2207	1793
Mass IN	kg/h	4000	Closure %	
Mass OUT	kg/h	4000	100	

SPECIFICATION OF MATERIAL BALANCE PROBLEMS

When you are working on a material balance problem with many variables and equations it can be easy to lose track of what is known and what is unknown. Process designers call this issue the "specification" of a process, and it is summarised in the following set of relations:

Number of unknown quantities > number of independent equations: UNDER-SPECIFIED
Number of unknown quantities = number of independent equations: FULLY-SPECIFIED
Number of unknown quantities < number of independent equations: OVER-SPECIFIED

***Degrees of freedom* (D of F)** = **number of unknown quantities** − **number of independent equations** *Equation 4.10*

An *under-specified* problem (D of F > 0) is essentially a *design problem*, in which the designer manipulates the values of the process variables to satisfy the design objective. Process design problems often become *optimisation* problems, where the design objective is to maximise a *figure of merit* such as the *return on investment.*

A *fully-specified* problem (D of F = 0) has no scope for manipulation. All values are uniquely fixed by the conditions specified in the problem statement, so the solution is just a matter of writing and solving the equations. This type of problem is usually stated in the form "given the inputs — find the single correct output" and is most familiar to undergraduate students in engineering and science (until they encounter a course in design).

An *over-specified* problem (D of F < 0) is sometimes caused by redundancy of correct information, but usually it is the result of an error. An over-specification error will give you the impossible task of finding results that are mutually incompatible. It is easy to inadvertently and incorrectly over-specify a complex material balance problem, then to go into a hopeless loop trying to find the solution. Try to avoid this type of error.

As you approach a material balance problem you should compare the number of unknown quantities with the number of *independent equations*[9] by the criteria listed above. When the problem is *fully-specified* then you can proceed to find a unique solution. When the problem is *under-specified* you will not be able to find a unique solution without more information. If insufficient data are available then you may treat the problem as a *design problem.* Design problems are beyond the scope of this text but you can read about them in *Refs. 8–11.*

When the problem is *over-specified* you should suspect an error and check the validity of each equation, then remove any invalid equations. If all independent equations of an over-specified problem are valid, then you can choose the set that yields the most efficient or the most accurate solution.

Example 4.08 illustrates the concept of process specification with a single process unit and *Table 4.02* summarises some rules of thumb for determining the degrees of freedom of multi-unit processes.

Unless indicated otherwise, all problems in this text are *fully-specified.*

[9] An *independent equation* is one that cannot be obtained by combining other equations in the set.

EXAMPLE 4.08 Specification of material balance problems (open system at steady-state).

A. Under-specification of a mixer

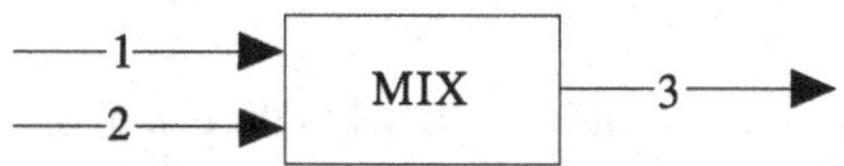

Stream Table		Mixer Specification		
Species	M	Stream		
		1	2	3
	kg/kmol	kmol/h		
A	40	8	?	14
B	60	5	?	?
Total	kg/h	620	?	?

Problem: Complete the material balance stream table.
Solution:

Mole balance on A: $0 = 8 + \dot{n}(2,A) - 14$ [1]

Mole balance on B: $0 = 5 + \dot{n}(2,B) - \dot{n}(3,B)$ [2]

$\overline{m}(2) = 40\dot{n}(2,A) + 60\dot{n}(2,B)$ [3]

$\overline{m}(3) = 40\dot{n}(14) + 60\dot{n}(3,B)$ [4]

Five unknowns. Four independent equations.

$D\ of\ F = 5 - 4 = 1$ $D\ of\ F > 0$ **UNDER-SPECIFIED**

or (*not counting stream totals*)

Number of stream variables = IJ + L= 6

Number (independent equations + specified values) = 5 $D\ of\ F = 6 - 5 = 1$

There is an infinite number of (potential) solutions to problem A.
A process design aims for the "optimum" solution.

B. Full-specification of a mixer

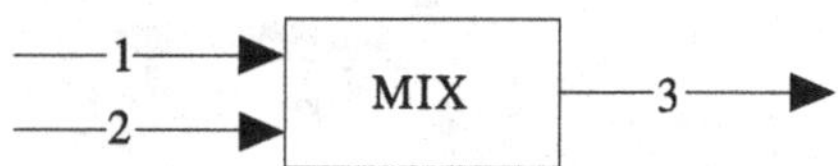

Stream Table		Mixer Specification		
Species	M	Stream		
		1	2	3
	kg/kmol	kmol/h		
A	40	8	?	14
B	60	5	2	?
Total	kg/h	620	?	?

Problem: Complete the material balance stream table.
Solution:

Mole balance on A: $0 = 8 + \dot{n}(2,A) - 14$ [1]

Mole balance on B: $0 = 5 + 2 - \dot{n}(3,B)$ [2]

$\overline{m}(3) = (40)\dot{n}(2,A) + 60\dot{n}(3,B)$ [3]

$\overline{m}(3) = (40)(14) + 60\dot{n}(3,B)$ [4]

Four unknowns. Four independent equations.

$D\ of\ F = 4 - 4 = 0$ $D\ of\ F = 0$ **FULLY-SPECIFIED**

or (*not counting stream totals*)

Number of stream variables = IJ + L = 6

Number (independent equations + specified values) = 6 $D\ of\ F = 6 - 6 = 0$

C. Over-specification of a mixer (redundant)

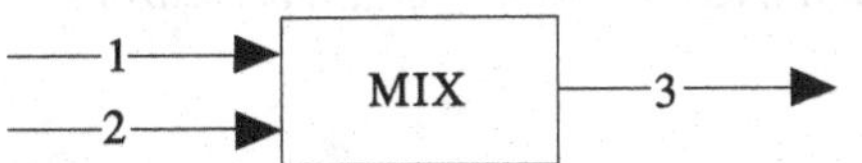

Stream 3 density = 870.2 kg/m³

Stream Table	Mixer Specification			
Species	M	Stream		
		1	2	3
	kg/kmol	kmol/h		
A	40	8	?	14
B	60	5	2	?
Total	kg/h	620	?	?

Density A = 900 kg/m³, B = 797 kg/m³
Ideal liquid mixtures.

Problem: Complete the material stream table.
Solution:

Mole balance on A: $0 = 8 + \dot{n}(2,A) - 14$ [1]

Mole balance on B: $0 = 5 + \dot{n}(2,B) - \dot{n}(3,B)$ [2]

$\overline{m}(2) = 40\dot{n}(2,A) + 60\dot{n}(2,B)$ [3]

$\overline{m}(3) = (40)(14) + 60\dot{n}(3,B)$ [4]

Stream 3 density: $\rho(3) = 870.2 \text{ kg/m}^3$ [5]

Four unknowns. Five "independent" equations.

D of F = 4 − 5 = −1 D of F < 0 **OVER-SPECIFIED and REDUNDANT**

or (*not counting stream totals*)

Number of stream variables = IJ + L = 6

Number (independent equations + specified values) = 7 D of F = 6 − 7 = −1

D. Over-Specification of a mixer (incorrect)

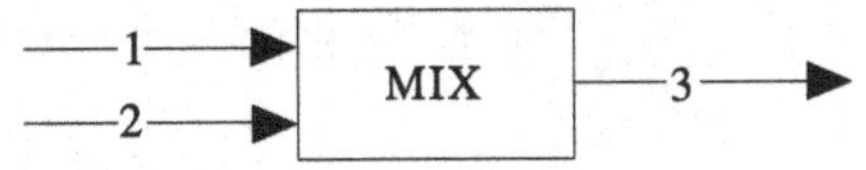

Stream 3 velocity = 2 m/s
Stream 3 pipe ID = 0.02 m

Stream Table	Mixer Specification			
Species	M	Stream		
		1	2	3
	kg/kmol	kmol/h		
A	40	8	?	14
B	60	5	2	?
Total	kg/h	620	?	?

Density A = 900 kg/m³, B = 797 kg/m³
Ideal liquid mixtures.

Problem: Complete the material stream table.
Solution:

Mole balance on A: $0 = 8 - \dot{n}(2,A) - 14$ [1]

Mole balance on B: $0 = 5 + 2 - \dot{n}(2,B)$ [2]

$\overline{m}(2) = 40\dot{n}(2,A) + 60\dot{n}(2,B)$ [3]

$\overline{m}(3) = (40)(14) + 60\dot{n}(3,B)$ [4]

Stream 3 velocity in 0.02 m internal diameter ID outlet pipe = $\tilde{u}(3) = 2$ m/s [5]

Four unknowns. Five independent equations.

D of F = 4 − 5 = −1 D of F < 0 **OVER-SPECIFIED and INCORRECT** (*a.k.a. inconsistent*)

or (*not counting stream totals*)

Number of stream variables = IJ + L = 6

Number (independent equations + specified values) = 7 D of F = 6 − 7 = −1

MULTIPLE PROCESS UNITS

Most chemical processes are systems with multiple process units joined in sequence through their process streams. Material balances on systems with multiple process units are obtained by simply adding the single unit balances, in the order of the process sequence. In these systems the output of one unit becomes the input of the next unit in the sequence, so individual units are *coupled* through their process streams.

There are two common methods used to solve the material balance equations for multiple process units:

- The method of *sequential modular solution*. Equations for each process unit are solved in sequence.
- The method of *simultaneous solution*. Equations describing the whole system are solved simultaneously.

Once it has been "set up" the method of simultaneous solution is faster and more efficient than the method of sequential solution. However the method of sequential solution is used for most examples in this text because it is easier to apply, more transparent and less error prone than the method of simultaneous solution. The method of sequential modular solution of multi-unit material balance problems is illustrated in *Examples 4.09A* and *4.10A*, while *Examples 4.09B* and *4.10B* illustrate the method of simultaneous solution.

Table 4.02. Specification (degrees of freedom) of multi-unit process material balance.

Process unit	Divide	Mix	Separate	Hex	Pump	React	OVERALL
Stream variables	IJ	IJ	IJ	IJ	IJ	IJ + L	IJ + L
Species balance equations	J	J	J	J	J	J	J K[10]
Element balance equations	A	A	A	A	A	< = A[10]	< = A K[10]
Subsidiary relations	Composition restrictions (D – 1)(J – 1) Flow ratios	None	Sep. efficiency Split fractions Flow ratios	None	None Yield	Conversion Extent Selectivity	Sum all units

Specified quantities	BASIS of calculations.[11]
Flows	Stream or component flows in mass, moles or volume per time
Compositions	Stream compositions in mass, mole or volume fraction (or %)

A = number of elements
D = number of streams out of a divider
I = number of streams
J = number of components (usually chemical species)
K = number of process units
L = number of independent chemical reactions[12]

For balances in multi-unit processes it is easiest to begin sequential calculations at a unit that is fully specified (i.e. D of F = 0). However, it is not necessary that there will be a unit with D of F = 0

The D of F for a material balance can be determined by either of two methods.

A. [Preferred method] Count the total number of stream variables and subtract the number of independent relations.
D of F = number of stream variables – number of independent relations
The number of stream variables for each process unit is tabulated above and includes the known component flows.
Independent relations ≡ Basis + known stream specifications + material balances + subsidiary relations
Stream specifications ≡ Total stream flows, component flows, stream compositions
Subsidiary relations ≡ Divider composition restrictions and flow ratios, separator split fractions and stream equilibria, reactant conversion, extent of reaction, selectivity, yield, etc.
In method A the independent relations include the basis of calculations and the known stream specifications.
When atom balances are used the number of stream variables *excludes* the number of reactions (L).

B. Count the unknown flows and subtract the number of independent equations.
D of F = number of unknowns – number of independent equations
Independent equations ≡ Stream compositions + material balances + subsidiary relations
In method B, if total stream flows are used then the independent equations include the stream summations:
$\bar{n}(i) = \Sigma[\dot{n}(i, j)]$ or $\bar{m}(i) = \Sigma\,[M(j)\dot{n}(i, j)]$
When atom balances are used the number of unknowns *excludes* the number of reactions (L).

[10] When atoms or molecules occur in fixed ratios, the individual species balances are not independent equations.
[11] Setting a *basis of calculations* means arbitrarily fixing the value of one unknown quantity, without over-specifying the problem. e.g. fixing the total mass, total flow rate, mass or flow of one species, etc.
[12] Independent reactions are those whose stoichiometry cannot be obtained by combining other reactions in the set.

Example 4.09 Material balances on a multi-unit process (open system at steady-state).

A. Sequential modular solution

Continuous process at steady-state

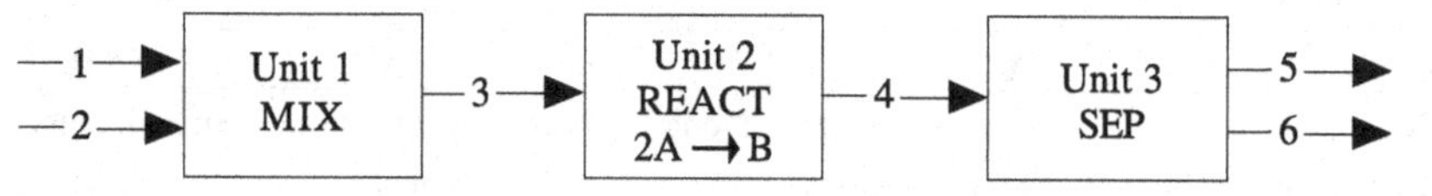

Specifications	*Stream*	
	1 (kmol/h)	*2* (kmol/h)
A (M=40)	10.0	5.0
B (M=80)	0.0	1.0
Conversion of A:X(A)		0.6
Split fractions	s(5,A) =	0.9
	s(5,B) =	0.2

Specifications	**Unit 1**	**Unit 2**	**Unit 3**	**Overall**
# stream variables	IJ = 6	IJ + L = 5	IJ = 6	IJ + L = 9
# specified values + subsidiary relations	4	1	2	4
# species balances	2	2	2	2
D of F	0	2	2	3

I = No. of streams, J = No. of components, L = No. of reactions

A. Total stream variables = (6)(2) + 1 = 13

Total independent relations (Units 1, 2, 3) = 13

D of F = 13 – 13 = 0

B. Total unknowns in the initial problem = (4)(2) + 1= 9

Total independent equations = 9 (includes split faction and conversion)

D of F = 9 – 9 = 0

Problem is FULLY-SPECIFIED

This procedure does not count the stream total mass or mole flows as unknowns, they are assumed known as the sum of the component flows.

Problem: Complete the stream table material balance.

Solution:

UNIT 1 (Mixer)

Mole balance on A

$$0 = \dot{n}(1,A) + \dot{n}(2,A) - \dot{n}(3,A) \quad [1]$$

Mole balance on B

$$0 = \dot{n}(1,B) + \dot{n}(2,B) - \dot{n}(3,B) \quad [2]$$

UNIT 2 (Reactor)

Mole balance on A

$$0 = \dot{n}(3,A) - \dot{n}(4,A) + 0 - X(A)\dot{n}(3,A) \quad [3]$$

Mole balance on B

$$0 = \dot{n}(3,B) - \dot{n}(4,B) + (1/2)X(A)\dot{n}(3,A) \quad [4]$$

UNIT 3 (Separator)

Mole balance on A

$$0 = \dot{n}(4,A) - \dot{n}(5,A) - \dot{n}(6,A) \quad [5]$$

Mole balance on B

$$0 = \dot{n}(4,B) - \dot{n}(5,B) - \dot{n}(6,B) \quad [6]$$

Specified values: $s(5,A) = 0.9 = \dot{n}(5,A)/\dot{n}(4,A)$ [7]

Split fractions: $s(5,B) = 0.2 = \dot{n}(5,B)/\dot{n}(4,B)$ [8]

Reactor conversion: $X(A) = 0.6$ [9]

Stream flows:

$\dot{n}(1,A) = 10$

$\dot{n}(1,B) = 0$

$\dot{n}(2,A) = 5$

$\dot{n}(2,B) = 1$ kmol/h

Solve the process units in sequence — beginning at Unit 1.

Stream Table	***Mixer-Reactor-Separator***			***Solution***			
Species	M	Stream					
		1	*2*	*3*	*4*	*5*	*6*
	kg/kmol	kmol/h					
A	40	10.0	5.0	**15.0**	**6.0**	**5.4**	**0.6**
B	80	0.0	1.0	**1.0**	**5.5**	**1.1**	**4.4**
Total	kg/h	400	280	680	680	304	376
Check.	UNIT 1		UNIT 2		UNIT 3		OVERALL
Mass IN	680 kg/h	Closure	680 kg/h	Closure	680 kg/h	Closure	680 kg/h
Mass OUT	680 kg/h	100%	680 kg/h	100%	680 kg/h	100%	680 kg/h
Overall yield of B from A =			2 (4.4) / 15	=	0.6 = 60%		

$\dot{n}(3, A) = 10 + 5 = 15$ kmol/h

$\dot{n}(4, A) = (15)(1 - 0.6) = 6$ kmol/h

$\dot{n}(5, A) = (0.9)(6) = 5.4$ kmol/h

$\dot{n}(6, A) = 6 - 5.4 = 0.6$ kmol/h

$\dot{n}(3, B) = 0 + 1 = 1$ kmol/h

$\dot{n}(4, B) = 1 + (1/2)(0.6)(15) = 5.5$ kmol/h

$\dot{n}(5, B) = (0.2)(5.5) = 1.1$ kmol/h

$\dot{n}(6, B) = 5.5 - 1.1 = 4.4$ kmol/h

B. Simultaneous equation method

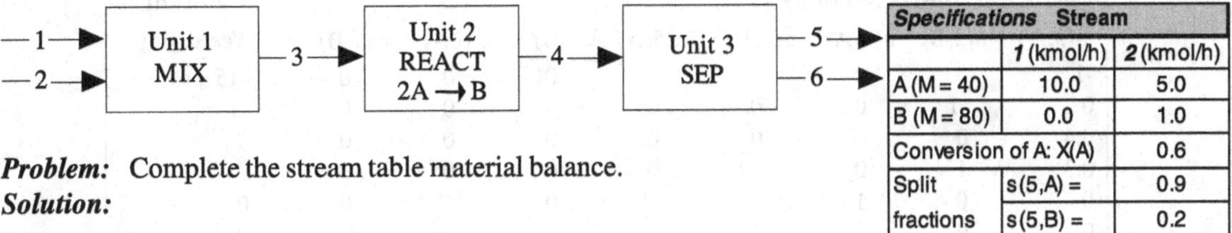

Specifications Stream		
	1 (kmol/h)	*2* (kmol/h)
A (M = 40)	10.0	5.0
B (M = 80)	0.0	1.0
Conversion of A: X(A)		0.6
Split	s(5,A) =	0.9
fractions	s(5,B) =	0.2

Problem: Complete the stream table material balance.
Solution:

UNIT 1 (Mixer)

Mole balance on A: $0 = \dot{n}(1, A) + \dot{n}(2, A) - \dot{n}(3, A)$ [1]

Mole balance on B: $0 = \dot{n}(1, B) + \dot{n}(2, B) - \dot{n}(3, B)$ [2]

UNIT 2 (Reactor)

Mole balance on A: $0 = \dot{n}(3, A) - \dot{n}(4, A) + 0 - X(A)\dot{n}(3, A)$ [3]

Mole balance on B: $0 = \dot{n}(3, B) - \dot{n}(4, B) + (1/2)X(A)\dot{n}(3, A)$ [4]

UNIT 3 (Separator)

Mole balance on A: $0 = \dot{n}(4, A) - \dot{n}(5, A) - \dot{n}(6, A)$ [5]

Mole balance on B: $0 = \dot{n}(4, B) - \dot{n}(5, B) - \dot{n}(6, B)$ [6]

Specified values: $s(5, A) = 0.9 = \dot{n}(5, A)/\dot{n}(4, A)$ [7]

Split fractions: $s(5, B) = 0.2 = \dot{n}(5, B)/\dot{n}(4, B)$ [8]

Reactor conversion: $X(A) = 0.6$ [9]

Stream flows: $\dot{n}(1, A) = 10 \quad \dot{n}(1, B) = 0 \quad \dot{n}(2, A) = 5 \quad \dot{n}(2, B) = 1$ kmol/h

Specifications same as *Problem 4.09A*. By method B, *Table 4.02*: **Nine unknowns. Nine independent equations.** Write the set of eight simultaneous linear equations (Equation [9] is included in the reactor mole balances). Solve the set by matrix algebra.[13]

$$-15 = -(1)\dot{n}(3,A) + (0)\dot{n}(3,B) + (0)\dot{n}(4,A) + (0)\dot{n}(4,B) + (0)\dot{n}(5,A) + (0)\dot{n}(5,B) + (0)\dot{n}(6,A) + (0)\dot{n}(6,B) \quad [1]$$

$$-1 = +(0)\dot{n}(3,A) - (1)\dot{n}(3,B) + (0)\dot{n}(4,A) + (0)\dot{n}(4,B) + (0)\dot{n}(5,A) + (0)\dot{n}(5,B) + (0)\dot{n}(6,A) + (0)\dot{n}(6,B) \quad [2]$$

$$0 = +(0.4)\dot{n}(3,A) + (0)\dot{n}(3,B) - (1)\dot{n}(4,A) + (0)\dot{n}(4,B) + (0)\dot{n}(5,A) + (0)\dot{n}(5,B) + (0)\dot{n}(6,A) + (0)\dot{n}(6,B) \quad [3]$$

---

[13] Solution vector = X = A^{-1} C where A^{-1} = inverse of coefficient matrix; C = constant vector

$$0 = +(0.3)\dot{n}(3,A) + (1)\dot{n}(3,B) + (0)\dot{n}(4,A) - (1)\dot{n}(4,B) + (0)\dot{n}(5,A) + (0)\dot{n}(5,B) + (0)\dot{n}(6,A) + (0)\dot{n}(6,B) \quad [4]$$

$$0 = +(0)\dot{n}(3,A) + (0)\dot{n}(3,B) + (1)\dot{n}(4,A) + (0)\dot{n}(4,B) - (1)\dot{n}(5,A) + (0)\dot{n}(5,B) - (1)\dot{n}(6,A) + (0)\dot{n}(6,B) \quad [5]$$

$$0 = +(0)\dot{n}(3,A) + (0)\dot{n}(3,B) + (0)\dot{n}(4,A) + (1)\dot{n}(4,B) + (0)\dot{n}(5,A) - (1)\dot{n}(5,B) + (0)\dot{n}(6,A) - (1)\dot{n}(6,B) \quad [6]$$

$$0 = +(0)\dot{n}(3,A) + (0)\dot{n}(3,B) + (0.9)\dot{n}(4,A) + (0)\dot{n}(4,B) - (1)\dot{n}(5,A) + (0)\dot{n}(5,B) + (0)\dot{n}(6,A) + (0)\dot{n}(6,B) \quad [7]$$

$$0 = +(0)\dot{n}(3,A) + (0)\dot{n}(3,B) + (0)\dot{n}(4,A) + (0.2)\dot{n}(4,B) + (0)\dot{n}(5,A) - (1)\dot{n}(5,B) + (0)\dot{n}(6,A) + (0)\dot{n}(6,B) \quad [8]$$

Coefficient Matrix [A] [8 by 8] / **Constant Vector[C]**

$\dot{n}(3,A)$	$\dot{n}(3,B)$	$\dot{n}(4,A)$	$\dot{n}(4,B)$	$\dot{n}(5,A)$	$\dot{n}(5,B)$	$\dot{n}(6,A)$	$\dot{n}(6,B)$	Constant Vector[C]
-1	0	0	0	0	0	0	0	-15
0	-1	0	0	0	0	0	0	-1
0.4	0	-1	0	0	0	0	0	0
0.3	1	0	-1	0	0	0	0	0
0	0	1	0	-1	0	-1	0	0
0	0	0	1	0	-1	0	-1	0
0	0	0.9	0	-1	0	0	0	0
0	0	0	0.2	0	-1	0	0	0

Inverse Matrix [A^{-1}] [8 by 8] / **Solution Vector [X]**

								Solution Vector [X]	
-1	0	0	0	0	0	0	0	15	$\dot{n}(3,A)$
0	-1	0	0	0	0	0	0	1	$\dot{n}(3,B)$
-0.4	0	-1	0	0	0	0	0	6	$\dot{n}(4,A)$
-0.3	-1	0	-1	0	0	0	0	5.5	$\dot{n}(4,B)$
-0.36	0	-0.9	0	0	0	-1	0	5.4	$\dot{n}(5,A)$
-0.06	-0.2	0	-0.2	0	0	0	-1	1.1	$\dot{n}(5,B)$
-0.04	0	-0.1	0	-1	0	1	0	0.6	$\dot{n}(6,A)$
-0.24	-0.8	0	-0.8	0	-1	0	1	4.4	$\dot{n}(6,B)$

Stream Table	*Mixer-Reactor-Separator*		*Solution*				
Species	M	Stream					
		1	2	3	4	5	6
	kg/kmol	kmol/h					
A	40	10.0	5.0	**15.0**	**6.0**	**5.4**	**0.6**
B	80	0.0	1.0	**1.0**	**5.5**	**1.1**	**4.4**
Total	kg/h	400	280	680	680	304	376
Check.	UNIT 1		UNIT 2		UNIT 3		OVERALL
Mass IN	680 kg/h	Closure	680 kg/h	Closure	680 kg/h	Closure	680 kg/h
Mass OUT	680 kg/h	100%	680 kg/h	100%	680 kg/h	100%	680 kg/h
Overall yield of B from A =			2 (4.4) /15	=	0.6=60%		

Recycling, Accumulation and Purging

Recycling is commonly used in chemical plants, both to conserve materials and for process control. The presence of a recycle stream in a process flowsheet complicates the material balance by coupling the output of one unit to the input of a previous unit in the sequence. Recycle balance problems can be solved indirectly by iterating the sequential modular method or directly by the method of simultaneous solution. Both methods are illustrated with a simple recycle flowsheet in *Example 4.10*.

The iterative sequential modular method is the basis of most commercial software for process modelling by computer. By this method one stream in the recycle loop is designated as the "tear stream" and given an arbitrary initial condition (e.g. zero flow). The material balance is then calculated in sequence and iterated through the tear point with each new condition of the tear stream. The iteration is terminated when the mass balance *closure* (*Equation 4.09*) converges to an acceptable value and/or shows little change with the number of iterations. Depending on the complexity of the problem the number of iterations required to reach acceptable closure (e.g. 99.9% or higher) can range from about 5 to 10,000. In bad cases the iteration may be unstable and never converge to a satisfactory point, though such cases can usually be resolved by numerical techniques such as *relaxation*.[14]

Recycling is a good way to increase the profitability and/or reduce the environmental impact of a chemical process. However the tendency of undesired materials to *accumulate* in recycle loops is a major problem in recycling. If they are not controlled, small amounts of undesired substances, such as inert "non-process elements",[15] catalyst poisons, etc. can build up in recycle loops to high levels that make the system inoperable. One way to control this accumulation is to *purge* a fraction of some stream(s) from the system to balance the rate of input of the undesired substance(s).

Example 4.11A shows how a small amount of inert substance (C) in a feed stream can accumulate in a recycle loop to virtually take over the process, then *Example 4.11B* shows that the accumulation is controlled by purging part of the recycle stream.

EXAMPLE 4.10 Material balances on a multi-unit recycle process (open system at steady-state).

A. Iterative sequential modular method

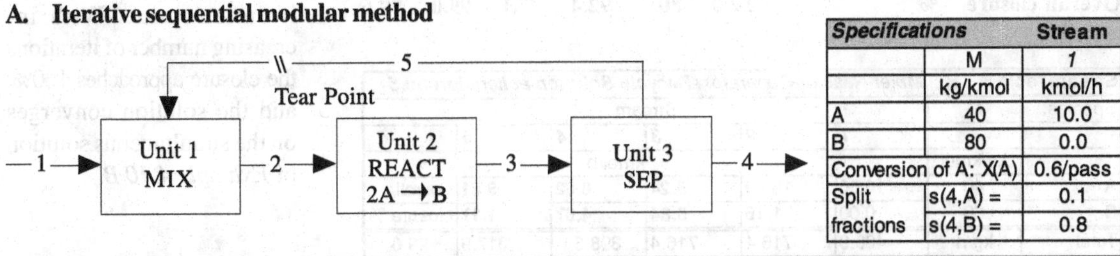

Specifications		Stream
	M	*1*
	kg/kmol	kmol/h
A	40	10.0
B	80	0.0
Conversion of A: X(A)		0.6/pass
Split	s(4,A) =	0.1
fractions	s(4,B) =	0.8

[14] *Relaxation* (specifically under-relaxation) means applying a damping factor to the progressive incremental change in an iterated value.

[15] *Non-process elements* are trace impurities that are not required for operation of the process, for example, *silicon* that is introduced with wood in the chemical pulping process.

Problem: Complete the stream table material balance.
Solution:

UNIT 1 (Mixer)	Mole balance on A:	$0 = 10 + \dot{n}(5,A) - \dot{n}(2,A)$	[1]
	Mole balance on B:	$0 = 0 + \dot{n}(5,B) - \dot{n}(2,B)$	[2]
UNIT 2 (Reactor)	Mole balance on A:	$0 = \dot{n}(2,A) - \dot{n}(3,A) - 0.6\dot{n}(2,A)$	[3]
	Mole balance on B:	$0 = \dot{n}(2,B) - \dot{n}(3,B) + (1/2)(0.6\dot{n}(2,A))$	[4]
UNIT 3 (Separator)	Mole balance on A:	$0 = \dot{n}(3,A) - \dot{n}(4,A) - \dot{n}(5,A)$	[5]
	Mole balance on B:	$0 = \dot{n}(3,B) - \dot{n}(4,B) - \dot{n}(5,B)$	[6]
Split fractions:	$0.1 = \dot{n}(4,A)/\dot{n}(3,A)$		[7]
	$0.8 = \dot{n}(4,B)/\dot{n}(3,B)$		[8]
Conversion:	Specified in the reactor balances		

FULLY-SPECIFIED. By method B, *Table 4.02*. Eight unknowns. Eight independent equations.

Equations are coupled through the recyle stream. Note also that the flows of A and B are coupled through the reactor. This procedure does not count the stream total mass or mole flows as unknowns, since they are assumed known as the sum of the component flows. Solve by the iterative sequential modular method shown below. Begin iterations with tear stream 5 "empty", i.e. $\dot{n}(5,A) = 0$

Sequenced explicit solution	**Iteration number** kmol/h					
	0	**1**	**2**	**3**	**4**	**5**
$\dot{n}(1,A) = 10$	10	10	10	10	10	10
$\dot{n}(1,B) = 0$	0	0	0	0	0	0
$\dot{n}(2,A) = 10 + \dot{n}(5,A)$	10.00	13.60	14.90	15.36	15.53	15.59
$\dot{n}(2,B) = 0 + \dot{n}(5,B)$	0.00	0.60	0.94	1.08	1.14	1.16
$\dot{n}(3,A) = \dot{n}(2,A) - 0.6\dot{n}(2,A)$	4.00	5.44	5.96	6.15	6.21	6.24
$\dot{n}(3,B) = (2,B) + (1/2)0.6\dot{n}(2,A)$	3.00	4.68	5.40	5.69	5.80	5.84
$\dot{n}(4,A) = 0.1\dot{n}(3,A)$	0.40	0.54	0.60	0.61	0.62	0.62
$\dot{n}(4,B) = 0.8\dot{n}(3,B)$	2.40	3.74	4.32	4.55	4.64	4.67
$\dot{n}(5,A) = \dot{n}(3,A) - \dot{n}(4,A)$	3.60	4.90	5.36	5.53	5.59	5.61
$\dot{n}(5,B) = \dot{n}(3,B) - \dot{n}(4,B)$	0.60	0.94	1.08	1.14	1.16	1.17
Overall closure %	**52.0**	**80.3**	**92.4**	**97.1**	**99.0**	**99.6**

NOTE: The iterations in *Examples 4.10A*, *4.11A* and *4.11B* are recorded here to only illustrate the sequential iterative method. You will normally do such iterative calculations by computer (e.g. spreadsheet) as shown later in this text. It is sometimes useful, but generally not necessary to record the iteration value. With increasing number of iterations the closure approaches 100%, and the solution converges on the simultaneous solution of *Example 4.10 B.*

StreamTable		*Mixer-Reactor-Separator+Recycle Solution at Iteration no.5*					
Species.	M	Stream					
		1	2	3	4	5	
	kg/kmol	kmol/h					
A	40	10.00	15.59	**6.24**	**0.62**	**5.61**	Overall
B	80	0.00	1.16	**5.84**	**4.67**	**1.17**	closure %
Total	kg/h	400.0	716.4	716.4	398.5	317.9	99.6
Check	UNIT 1		UNIT 2		UNIT 3		OVERALL
Mass IN	717.9	Closure %	716.4	Closure %	716.4	Closure %	400.0
Mass OUT	716.4	99.8	716.4	100.0	716.4	100.0	398.5

Overall yield of B from A = 2(4.67)/10 = 0.94 = 94%

Overall yield of B from A
= 2 (4.67)/10 = 0.94 = 94%

B. Simultaneous equation method

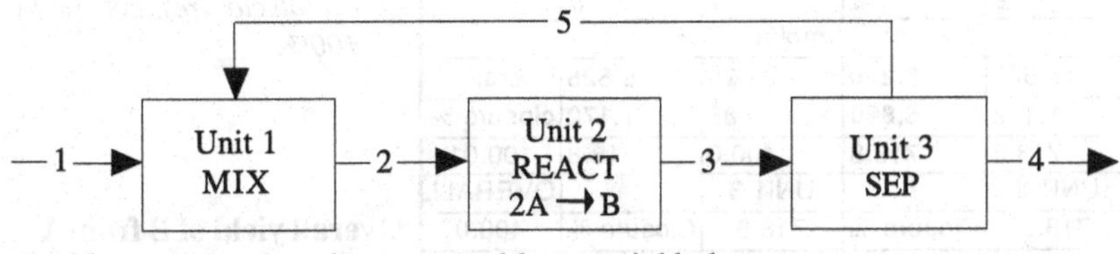

Specifications		Stream
	M	*1*
	kg/kmol	kmol/h
A	40	10.0
B	80	0.0
Conversion of A: X(A)		0.6/pass
Split fractions	s(4, A) =	0.1
	s(4, B) =	0.8

Problem: Complete the stream table material balance.

Solution:

UNIT 1 (Mixer)	Mole balance on A:	$0 = 10 + \dot{n}(5,A) - \dot{n}(2,A)$	[1]
	Mole balance on B:	$0 = 0 + \dot{n}(5,B) - \dot{n}(2,B)$	[2]
UNIT 2 (Reactor)	Mole balance on A:	$0 = \dot{n}(2,A) - \dot{n}(3,A) - 0.6\dot{n}(2,A)$	[3]
	Mole balance on B:	$0 = \dot{n}(2,B) - \dot{n}(3,B) + (1/2)(0.6\dot{n}(2,A))$	[4]
UNIT 3 (Separator)	Mole balance on A:	$0 = \dot{n}(3,A) - \dot{n}(4,A) - \dot{n}(5,A)$	[5]
	Mole balance on B:	$0 = \dot{n}(3,B) - \dot{n}(4,B) - \dot{n}(5,B)$	[6]
Split fractions:	$0.1 = \dot{n}(4,A)/\dot{n}(3,A)$		[7]
	$0.8 = \dot{n}(4,B)/\dot{n}(3,B)$		[8]
Conversion:	Specified in the reactor balances		

FULLY-SPECIFIED. By method B, *Table 4.02*. Eight unknowns. Eight independent equations.

Equations are coupled through the recycle stream. Note also the the flows of A and B are coupled through the reactor. This procedure does not count the stream total mass or mole flows as unknowns, since they are easily found as the sum of the component flows.

Simultaneous solution:

Combine equations [7 and 8] with [5 and 6]: $0 = \dot{n}(3,A) - (0.1)\dot{n}(3,A) - \dot{n}(5,A)$ [9]

$0 = \dot{n}(3,B) - (0.8)\dot{n}(3,B) - \dot{n}(5,B)$ [10]

Combine equations [9 and 10] with [3 and 4]: $0 = \dot{n}(2,A) - \dot{n}(5,A)/0.9 - 0.6\dot{n}(2,A)$ [11]

$0 = \dot{n}(2,B) - \dot{n}(5,B)/0.2 + (1/2)0.6\dot{n}(2,A)$ [12]

Combine equations [11 and 12] with [1 and 2]: $0 = 10 + \dot{n}(5,A) - \dot{n}(5,A)/(0.36)$ [13]

$0 = 0 - 4\dot{n}(5,B) + 833\dot{n}(5,A)$ [14]

Yields the solution: $\dot{n}(5,A) = 5.625$ $\quad \dot{n}(2,A) = 15.625$ $\quad \dot{n}(3,A) = 6.250$ $\quad \dot{n}(4,A) = 0.625$

$\dot{n}(5,B) = 1.172$ $\quad \dot{n}(2,B) = 1.172$ $\quad \dot{n}(3,B) = 5.860$ $\quad \dot{n}(4,B) = 4.688$ kmol/h

StreamTable		*Mixer-Reactor-Separator + Recycle Simultaneous Solution*					
Species	M	Stream					
		1	*2*	*3*	*4*	*5*	
	kg/kmol	kmol/h					
A	40	10.000	15.625	**6.250**	**0.625**	**5.625**	Overall*
B	80	0.000	1.172	**5.860**	**4.688**	**1.172**	closure %
Total	kg/h	400.0	718.8	718.8	400.0	318.8	100.01
Check	UNIT 1		UNIT 2		UNIT 3		OVERALL
Mass IN	718.8	Closure %	718.8	Closure %	718.8	Closure %	400.0
Mass OUT	718.8	100.00	718.8	100.01	718.8	100.00	400.0

** Due to rounding of the stream flow values the overall closure is not exactly 100%.*

Overall yield of B from A = 2(4.688/10) = 0.94 = 94%

EXAMPLE 4.11 Material balance on a multi-unit recycle process (open system at steady-state).

A. Accumulation of non-process species

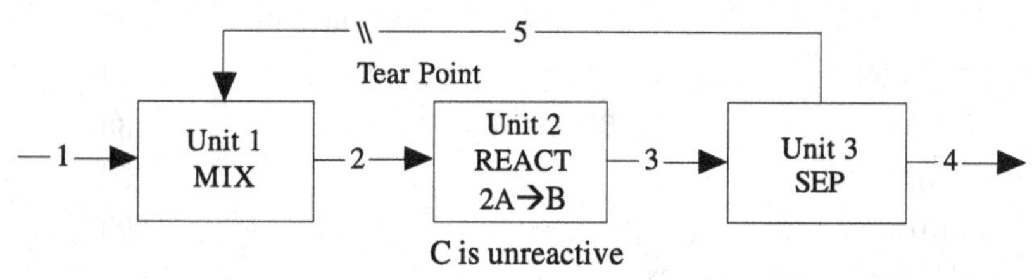

Specifications		Stream
	M	*1*
	kg/kmol	kmol/h
A	40	10.0
B	80	0.0
C	50	0.1
Conversion of A: X(A)		0.6/pass
Split fractions	s(4,A) =	0.1
	s(4,B) =	0.8
	s(4,C) =	0.05

Problem: Complete the stream table material balance.

Solution:

UNIT 1 (Mixer) Mole balance on A: $0 = 10 + \dot{n}(5,A) - \dot{n}(2,A)$ [1]

Mole balance on B: $0 = 0 + \dot{n}(5,B) - \dot{n}(2,B)$ [2]

Mole balance on C: $0 = 0.1 + \dot{n}(5,C) - \dot{n}(2,C)$ [3]

UNIT 2 (Reactor) Mole balance on A: $0 = \dot{n}(2,A) - \dot{n}(3,A) - 0.6\dot{n}(2,A)$ [4]

Mole balance on B: $0 = \dot{n}(2,B) - \dot{n}(3,B) + (1/2)(0.6\dot{n}(2,A))$ [5]

Mole balance on C: $0 = \dot{n}(2,C) - \dot{n}(3,C)$ [Unreactive] [6]

UNIT 3 (Separator) Mole balance on A: $0 = \dot{n}(3,A) - \dot{n}(4,A) - \dot{n}(5,A)$ [7]

Mole balance on B: $0 = \dot{n}(3,B) - \dot{n}(4,B) - \dot{n}(5,B)$ [8]

Mole balance on C: $0 = \dot{n}(3,C) - \dot{n}(4,C) - \dot{n}(5,C)$ [9]

Split fractions: $0.1 = \dot{n}(4,A)/\dot{n}(3,A)$ [10]

$0.8 = \dot{n}(4,B)/\dot{n}(3,B)$ [11]

$0.05 = \dot{n}(4,C)/\dot{n}(3,C)$ [12]

Conversion: Specified in reactor balances

FULLY-SPECIFIED. By method B, *Table 4.02*. 12 independent equations. 12 unknowns.

Solve by the iterative sequential modular method. Begin iterations with tear stream 5 "empty".

Sequenced explicit solutions	**Iteration number** kmol/h					
	0	**1**	**2**	**3**	**4**	**100**
$\dot{n}(1,A) = 10$	10	10	10	10	10	10
$\dot{n}(1,B) = 0$	0	0	0	0	0	0
$\dot{n}(1,C) = 0.1$	0.1	0.1	0.1	0.1	0.1	0.1
$\dot{n}(2,A) = 10 + \dot{n}(5,A)$	10.00	13.60	14.90	15.36	15.53	15.63
$\dot{n}(2,B) = 0 + \dot{n}(5,B)$	0.00	0.60	0.94	1.08	1.14	1.17
$\dot{n}(2,C) = 0.1 + \dot{n}(5,C)$	0.10	0.20	0.29	0.37	0.45	2.00
$\dot{n}(3,A) = \dot{n}(2,A) - 0.6\dot{n}(2,A)$	4.00	5.44	5.96	6.15	6.21	6.25
$\dot{n}(3,B) = \dot{n}(2,B) + (1/2)0.6\dot{n}(2,A)$	3.00	4.68	5.40	5.69	5.80	5.85
$\dot{n}(3,C) = \dot{n}(2,C)$	0.10	0.20	0.29	0.37	0.45	2.00
$\dot{n}(4,A) = 0.1\dot{n}(3,A)$	0.40	0.54	0.60	0.61	0.62	0.63
$\dot{n}(4,B) = 0.8\dot{n}(3,B)$	2.40	3.74	4.32	4.55	4.64	4.68
$\dot{n}(4,C) = 0.05\dot{n}(3,C)$	0.005	0.010	0.014	0.019	0.023	0.100
$\dot{n}(5,A) = \dot{n}(3,A) - \dot{n}(4,A)$	3.60	4.90	5.36	5.53	5.59	5.63
$\dot{n}(5,B) = \dot{n}(3,B) - \dot{n}(4,B)$	0.60	0.94	1.08	1.14	1.16	1.17
$\dot{n}(5,C) = \dot{n}(3,C) - \dot{n}(4,C)$	0.095	0.185	0.271	0.352	0.430	1.900
Overall closure %	**51.42**	**79.45**	**91.47**	**96.21**	**98.02**	**99.93**

Stream Table *Mixer-Reactor-Separator+Recycle+Accumulation Solution at itn100*							
Species	M		Stream				
		1	*2*	*3*	*4*	*5*	
	kg/kmol			kmol/h			
A	40	10.00	15.63	6.25	0.63	5.63	
B	80	0.00	1.17	5.85	4.68	1.17	Overall
C	50	0.10	2.00	2.00	0.10	1.90	closure %
Total	kg/h	405.0	818.4	818.4	404.7	413.4	99.93
	UNIT 1		UNIT 2		UNIT 3		OVERALL
Mass IN	818.4	Closure %	818.4	Closure %	818.4	Closure %	405.00
Mass OUT	818.4	100.0	818.4	100.0	818.1	100.0	404.71
Overall yield of B from A=			2(4.68)/10 = 0.94 = 94%				
Note the accumulation of C in the recycle stream 5.							

B. Accumulation and purge of non-process species

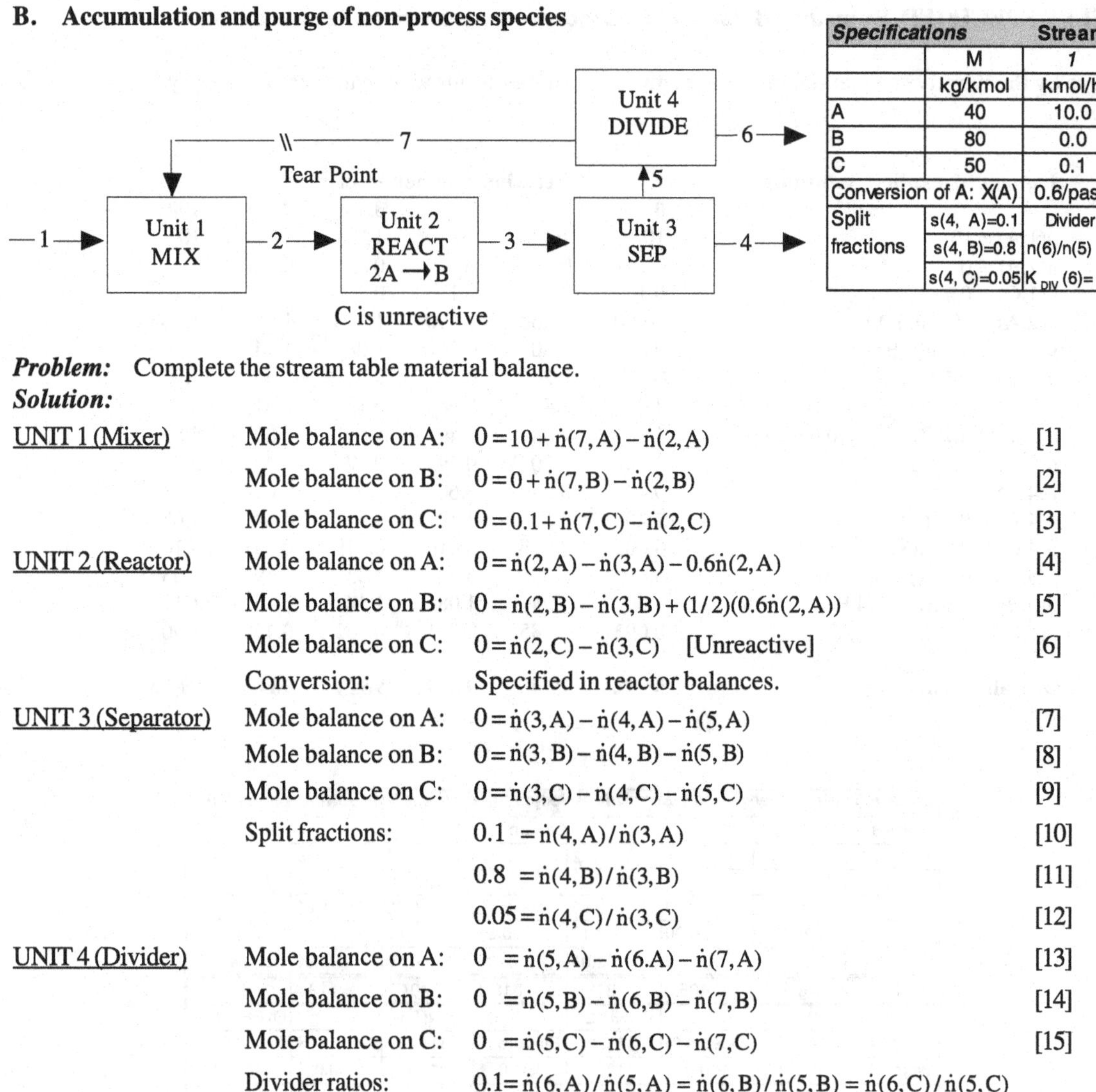

Specifications		Stream
	M	1
	kg/kmol	kmol/h
A	40	10.0
B	80	0.0
C	50	0.1
Conversion of A: X(A)		0.6/pass
Split fractions	s(4, A)=0.1	Divider
	s(4, B)=0.8	n(6)/n(5) =
	s(4, C)=0.05	K_{DIV}(6)= 0.1

Problem: Complete the stream table material balance.

Solution:

UNIT 1 (Mixer)	Mole balance on A:	$0 = 10 + \dot{n}(7,A) - \dot{n}(2,A)$	[1]
	Mole balance on B:	$0 = 0 + \dot{n}(7,B) - \dot{n}(2,B)$	[2]
	Mole balance on C:	$0 = 0.1 + \dot{n}(7,C) - \dot{n}(2,C)$	[3]
UNIT 2 (Reactor)	Mole balance on A:	$0 = \dot{n}(2,A) - \dot{n}(3,A) - 0.6\dot{n}(2,A)$	[4]
	Mole balance on B:	$0 = \dot{n}(2,B) - \dot{n}(3,B) + (1/2)(0.6\dot{n}(2,A))$	[5]
	Mole balance on C:	$0 = \dot{n}(2,C) - \dot{n}(3,C)$ [Unreactive]	[6]
	Conversion:	Specified in reactor balances.	
UNIT 3 (Separator)	Mole balance on A:	$0 = \dot{n}(3,A) - \dot{n}(4,A) - \dot{n}(5,A)$	[7]
	Mole balance on B:	$0 = \dot{n}(3,B) - \dot{n}(4,B) - \dot{n}(5,B)$	[8]
	Mole balance on C:	$0 = \dot{n}(3,C) - \dot{n}(4,C) - \dot{n}(5,C)$	[9]
	Split fractions:	$0.1 = \dot{n}(4,A)/\dot{n}(3,A)$	[10]
		$0.8 = \dot{n}(4,B)/\dot{n}(3,B)$	[11]
		$0.05 = \dot{n}(4,C)/\dot{n}(3,C)$	[12]
UNIT 4 (Divider)	Mole balance on A:	$0 = \dot{n}(5,A) - \dot{n}(6.A) - \dot{n}(7,A)$	[13]
	Mole balance on B:	$0 = \dot{n}(5,B) - \dot{n}(6,B) - \dot{n}(7,B)$	[14]
	Mole balance on C:	$0 = \dot{n}(5,C) - \dot{n}(6,C) - \dot{n}(7,C)$	[15]
	Divider ratios:	$0.1 = \dot{n}(6,A)/\dot{n}(5,A) = \dot{n}(6,B)/\dot{n}(5,B) = \dot{n}(6,C)/\dot{n}(5,C)$	[16][17][18]

FULLY-SPECIFIED. By method B, *Table 4.02*. 18 unknowns. 18 independent equations.

Solve by the ***iterative sequential modular method*** shown on next page.
Begin iterations with tear stream "empty".

Sequenced explicit solutions	**Iteration number** kmol/h					
	0	**1**	**2**	**3**	**4**	**100**
$\dot{n}(1,A)$	10	10	10	10	10	10
$\dot{n}(1,B)$	0	0	0	0	0	0
$\dot{n}(1,C) = 0.1$	0.1	0.1	0.1	0.1	0.1	0
$\dot{n}(2,A) = 10 + \dot{n}(7,A)$	10.00	13.24	14.29	14.63	14.74	14.79
$\dot{n}(2,B) = 0 + \dot{n}(7,B)$	0.00	0.54	0.81	0.92	0.96	0.97
$\dot{n}(2,C) = 0.1 + \dot{n}(7,C)$	0.10	0.19	0.26	0.32	0.37	0.69
$\dot{n}(3,A) = \dot{n}(2,A) - 0.6\dot{n}(2,A)$	4.00	5.30	5.72	5.85	5.90	5.92
$\dot{n}(3,B) = \dot{n}(2,B) + (1/2)0.6\dot{n}(2,A)$	3.00	4.51	5.10	5.31	5.38	5.41
$\dot{n}(3,C) = \dot{n}(2,C)$	0.10	0.19	0.26	0.32	0.37	0.69
$\dot{n}(4,A) = 0.1\dot{n}(3,A)$	0.40	0.53	0.57	0.59	0.59	0.59
$\dot{n}(4,B) = 0.8\dot{n}(3,B)$	2.40	3.61	4.08	4.25	4.30	4.33
$\dot{n}(4,C) = 0.05\dot{n}(3,C)$	0.005	0.009	0.013	0.016	0.019	0.034
$\dot{n}(5,A) = \dot{n}(3,A) - \dot{n}(4,A)$	3.60	4.77	5.14	5.27	5.31	5.33
$\dot{n}(5,B) = \dot{n}(3,B) - \dot{n}(4,B)$	0.60	0.90	1.02	1.06	1.08	1.08
$\dot{n}(5,C) = \dot{n}(3,C) - \dot{n}(4,C)$	0.095	0.176	0.246	0.305	0.356	0.655
$\dot{n}(6,A) = 0.1\dot{n}(5,A)$	0.36	0.48	0.51	0.53	0.53	0.53
$\dot{n}(6,B) = 0.1\dot{n}(5,B)$	0.060	0.090	0.102	0.106	0.108	0.108
$\dot{n}(6,C) = 0.1\dot{n}(5,C)$	0.010	0.018	0.025	0.031	0.036	0.066
$\dot{n}(7,A) = \dot{n}(5,A) - \dot{n}(6,A)$	3.24	4.29	4.63	4.74	4.78	4.79
$\dot{n}(7,B) = \dot{n}(5,B) - \dot{n}(6,B)$	0.54	0.81	0.92	0.96	0.97	0.97
$\dot{n}(7,C) = \dot{n}(5,C) - \dot{n}(6,C)$	0.086	0.159	0.221	0.275	0.320	0.59
Overall closure %	**56.28**	**83.35**	**93.78**	**97.51**	**98.83**	**100.00**

Stream Table ***Mixer-Reactor-Separator+Recycle+Accumulation+Purge. Solution at itn100***								
Species	M	Stream						
		1	*2*	*3*	*4*	*5*	*6*	*7*
	kg/kmol	kmol/h						
A	40	10.00	14.79	5.92	0.59	5.33	0.53	4.79
B	80	0.00	0.97	5.41	4.33	1.08	0.11	0.97 **Overall**
C	50	0.10	0.69	0.69	0.03	0.66	0.07	0.59 **closure %**
Total	kg/h	405.0	704.1	704.1	371.8	332.4	33.2	299.1 **100.0**
Check	UNIT 1		UNIT 2		UNIT 3		UNIT 4		OVERALL
Mass IN	704.1	Closure %	704.1	Closure %	704.1	Closure %	332.4	Closure %	405.0
Mass OUT	704.1	100.0	704.1	100.0	704.1	100.0	332.4	100.0	405.0

Overall yield of B from A + (2)(4.33)/10 = 0.87 = 87%

MATERIAL BALANCE CALCULATIONS BY COMPUTER (SPREADSHEET)

Material balance problems often require a lot of calculations. The most efficient way to do these calculations is by computer. There are three ways that you can use a computer to help solve material balance problems:

1. By spreadsheet (e.g. Excel™, Quattro Pro™)
2. By writing your own "in house" code in a high level language (e.g. Basic, C, Fortran, MATLAB™)
3. By commercial "process simulation" software (e.g. ASPENPLUS™, HYSYS™, PRO/II™, CHEMCAD™)

Process engineers use all of the above three methods, but this text focuses only on method 1, i.e. solving material (and energy) balances by *spreadsheet.*

As outlined in Chapter 3 there are two ways to set up a spreadsheet for material balances:

A. As a separate *stream table* attached to the labelled flowsheet
B. As sets of data adjacent to each stream in the flowsheet

In option A the *spreadsheet* is used as a two-dimensional matrix of cells in which the column and row values correspond to the stream and quantity entries in a process *stream table.* In option B the stream quantities are calculated and displayed in cells adjacent to each process stream.

For both options A and B the values in each cell are assigned or calculated from the appropriate balance equations. In simple cases each equation is solved explicitly for the corresponding cell value. In more complex cases (e.g. non-linear equations) the cell value may be found by a "GoalSeek" or "Solver" routine,[16] or calculated through a *macro* via Visual Basic code (see *Ref.* 7).

Iterative calculations in the spreadsheet first appear as "circular references". These are handled in Excel for example, by ticking the "iterations" box in the "tools-options-calculation" window. On the "calculate" command the computer will then run through the set number of iterations, or it can be programed to terminate the iterations if the balance reaches acceptable *closure,* where the closure may be defined as in *Equation 4.09.* The acceptable closure value can be set as the *convergence criterion* in a conditional command (e.g. IF $e_M > 99.9\%$ THEN stop iterating).

Example 4.12 shows the spreadsheet solution of the recycle material balance problem in *Example 4.10* by both the method of option A and the method of option B. In this text option A is preferred for its superior transparency, and will be used in all subsequent spreadsheet calculation of steady-state material and energy balances.

---

[16] "GoalSeek" can solve single linear and non-linear equations. The "Solver" is a more powerful tool that can solve both single and simultaneous linear and non-linear equations (see *Ref.* 7), although non-linear sets often give difficulties.

Examples 4.13 to *4.16* illustrate the set-up and solution of some more complex material balance problems. Each of these problems is solved by spreadsheet, using the iterative sequential modular method to produce a stream table attached to the process flowsheet.

Examples 4.13, *4.14* and *4.15* show examples in process engineering, being respectively a biochemical process, an electrochemical process and a thermochemical process, each in continuous operation at steady-state. *Example 4.16* presents a case related to the Earth's environment and sustainable development.

Example 4.17 shows how the concepts of conversion, extent of reaction and selectivity, together with atom and mole balances, can be used in different methods to calculate the material balance on a continuous combustion process at steady-state.

The best way to become competent with material balance calculations is to practice solving problems. The illustrative material balance problems on the next pages will get you started. You will find many more practice problems in *Refs. 2–6*.

EXAMPLE 4.12 Spreadsheet calculation of a material balance (open system at steady-state).

A: Flowsheet and stream table

	A	B	C	D	E	F	G	H
1	**Example 4.12**					Specifications		Stream
2	**Option A: Flowsheet and stream table - set up**						M	1
3						A kmol/h	40	10
4	**Problem**	Complete the material balance.				B kmol/h	80	0
5						Conversion of A: X(A)		0.6
6						Split	s(4,A)	0.1
7	**Solution**					fractions	s(4,B)	0.8
8	Option A	Flowsheet and stream table						
9			//	5				
10			Tear Point (Iterate)					
11		Unit 1		Unit 2		Unit 3		
12	1	**MIX**	2	**REACT**	3	**SEPARATE**	4	
13				2A>B				
14								
15	**Stream Table**				**Mixer-Reactor-Separator**			
16	**Species**	**M**			**Stream**			
17			***1***	***2***	***3***	***4***	***5***	
18		kg/kmol			kmol/h			
19	A	40	10	**=C19+G19**	**=D19*(1-$H$5)**	**=$H$6*E19**	**=E19-F19**	Overall
20	B	80	0	**=C20+G20**	**=D20+(1/2)*$H$5*D19**	**=$H$7*E20**	**=E20-F20**	Closure %
21	**Total**	**kg/h**	=$B$19*C19+$B$20*C20	=$B$19*D19+$B$20*E20	=$B$19*E19+$B$20*E20	=$B$19*F19+$B$20*F20	=$B$19*G19+$B$20*G20	=100*H24/H23
22		UNIT 1		Unit 2		UNIT 3		OVERALL
23	Mass IN	=C21+G21	Closure %	=D21	Closure %	=E21	Closure %	=C21
24	Mass OUT	=D21	=100*B24/B23	=E21	=100*D24/D23	=F21+G21	=100*F24/F23	=F21

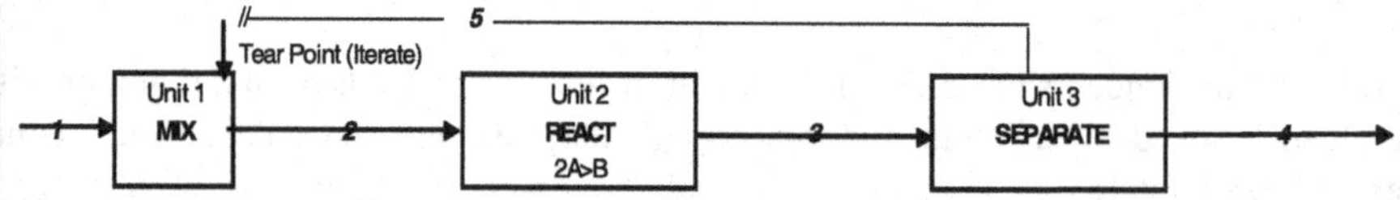

B: Flowsheet with stream data

	A	B	C	D	E	F	G	H	I
1	**Example 4.12**					Specifications		Stream	
2	**Option B: Flowsheet with stream data - set up**						M	1	
3						A kmol/h	40	10	
4	**Problem**	Complete the material balance.				B kmol/h	80	0	
5						Conversion of A: X(A)		0.6	
6	**Solution**					Split	s(4,A)	0.1	
7	Option B	Flowsheet with stream data				fractions	s(4,B)	0.8	
8									
9									
10					5	\\			
11				=F15-H15	A kmol/h	Tear point (iterate)			
12				=F16-H16	B kmol/h				
13				=40*D11+80*D12	kg/h				
14									
15	10	A kmol/h		=A15+D11	A kmol/h	=D15*(1-$H$5)	A kmol/h	=$H$6*F15	A kmol/h
16	0	B kmol/h		=A16+D12	B kmol/h	=D16+(1/2)*$H$5*D15	B kmol/h	=$H$7*F16	B kmol/h
17	=40*A15+80*A16	kg/h		=40*D15+80*D16	kg/h	=40*F15+80*F16	kg/h	=40*H15+80*H16	kg/h
18			Unit 1		Unit 2		Unit 3		
19		1	**MIX**	2	**REACT**	3	**SEPARATE**	4	
20					2A>B				
21	Mass balance check								
22	Mass IN	kg/h	=A17+D13		=D17		=F17		
23	Mass OUT	kg/h	=D17		=F17		=H17+D13		
24	Closure	%	=100*C23/C22		=100*E23/E22		=100*G23/G22		
25									
26	Overall								
27	Mass IN	kg/h	=A17						
28	Mass OUT	kg/h	=H17						
29	Closure	%	=100*C28/C27						

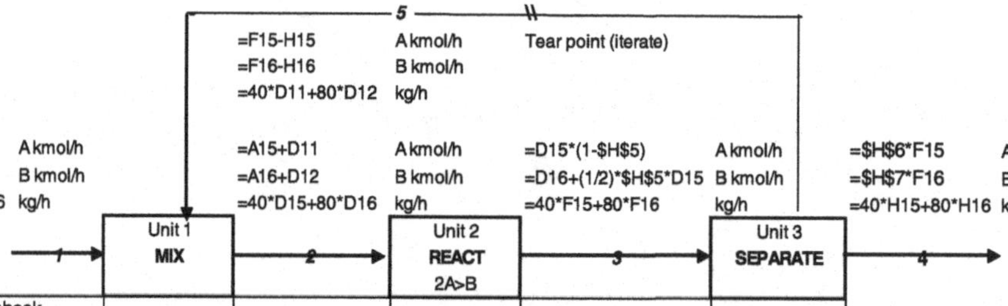

A: Flowsheet and stream table — solution

	A	B	C	D	E	F	G	H
1	**Example 4.12**					Specifications		Stream
2	**Option A: Flowsheet and stream table - solution**						M	1
3						A kmol/h	40	10.0
4						B kmol/h	80	0.0
5	**Problem**	Complete the material balance.				Conversion of A: X(A)		0.6
6						Split	s(4,A)	0.1
7	**Solution**					fractions	s(4,B)	0.8
8	Option A	Flowsheet and stream table						
9			// 5					
10			Tear Point (Iterate)					
11		Unit 1		Unit 2		Unit 3		
12	1	**MIX**	2	**REACT**	3	**SEPARATE**	4	
13				2A>B				
14								
15	**Stream Table**		**Mixer-Reactor-Separator**					
16	**Species**	**M**			**Stream**			
17			***1***	***2***	***3***	***4***	***5***	
18		kg/kmol			kmol/h			
19	A	40	10.000	**15.625**	**6.250**	**0.625**	**5.625**	Overall
20	B	80	0.000	**1.172**	**5.859**	**4.688**	**1.172**	Closure %
21	**Total**	**kg/h**	**400.0**	**718.8**	**718.2**	**400.0**	**318.8**	**100.00**
22		UNIT 1		Unit 2		UNIT 3		OVERALL
23	Mass IN	718.8	Closure %	718.8	Closure %	718.8	Closure %	400.00
24	Mass OUT	718.8	100.00	718.8	100.00	718.8	100.00	400.00

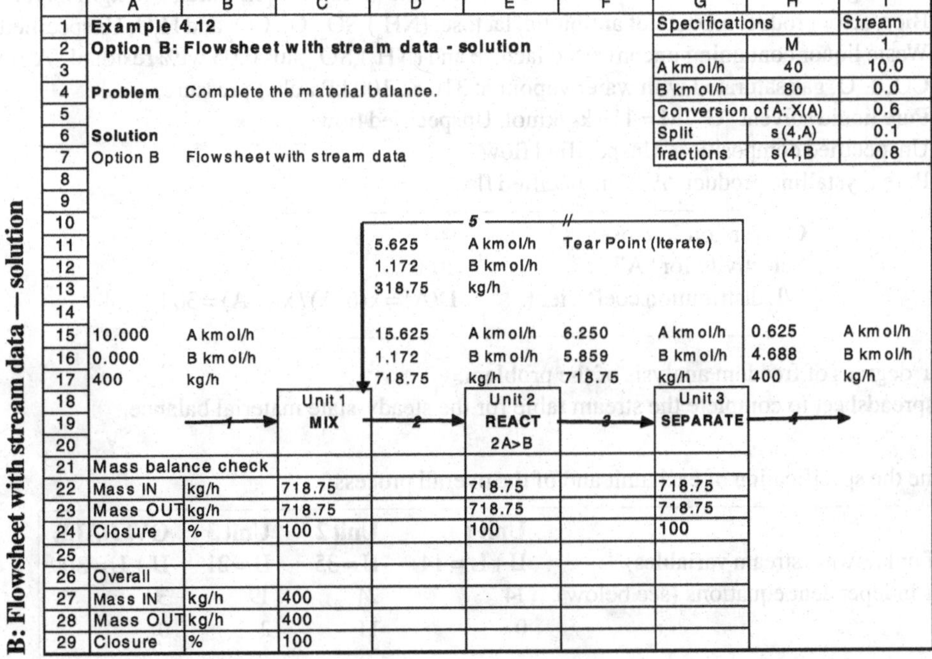

	A	B	C	D	E	F	G	H	I
1	**Example 4.12**						Specifications		Stream
2	**Option B: Flowsheet with stream data - solution**							M	1
3							A kmol/h	40	10.0
4	**Problem**	Complete the material balance.					B kmol/h	80	0.0
5							Conversion of A: X(A)		0.6
6	**Solution**						Split	s(4,A)	0.1
7	Option B	Flowsheet with stream data					fractions	s(4,B	0.8
8									
9									
10					5 //				
11				5.625	A kmol/h	Tear Point (Iterate)			
12				1.172	B kmol/h				
13				318.75	kg/h				
14									
15	10.000	A kmol/h		15.625	A kmol/h	6.250	A kmol/h	0.625	A kmol/h
16	0.000	B kmol/h		1.172	B kmol/h	5.859	B kmol/h	4.688	B kmol/h
17	400	kg/h		718.75	kg/h	718.75	kg/h	400	kg/h
18			Unit 1		Unit 2		Unit 3		
19		1	MIX	2	REACT	3	SEPARATE	4	
20					2A>B				
21	Mass balance check								
22	Mass IN	kg/h	718.75		718.75		718.75		
23	Mass OUT	kg/h	718.75		718.75		718.75		
24	Closure	%	100		100		100		
25									
26	Overall								
27	Mass IN	kg/h	400						
28	Mass OUT	kg/h	400						
29	Closure	%	100						

B: Flowsheet with stream data — solution

EXAMPLE 4.13 Material balance on a biochemical process (production of antibiotic).

This figure shows a simplified flowsheet of a continuous biochemical process for production of an antibiotic. "A" = $C_{16}H_{18}O_5N_2S$ by fermentation from the substrate lactose, $C_{12}H_{22}O_{11}$.
This is a continuous process at steady-state.

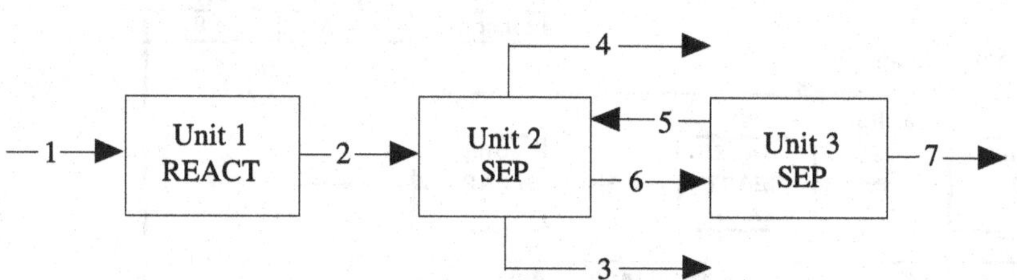

In this process, an aqueous solution of lactose and ammonium sulphate is fed with oxygen (Str. 1) and a small amount of active mold to a bioreactor for partial conversion to "A". The reaction product (Str. 2) is separated in two stages. In stage 1 (Unit 2) CO_2 and excess O_2 are removed as gas (Str. 4), "A" is recovered from the water solution by liquid–liquid extraction into a solvent "G", and the residual waste liquor is discharged (Str. 3). In stage 2 (Unit 3) the solvent is evaporated from the rich A + G solution (Str. 6), then condensed and recycled to stage 1 (Str. 5). Pure product "A" is crystallised in Unit 3 and recovered in Str. 7.

Assume negligible contribution of the mold to the material balance. The streams are specified as follows:

Stream 1: 3420 kg/h liquid solution of 5 wt% lactose + 0.4 wt% $(NH_4)_2SO_4$ in water, 448 kg/h oxygen gas.
Stream 2: Bioreactor product mixture of antibiotic, lactose, $(NH_4)_2SO_4$, O_2, CO_2 and H_2O. Unspecified flow.
Stream 3: Waste liquor containing unconverted lactose and $(NH_4)_2SO_4$ plus 0.05 wt% residual "A" in water.
Stream 4: CO_2 + O_2 gas saturated with water vapour at 310 K, 100 kPa(abs) pressure.
Stream 5: Pure liquid solvent "G". M = 119 kg/kmol. Unspecified flow.
Stream 6: Unspecified composition. Unspecified flow.
Stream 7: Pure crystalline product "A". Unspecified flow.

Conversion Lactose	90%
Selectivity for "A"	10%
L/L distribution coefficient	D(A) = x(6, A) / x(3, A) = 30

Problem:

A. Make a degrees of freedom analysis of the problem.

B. Use a spreadsheet to complete the stream table for the steady-state material balance.

Solution:

A. Examine the specification of each unit and of the overall process:

	Unit 1	Unit 2	Unit 3	OVERALL
Number of unknowns (stream variables)	IJ + L = 14	IJ = 35	IJ = 21	IJ + L = 30
Number of independent equations (see below)	14	24	19	30
D of F	0	11	2	0

B. Write the equations. Define the species as shown in the stream table, column 1:

UNIT 1 (React) **[Six species, excluding the solvent, two streams, two reactions]**

Atom balance on:

C: $0 = 16\dot{n}(1,A) + 12\dot{n}(1,B) + 1\dot{n}(1,E) - 16\dot{n}(2,A) - 12\dot{n}(2,B) - 1\dot{n}(2,E)$ [1]

H: $0 = 18\dot{n}(1,A) + 22\dot{n}(1,B) + 8\dot{n}(1,C) + 2\dot{n}(1,F) - 18\dot{n}(2,A) - 22\dot{n}(2,B) - 1\dot{n}(2,E)$ [2]

O: $0 = 5\dot{n}(1,A) + 11\dot{n}(1,B) + 4\dot{n}(1,C) - 2\dot{n}(1,D) + 2\dot{n}(1,E) + 1\dot{n}(1,F)$
$-5\dot{n}(2,A) - 1\dot{n}(2,B) - 4\dot{n}(2,C) - 2\dot{n}(2,E) - 1\dot{n}(2,F)$ [3]

N: $0 = 2\dot{n}(1,A) + 2\dot{n}(2,C) - 2\dot{n}(2,A) - 2\dot{n}(2,C)$ [4]

S: $0 = 1\dot{n}(1,A) + 1\dot{n}(1,C) - 1\dot{n}(2,A) - 1\dot{n}(2,C)$ [5]

NOTE: Equations [4 and 5] make only 1 independent equation.

Stream compositions:

Stream 1: $\dot{n}(1,A) = 0$ [6]

$\dot{n}(1,B) = (0.05)(3420/342) = 0.5$ [7]

$\dot{n}(1,C) = (0.004)(3420/132) = 0.104$ [8]

$\dot{n}(1,D) = 448/32 = 14$ [9][10]

$\dot{n}(1,F) = (1-0.05-0.004)(4320/18) = 179.7$ [11]

$\dot{n}(1,G) = 0$ [12]

Stream 2: $\dot{n}(2,G) = 0$ $\qquad \dot{n}(1,E) = 0$ [13]

Conversion: $X(B) = 0.90 = (\dot{n}(1,B) - \dot{n}(2,B))/(\dot{n}(1,B)$ [14]

Selectivity [based on carbon] $S(A) = 0.10 = (16/12)(\dot{n}(2,A) - \dot{n}(1,A))/(\dot{n}(1,B) - \dot{n}(2,B))$ [15]

UNIT 2 (Separate) **[Seven species, including the solvent, five streams, no reactions]**

Mole balance on:

Antibiotic: $0 = \dot{n}(2,A) + \dot{n}(5,A) - \dot{n}(3,A) - \dot{n}(4,A) - \dot{n}(6,A)$ [16]

Lactose: $0 = \dot{n}(2,B) + \dot{n}(5,B) - \dot{n}(3,B) - \dot{n}(4,B) - \dot{n}(6,B)$ [17]

$(NH_4)_2SO_4$: $0 = \dot{n}(2,C) + \dot{n}(5,C) - \dot{n}(3,C) - \dot{n}(4,C) - \dot{n}(6,C)$ [18]

O_2: $0 = \dot{n}(2,D) + \dot{n}(5,D) - \dot{n}(3,D) - \dot{n}(4,D) - \dot{n}(6,D)$ [19]

CO_2: $0 = \dot{n}(2,E) + \dot{n}(5,E) - \dot{n}(3,E) - \dot{n}(4,E) - \dot{n}(6,E)$ [20]

H_2O: $0 = \dot{n}(2,F) + \dot{n}(5,F) - \dot{n}(3,F) - \dot{n}(4,F) - \dot{n}(6,F)$ [21]

Solvent: $0 = \dot{n}(2,G) + \dot{n}(5,G) - \dot{n}(3,G) - \dot{n}(6,G)$ [22]

Stream compositions:

Stream 3: $0.0005 = 350\dot{n}(3,A)/350\dot{n}(3,A) + 342\dot{n}(3,B) + 132(3,C) + 18\dot{n}(3,F))$ [23]

$\dot{n}(3,D) = 0 \quad \dot{n}(3,E) = 0 \quad \dot{n}(3,G) = 0$ 24][25][26]

Stream 4: $\dot{n}(4,A) = 0 \quad \dot{n}(4,B) = 0$ [27][28]

$\dot{n}(4,C) = 0 \quad \dot{n}(4,G) = 0$

[29][30]

Vapour pressure: H_2O at 310 K = 6.2 kPa(abs) see *Equation 2.32* or steam table

$\dot{n}(4,F)/(\dot{n}(4,D) + \dot{n}(4,E) + \dot{n}(4,F)) = 6.23/100 = 0.062$ [31]

Stream 5: $\dot{n}(5,A) = 0 \quad \dot{n}(5,B) = 0 \quad \dot{n}(5,C) = 0$ [32][33][34]

$\dot{n}(5,D) = 0 \quad \dot{n}(5,E) = 0 \quad \dot{n}(5,F) = 0$ [35][36][37]

L/L distribution coefficient: $[\dot{n}(6,A)/(\dot{n}(6,A) + \dot{n}(6,G))]/[\dot{n}(3,A)/(\dot{n}(3,A) + \dot{n}(3,B) + (3,C) + \dot{n}(3,F))] = 30$ [38]

UNIT 3 (Separate) **[Seven species, including the solvent, three streams, no reactions]**

Mole balance on:

Antibiotic: $0 = \dot{n}(6,A) - \dot{n}(5,A) - \dot{n}(7,A)$ [39]

Lactose: $0 = \dot{n}(6,B) - \dot{n}(5,B) - \dot{n}(7,B)$ [40]

$(NH_4)_2SO_4$: $0 = \dot{n}(6,C) - \dot{n}(5,C) - \dot{n}(7,C)$ [41]

O_2: $0 = \dot{n}(6,D) - \dot{n}(5,D) - \dot{n}(7,D)$ [42]

CO_2: $0 = \dot{n}(6,E) - \dot{n}(5,E) - \dot{n}(7,E)$ [43]

H_2O: $0 = \dot{n}(6,F) - \dot{n}(5,F) - \dot{n}(7,F)$ [44]

Solvent: $0 = \dot{n}(6,G) - \dot{n}(5,G) - \dot{n}(7,G)$ [45]

FULLY-SPECIFIED. By method A, *Table 4.02*. 45 independent equations. 45 stream variables.
The problem is fully-specified and can be solved in a spreadsheet by the sequential modular method.

StreamTable	***Example 4.13***	***Biosynthesis of an Antibiotic from Lactose***					***Solution***
Species	M	Stream					
		1	**2**	**3**	**4**	**5**	**6**	**7**
	kg/kmol	kmol/h						
[A] $C_{16}H_{18}O_5N_2S$	350	0.000	0.034	0.005	0.000	0.000	0.029	0.029
[B] $C_{12}H_{22}O_{11}$	342	0.500	0.050	0.050	0.000	0.000	0.000	0.000
[C] $(NH_4)_2SO_4$	132	0.104	0.070	0.070	0.000	0.000	0.000	0.000
[D] O_2	32	14.000	9.208	0.000	9.208	0.000	0.000	0.000
[E] CO_2	44	0.000	4.860	0.000	4.860	0.000	0.000	0.000
[F] H_2O	18	179.700	184.481	183.551	0.930	0.000	0.000	0.000
[G] Solvent	119	0.000	0.000	0.000	0.000	37.064	37.064	0.000
Total	kg/h	3867	3867	3332	525	4411	4421	10
Mass balance checks	Unit 1		Unit 2		Unit 3		Overall	
Mass IN kg/h	3867	Closure %	8278	Closure %	4421	Closure %	3867	Closure %
Mass OUT kg/h	3867	100.0	8278	100.0	4421	100.0	3867	100.0

EXAMPLE 4.14 Material balance on an electrochemical process (production of sodium chlorate).

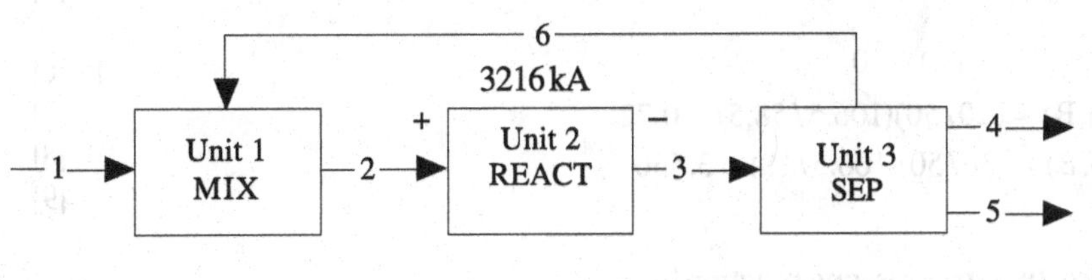

This figure, shows a simplified flowsheet of a continuous electrochemical process for the production of sodium chlorate from salt by the overall cell *Reaction 1*, assumed to occur at 100% current efficiency. This is a continuous process at steady-state.

$$NaCl + 3H_2O \xrightarrow{6F} NaClO_3 + 3H_2 \qquad [3216\,kA] \qquad \textit{Reaction 1}$$

In this process a fresh feed of NaCl plus water (Str. 1) is mixed with a recycle mother liquor (Str. 6) and fed to electrochemical reactors where NaCl undergoes 20% conversion to $NaClO_3$ by Reaction 1, using 3216 kA. The reaction product mix (Str. 3) is separated to give H_2 gas (Str. 4), $NaClO_3$ crystals (Str. 5) and a mother liquor (Str. 6) which is recycled to the process. The streams are specified as follows:

Stream 1:	$NaCl + H_2O$	Unspecified flow
Stream 2:	Unspecified composition	Unspecified flow
Stream 3:	Unspecified composition	Unspecified flow
Stream 4:	H_2 gas saturated with water vapour at 330 K, total pressure 100 kPa(abs)	Unspecified flow
Stream 5:	Pure $NaClO_3$ crystals	Unspecified flow
Stream 6:	Mother liquor = liquid solution of 20 wt% NaCl + 30 wt% $NaClO_3$ in H_2O	Unspecified flow

Problem:

A. Make a degrees of freedom analysis of the problem.

B. Use a spreadsheet to complete the stream table for the steady-state material balance.

Solution:

A. Examine the specification of each unit and of the overall process.

	Unit 1	**Unit 2**	**Unit 3**	**OVERALL**
Number of unknowns (stream variables)	IJ = 12	IJ + L = 9	IJ = 16	IJ + L = 13
Number of independent equations (see below)	9	6	10	13
D of F	3	3	6	0

B. Write the equations. Define the species as shown in the stream table, column 1.

UNIT 1 (Mix) **[Four species, three streams, no reactions]**

Mole balance on:

NaCl: $0 = \dot{n}(1,A) + \dot{n}(6,A) - \dot{n}(2,A)$ [1]

$NaClO_3$: $0 = \dot{n}(1,B) + \dot{n}(6,B) - \dot{n}(2,B)$ [2]

H_2O: $0 = \dot{n}(1,C) + \dot{n}(6,C) - \dot{n}(2,C)$ [3]
H_2: $0 = \dot{n}(1,D) + \dot{n}(6,D) - \dot{n}(2,D)$ [4]
Stream compositions:
Stream 1: $\dot{n}(1,B) = 0$ $\dot{n}(1,D) = 0$ [5][6]
Stream 6: $\dot{n}(6,A)/\dot{n}(6,B) = (20/50)(106.5/58.5) = 0.728$ [7]
$\dot{n}(6,C)/\dot{n}(6,B) = (30/50)(106.5/18) = 3.550$ [8]
$\dot{n}(6,D) = 0$ [9]

Unit 2 (REACT) **[Four species, two streams, one reaction]**

Let $I' = \text{current} = 3216 \text{ kA} = \text{kC/s}$
Mole balance on:
NaCl: $0 = \dot{n}(2,A) - \dot{n}(3,A) + 0 - X(A)\dot{n}(2,A)$ [10]
$NaClO_3$: $0 = \dot{n}(2,B) - \dot{n}(3,B) + X(A)\dot{n}(2,A) - 0$ [11]
H_2O: $0 = \dot{n}(2,C) - \dot{n}(3,C) + 0 - 3X(A)\dot{n}(2,A)$ [12]
H_2: $0 = \dot{n}(2,D) - \dot{n}(3,D) + 3X(A)\dot{n}(2,A) - 0$ [13]
Conversion: $X(A) = 0.2$ [14]
Faraday's law: $X(A)\,\dot{n}(2,A) = (3600)(I')/(6F)$ [15]
$F = \text{Faraday's number} = 96480 \text{ kC/kmol}$

Unit 3 (SEPARATE) **[Four species, four streams, no reactions]**

Mole balance on:
NaCl: $0 = \dot{n}(3,A) - \dot{n}(4,A) - \dot{n}(5,A) - \dot{n}(6,A) + 0 - 0$ [16]
$NaClO_3$: $0 = \dot{n}(3,B) - \dot{n}(4,B) - \dot{n}(5,B) - \dot{n}(6,B) + 0 - 0$ [17]
H_2O: $0 = \dot{n}(3,C) - \dot{n}(4,C) - \dot{n}(5,C) - \dot{n}(6,C) + 0 - 0$ [18]
H_2: $0 = \dot{n}(3,D) - \dot{n}(4,D) - \dot{n}(5,D) - \dot{n}(6,D) + 0 - 0$ [19]
Stream compositions:
Stream 4: $\dot{n}(4,A) = 0$ $\dot{n}(4,B) = 0$ [20][21]
Vapour pressure: H_2O at 330K = 17.2 kPa(abs) see *Equation 2.32* or steam table
$\dot{n}(4,C)/\dot{n}(4,D) = 0.172/(1-0.172) = 0.208$ see *Equation 2.07* [22]
Stream 5: $\dot{n}(5,A) = 0$ $\dot{n}(5,C) = 0$ $\dot{n}(5,D) = 0$ [23][24][25]

FULLY-SPECIFIED. By method A, *Table 4.02*. 25 independent equations. 25 stream variables.
The problem is fully-specified and can be solved by the sequential modular method.

StreamTable		Example 4.14	Electro-synthesis of Sodium Chlorate				Solution
Species	M	Stream					
		1	2	3	4	5	6
	kg/kmol	kmol/h					
[A] NaCl	58.5	20.0	100.0	80.0	0.0	0.0	80.0
[B] $NaClO_3$	106.5	0.0	109.9	129.9	0.0	20.0	109.9
[C] H_2O	18	72.5	462.6	402.6	12.5	0.0	390.1
[D] H_2	2	0.0	0.0	60.0	60.0	0.0	0.0
Total	kg/h	2475	25880	25880	345	2130	23405
Checks	Unit 1		Unit 2		Unit 3		Overall
Mass IN	25880	Closure %	25880	Closure %	25880	Closure %	2475
Mass OUT	25880	100	25880	100	25880	100	2475

EXAMPLE 4.15 Material balance on a thermochemical process (gas sweetening).

This figure shows a simplified flowsheet of a continuous thermochemical process for the removal of hydrogen sulphide gas from natural gas by a cyclic absorption-desorption procedure. In this process the contaminated natural gas (Str. 1) is treated in an absorption column (Unit 1) where the H_2S is absorbed into a countercurrent stream of a liquid amine absorbent (M = 111 kg/kmol) (Str. 4). Sweetened natural gas leaves the column as Str. 2 and the H_2S rich absorbent (Str. 3) is passed to a stripping column (Unit 2) where the H_2S is desorbed into steam (Str. 5) and leaves the process (Str. 6). The lean absorbent (Str. 4) then recycles to the absorber. The streams are specified as follows *[this is a continuous process at steady-state]*:

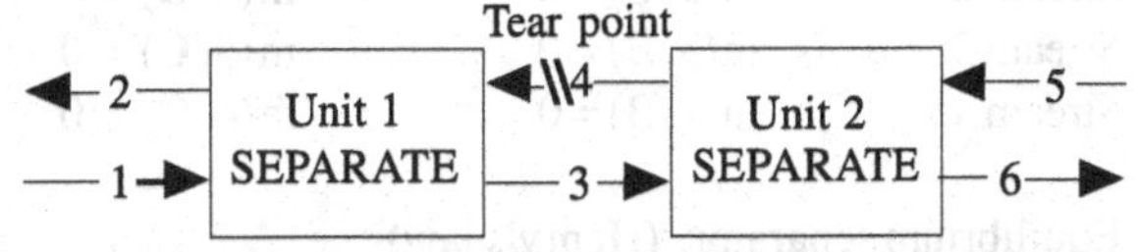

Stream 1: 10,000 kg/h of natural gas containing 95 vol% CH_4 + 5 vol% H_2S
Stream 2: Sweetened natural gas with H_2S in equilibrium[17] with stream 4 at 120 kPa(abs) Unspecified flow
Stream 3: Rich amine containing dissolved H_2S Unspecified flow
Stream 4: Lean amine containing dissolved H_2S Unspecified flow
Stream 5: Steam Unspecified flow
Stream 6: Steam containing 6 wt% H_2S Unspecified flow
Henry's constant H_2S in absorbent $K_H(A)$ = 20 kPa. Assume gas streams behave as ideal gases.

Problem:

A. Make a degrees of freedom analysis of the problem.
B. Use a spreadsheet to complete the stream table for the steady-state material balance.

Solution:

A. Examine the specification of each unit and of the overall process.

	Unit 1	Unit 2	OVERALL
Number of unknowns (stream variables)	IJ=16	IJ = 16	IJ = 16
Number of independent equations (see below)	15	14	16
D of F	1	2	0

[17] That is, an equilibrium separator.

B. Write the equations. Define the species as shown in the stream table, column 1.

UNIT 1 (Separate)

[Four species including H_2O, four streams, no reactions]

Mass balance on:

H_2S: $0 = \dot{m}(1,A) + \dot{m}(4,A) - \dot{m}(2,A) - \dot{m}(3,A)$ [1]

CH_4: $0 = \dot{m}(1,B) + \dot{m}(4,B) - \dot{m}(2,B) - \dot{m}(3,B)$ [2]

H_2O: $0 = \dot{m}(1,C) + \dot{m}(4,C) - \dot{m}(2,C) - \dot{m}(3,C)$ [3]

Absorbent: $0 = \dot{m}(1,D) + \dot{m}(4,D) - \dot{m}(2,D) - \dot{m}(3,D)$ [4]

Stream compositions:

Stream 1: $y(1,A) = 0.05$ $\overline{m}(1) = 10000$ [5][6]

$\dot{m}(1,C) = 0$ $\dot{m}(1,D) = 0$ [7][8]

Stream 2: $\dot{m}(2,C) = 0$ $\dot{m}(2,D) = 0$ [9][10]

Stream 3: $\dot{m}(3,B) = 0$ $\dot{m}(3,C) = 0$ 11][12]

Stream 4: $\dot{m}(4,B) = 0$ $\dot{m}(4,C) = 0$ [13][14]

Equilibrium separator: (Henry's law)

$y(2,A) = K_H(A)x(4,A)/P(2) = 20x(4,A)/120$ see *Equation 2.34* [15]

y = mole fraction in gas

x = mole fraction in liquid

UNIT 2 (Separate)

[Four species, four streams, no reactions]

Mass balance on:

H_2S: $0 = \dot{m}(3,A) + \dot{m}(5,A) - \dot{m}(4,A) - \dot{m}(6,A)$ [16]

CH_4: $0 = \dot{m}(3,B) + \dot{m}(5,B) - \dot{m}(4,B) - \dot{m}(6,B)$ [17]

H_2O: $0 = \dot{m}(3,C) + \dot{m}(5,C) - \dot{m}(4,C) - \dot{m}(6,C)$ [18]

Absorbent: $0 = \dot{m}(3,D) + \dot{m}(5,D) - \dot{m}(4,D) - \dot{m}(6,D)$ [19]

Stream compositions:

Stream 5: $\dot{m}(5,A) = 0$ $\dot{m}(5,B) = 0$ $\dot{m}(5,D) = 0$ [20][21][22]

Stream 6: $\dot{m}(6,A)/\overline{m}(6,A) = 0.06$ $\dot{m}(6,C) = 0$ [23][24]

FULLY-SPECIFIED. By method A, *Table 4.02*. 24 independent equations. 24 stream variables.

The problem is fully-specified and can be solved by the sequential modular method.

Convert vol% H_2S (i.e. mol%) in Str.1 to mass fraction:

w(1, A) = (0.05)(34)/[(0.05)(34) + (0.95)(16)] = 0.10
see *Equation 2.28*

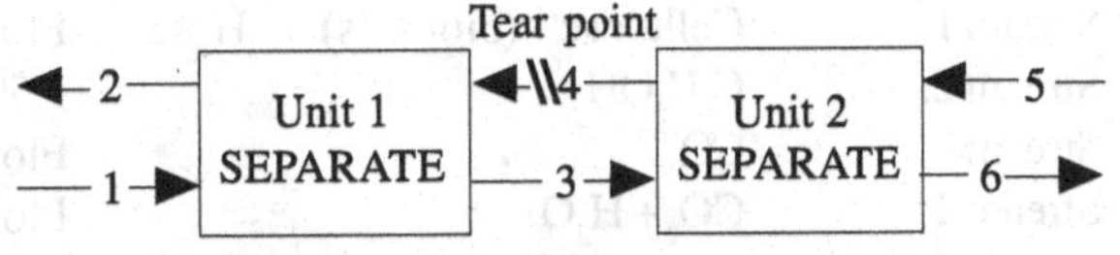

Convert wt% H_2S in Str. 4 to mole fraction:
x(4, A) = 0.0007
see *Equation 2.27*

Henry's law:
y(2, A) = 0.0001
see *Equation 2.34*

StreamTable		*Example 4.15*	*Gas Sweetening Process Solution*				
Species	M	Stream					
		1	2	3	4	5	6
	kg/kmol	kg/h					
[A] H_2S	34	1000	2	999	1	0	998
[B] CH_4	16	9000	9000	0	0	0	0
[C] H_2O	18	0	0	0	0	15634	15634
[D] Absorbent	111	0	0	5000	5000	0	0
Total	kg/h	10000	9002	5999	5001	15634	16632
Mass balance check		Unit 1		Unit 2		Overall	
Mass IN	kg/h	15001	Closure %	21633	Closure %	25634	Closure %
Mass OUT	kg/h	15001	100.0	21633	100.0	25634	100.0

EXAMPLE 4.16 Material balance on an environmental system (ethanol from biomass).

This figure shows a conceptual flowsheet for a "greenhouse gas neutral" closed carbon cycle fuel production system in which gasoline is replaced by ethanol from biomass as the fuel for the world's motor vehicles. In this proposed system energy from the Sun is used in the photosynthesis of selected plants (Unit 1) to produce cellulose biomass (Str. 1) which is converted to ethanol (Str. 2) and CO_2 (Str. 3) by thermal hydrolysis and fermentation (Unit 2). The ethanol is then used in automobiles (Unit 3) where it burns with oxygen from the air (Str. 10) to produce CO_2 and H_2O (Str. 4). The auto-exhaust gas (Str. 4) mixes with CO_2 from Unit 2 in the atmosphere (Unit 4, Str. 5). Water separates as rain (Unit 5) and is divided (Unit 6) to recycle to Unit 1 (Str. 8) and Unit 2 (Str. 9), while the CO_2 recycles (Str. 6) to support photosynthesis in Unit 1. Oxygen from the photosynthesis recycles (Str. 10) to support the combustion in Unit 3. The reactions and process specifications are as follows: *[This is assumed to be a continuous process at steady-state with 100% conversion of stoichiometric reactants in all reactors and 100% separator efficiency]*

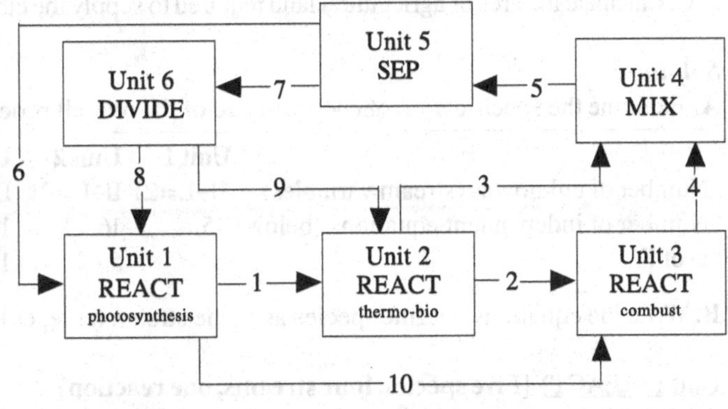

Unit 1:	Reaction 1	$6CO_2 + 5H_2O \rightarrow C_6H_{10}O_5 + 6O_2$	photosynthesis
Unit 2:	Reaction 2	$C_6H_{10}O_5 + H_2O \rightarrow (C_6H_{12}O_6) \rightarrow 2C_2H_5OH + 2CO_2$	hydrolysis + fermentation
Unit 3:	Reaction 3	$C_2H_5OH + 3O_2 \rightarrow 2CO_2 + 3H_2O$	combustion

Stream 1:	Cellulose (biomass) $C_6H_{10}O_5$	Flow unspecified
Stream 2:	C_2H_5OH	3.50E+12 kg/year
Stream 3:	CO_2	Flow unspecified
Stream 4:	$CO_2 + H_2O$	Flow unspecified
Stream 5:	Composition unspecified	Flow unspecified
Stream 6:	CO_2	Flow unspecified
Stream 7:	H_2O	Flow unspecified
Stream 8:	Composition unspecified	Flow unspecified
Stream 9:	Composition unspecified	Flow unspecified
Stream 10:	O_2	Flow unspecified

Unit 1: Converts 600 tonne CO_2/(km^2.year) to cellulose by photosynthesis.
Unit 3: Contains 7.00E+08 motor vehicles, each consuming 5000 kg/year ethanol.

NOTE: Nitrogen (from air) is assumed unreactive and is not considered in this balance.

Problem:

A. Make a degrees of freedom analysis of the problem.
B. Use a spreadsheet to complete the stream table for the steady-state material balance.
C. Calculate the area of agricultural land required to supply the ethanol for 700 million (=7.00E8) automobiles.

Solution:

A. Examine the specification of each unit and of the overall process.

	Unit 1	Unit 2	Unit 3	Unit 4	Unit 5	Unit 6	OVERALL
Number of unknowns (stream variables)	IJ+L=21	IJ+L=21	IJ+L=16	IJ=15	IJ=15	IJ=15	IJ+L=3
Number of independent equations (below)	15	16	15	10	11	11	3
D of F	6	5	1	5	4	4	0

B. Write the equations. Define species as in the stream table, column 1.

Unit 1 (REACT) **[Five species, four streams, one reaction]**

Atom balance on:

C: $0 = \dot{n}(6,A) - 6\dot{n}(1,D)$ [1]

H: $0 = 2\dot{n}(8,B) - 10\dot{n}(1,D)$ [2]

O: $0 = 2\dot{n}(6,A) + \dot{n}(8,B) - 5\dot{n}(1,D) - 2\dot{n}(10,C)$ [3]

Stream compositions:

Stream 1: $\dot{n}(1,A) = 0$ $\dot{n}(1,B) = 0$ $\dot{n}(1,C) = 0$ $\dot{n}(1,E) = 0$ [4][5][6][7]

Stream 6: $\dot{n}(6,B) = 0$ $\dot{n}(6,C) = 0$ $\dot{n}(6,D) = 0$ $\dot{n}(6,E) = 0$ [8][9][10][11]

Stream 10: $\dot{n}(10,A) = 0$ $\dot{n}(10,B) = 0$ $\dot{n}(10,D) = 0$ $\dot{n}(10,E) = 0$ [12][13][14][15]

UNIT 2 (React) **[Five species, four streams, one reaction]**

Atom balance on:

C: $0 = 6\dot{n}(1,D) - 2\dot{n}(2,E) - \dot{n}(3,A)$ [16]

H: $0 = 10\dot{n}(1,D) + 2\dot{n}(9,B) - 6\dot{n}(2,E)$ [17]

O: $0 = 5\dot{n}(1,D) + \dot{n}(9,B) - \dot{n}(2,E)$ [18]

Stream compositions:

Stream 2: $\dot{n}(2,A) = 0$ $\dot{n}(2,B) = 0$ $\dot{n}(2,C) = 0$ $\dot{n}(2,D) = 0$ [19][20][21][22]

$\dot{n}(2,E) = 9.11E{+}06\text{kmol/h}$ [23]

Stream 3: $\dot{n}(3,B) = 0$ $\dot{n}(3,C) = 0$ $\dot{n}(3,D) = 0$ $\dot{n}(3,E) = 0$ [24][25][26][27]

UNIT 3 (React) **[Five species, three streams, one reaction]**

Atom balance on:

C: $0 = 2\dot{n}(2,E) - \dot{n}(4,A)$ [28]

H: $0 = 6\dot{n}(2,E) - 2\dot{n}(4,B)$ [29]

O: $0 = \dot{n}(2,E) + 2\dot{n}(10,C) - 2\dot{n}(4,A) - \dot{n}(4,B)$ [30]

Stream compositions

Stream 4: $\dot{n}(4,C) = 0$ $\dot{n}(4,D) = 0$ $\dot{n}(4,E) = 0$ [31][32][33]

UNIT 4 (Mix) **[Five species, three streams, zero reactions]**

Atom balance on:

C: $0 = \dot{n}(3,A) + \dot{n}(4,A) - \dot{n}(5,A)$ [34]

H: $0 = 2\dot{n}(2,B) - 2\dot{n}(4,B)$ [35]

O: $0 = 2\dot{n}(3,A) + 2\dot{n}(4,A) + \dot{n}(4,B) - 2\dot{n}(5,A) - \dot{n}(5,B)$ [36]

UNIT 5 (Separate) **[Five species, three streams, zero reactions]**

Atom balance on:

C: $0 = \dot{n}(5,A) + 6\dot{n}(5,D) + 2\dot{n}(5,E) - \dot{n}(6,A) - \dot{n}(7,A)$ [37]

H: $0 = 2\dot{n}(5,B) + 10\dot{n}(6,D) + 6\dot{n}(5,E) - 2\dot{n}(6,B) - 2\dot{n}(7,B)$ [38]

O: $0 = 2\dot{n}(5,A) + \dot{n}(5,B) + 2\dot{n}(5,C) + 5\dot{n}(5,D) + \dot{n}(5,E) - 2\dot{n}(6,A) - \dot{n}(7,A) - \dot{n}(7,B)$ [39]

Stream compositions

Stream 7 $\dot{n}(7,A) = 0$ $\dot{n}(7,C) = 0$ $\dot{n}(7,D) = 0$ $\dot{n}(7,E) = 0$ [40][41][42][43]

UNIT 6 (Divide) **[Five species, three streams, zero reactions]**

Atom balance on:

C: $0 = \dot{n}(7,A) - \dot{n}(8,A) - 6\dot{n}(8,D) - 2\dot{n}(8,E) - \dot{n}(9,A) - 6\dot{n}(9,D) - 2\dot{n}(9,E)$ [44]

H: $0 = 2\dot{n}(7,B) - 2\dot{n}(8,B) - 10\dot{n}(8,D) - 6\dot{n}(8,E) - 2\dot{n}(9,B) - 10\dot{n}(9,D) - 6\dot{n}(9,E)$ [45]

O: $0 = 2\dot{n}(7,A) + \dot{n}(7,B) - 2\dot{n}(8,A) - \dot{n}(8,B) - 2\dot{n}(2,C) - 5\dot{n}(8,D)$ [46]

$- \dot{n}(8,E) - 2\dot{n}(9,A) - \dot{n}(9,B) - 2\dot{n}(9,C) - 5\dot{n}(9,D) - \dot{n}(9,E)$

Divider relations: $\dot{n}(8,A)/\overline{n}(8) = \dot{n}(7,A)/\overline{n}(7)$ $\quad$ $\dot{n}(8,C)/\overline{n}(8) = \dot{n}(7,C)/\overline{n}(7)$ [47][48]

$\dot{n}(8,D)/\overline{n}(8) = \dot{n}(7,D)/\overline{n}(7)$ $\quad$ $\dot{n}(8,E)/\overline{n}(8) = \dot{n}(7,E)/\overline{n}(7)$ [49][50]

FULLY-SPECIFIED. By method A, *Table 4.02*. 50 independent equations. 50 stream variables.

Stream Table		*Example 4.16 Material Balance on Carbon Cycle for Ethanol Fuel from Biomass*								*Solution*	
Species	M	Stream									
		1	2	3	4	5	6	7	8	9	10
	kg/kmol	kmol/hr									
[A]CO_2	44	0	0	***9.1E+06***	***1.8E+07***	***2.7E+07***	***2.7E+07***	0	**0**	**0**	0
[B]H_2O	18	0	0	0	***2.73E+07***	***2.73E+07***	0	***2.7E+07***	***2.3E+07***	***4.6E+06***	0
[C]O_2	32	0	0	0	0	0	0	0	**0**	**0**	***2.7E+07***
[D]$C_6H_{10}O_5$	162	***4.6E+06***	0	0	0	0	0	0	**0**	**0**	0
[E]C_2H_5OH	46	0	9.1E+06	0	0	0	0	0	**0**	**0**	0
Total	kg/h	7.38E+08	4.19E+08	4.01E+08	1.29E+09	1.69E+09	1.20E+09	4.92E+08	4.10E+08	8.19E+07	8.74E+08
Mass balance check		Unit 1	Unit 2	Unit 3	Unit 4	Unit 5	Unit 6	Overall			
Mass IN		1.61E+09	8.19E+08	1.29E+09	1.69E+09	1.69E+09	4.92E+08	0			
Mass OUT		1.61E+09	8.19E+08	1.29E+09	1.69E+09	1.69E+09	4.92E+08	0			
Closure %		100.0	100.0	100.0	100.0	100.0	100.0	100.0			

C. Land area to supply ethanol for 7.00E+08 vehicles = **<u>1.75E+07 km²</u>**

Fraction of land area of Earth = **<u>11.7%</u>**

EXAMPLE 4.17 Material balance on a combustion process, using various methods of solution.

Problem: 100 kmol/h of the fuel ethane (C_2H_6) undergoes partial combustion in excess dry air to give a product gas with 0.501 vol% C_2H_6, 1.803 vol% CO and a dew-point of 325 K at 100 kPa(abs). Calculate the complete material balance stream table for this continuous process at steady-state. Assume negligible production of nitrogen oxides.

Reactions: $C_2H_6 + 3.5O_2 \rightarrow 2CO_2 + 3H_2O$ [1] $\quad$ $C_2H_6 + 2.5O_2 \rightarrow 2CO + 3H_2O$ [2]

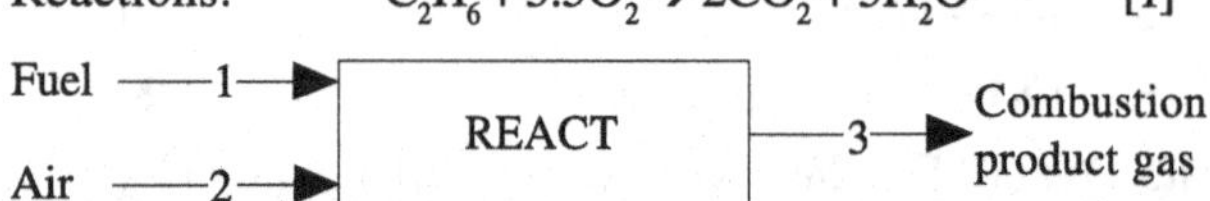

Component	C_2H_6	O_2	N_2	CO_2	CO	H_2O
Symbol	A	B	C	D	E	F

Solution:

Rate ACC = Rate IN – Rate OUT + Rate GEN – Rate CON

$\dot{n}(1,A) = 100$ kmol/h. Vapour pressure of water at 325 K = 13.53 kPa(abs) (see *Equation 2.32*):

Define: X = overall conversion of C_2H_6 $\quad$ S = selectivity for CO_2 from C_2H_6

Define: $\varepsilon(1)$ = extent of reaction 1 $\quad$ $\varepsilon(2)$ = extent of reaction 2 (kmol/h)

<u>Method A</u>[18]

Atom balances.

H_1: $0 = 6\dot{n}(1,A) - [6\dot{n}(3,A) + 2\dot{n}(3,F)]$

O_1: $0 = 2\dot{n}(2,B) - [2\dot{n}(3,B) + 2\dot{n}(3,D) + 1\dot{n}(3,E) + 1\dot{n}(3,F)]$

<u>Method B</u>

Atom balances using extents of reaction.

$0 = 6\dot{n}(1,A) - [6\dot{n}(3,B) + 6\varepsilon(1) + 6\varepsilon(2)]$ [1]

$0 = 2\dot{n}(2,B) - [2\dot{n}(3,B) + 7\varepsilon(1) + 5\varepsilon(2)]$ [2]

[18] Equation numbers appear on the right of the page.

C_1: $0=2\dot{n}(1,A)-[2\dot{n}(3,A)+1\dot{n}(3,D)+1\dot{n}(3,E)$

N_1: $0=2\dot{n}(2,C)-2\dot{n}(3,C)$

Stream composition:

Stream 2: $\dot{n}(2,B)/\dot{n}(2,C)=21/79$

Stream 3: $y(3,A)=\dot{n}(3,A)/\bar{n}(3)=0.00501$

$y(3,E)=\dot{n}(3,E)/\bar{n}(3)=0.01803$

$y(3,F)=\dot{n}(3,F)/\bar{n}(3)=p(3,F)/P(3)=0.1353$

<u>Method C</u>

Mole balances using conversion and selectivity.

C_2H_6 $\quad 0=\dot{n}(1,A)-\dot{n}(3,A)-X\dot{n}(1,A)$

O_2: $\quad 0=\dot{n}(2,B)-\dot{n}(3,B)-X[3.5S+2.5(1-S)]\dot{n}(1,A)$

N_2: $\quad 0=\dot{n}(2,C)-\dot{n}(3,C)$

CO_2: $\quad 0=0-\dot{n}(3,D)+2XS\dot{n}(1,A)$

CO: $\quad 0=0-\dot{n}(3,E)+2X(1-S)\dot{n}(1,A)$

H_2O: $\quad 0=0-\dot{n}(3,F)+3X\dot{n}(1,A)$

Stream composition:

Stream 2: $\dot{n}(2,B)/\dot{n}(2,C)=21/79$

Stream 3: $y(3,A)=\dot{n}(3,A)/\bar{n}(3)=0.00501$

$y(3,E)=\dot{n}(3,E)/\bar{n}(3)=0.01803$

$y(3,F)=\dot{n}(3,F)/\bar{n}(3)=p(3,F)/P(3)=0.1353$

$0=2\dot{n}(1,A)-[2\dot{n}(3,A)+2\varepsilon(1)+2\varepsilon(2)]$ [3]

$0=2\dot{n}(2,C)-2\dot{n}(3,C)$ [4]

Stream composition:

$\dot{n}(2,B)/\dot{n}(2,C)=21/79$ [5]

$y(3,A)=\dot{n}(3,A)/\bar{n}(3)=0.00501$ [6]

$y(3,E)=\dot{n}(3,E)/\bar{n}(3)=0.01803$ [7]

$y(3,F)=\dot{n}(3,F)/\bar{n}(3)=p(3,F)/P(3)=0.1353$ [8]

<u>Method D</u>

Mole balances using extents of reaction.

$0=\dot{n}(1,A)-\dot{n}(3,A)-\varepsilon(1)-\varepsilon(2)$ [1]

$0=\dot{n}(2,B)-\dot{n}(3,B)-3.5\varepsilon(1)-2.5\varepsilon(2)$ [2]

$0=\dot{n}(2,C)-\dot{n}(3,C)$ [3]

$0=0-\dot{n}(3,D)+2\varepsilon(1)$ [4]

$0=0-\dot{n}(3,E)+2\varepsilon(2)$ [5]

$0=0-\dot{n}(3,F)+3\varepsilon(1)+3\varepsilon(2)$ [6]

Stream composition:

$\dot{n}(2,B)/\dot{n}(2,C)=21/79$ [7]

$y(3,A)=\dot{n}(3,A)/\bar{n}(3)=0.00501$ [8]

$y(3,E)=\dot{n}(3,E)/\bar{n}(3)=0.01803$ [9]

$y(3,F)=\dot{n}(3,F)/\bar{n}(3)=p(3,F)/P(3)=0.135$ [10]

In all methods:

$$\bar{n}(3)=\dot{n}(3,A)+\dot{n}(3,B)+\dot{n}(3,C)+\dot{n}(3,D)+\dot{n}(3,E)+\dot{n}(3,F) \quad [11]$$

Method A: From [1, 6 and 8], $\dot{n}(3)$ = 1997 kmol/h . Equations [6, 7, 8] then [2, 3, 4, 5 and 11] give the rest.

Method B: Equations [1 and 3] are identical. Seven equations with eight unknowns. No solution.

Method C: From [1 and 6], $X=0.9$, $\dot{n}(3)$ = 1997 kmol/h.

From [5], $S=0.8$.

Equations [2, 3, 4, 7 and 11] give the rest.

Method D: From [1 and 6], $[\varepsilon(1)+\varepsilon(2)]=90$ kmol/h,

$\dot{n}(3)$ = 1997 kmol/h

From [5], $\varepsilon(2)=18$, $\varepsilon(1)=72$ kmol/h.

Equations [2, 3, 4, 7 and 11] give the rest.

Stream Table		Combustion Solution			
Component	M	Stream [kmol/h]			
	kg/kmol	1	2	3	y mol %
[A] C_2H_6	30	100	0	10	0.50
[B] O_2	32	0	385	88	4.41
[C] N_2	28	0	1448	1448	72.55
[D] CO_2	44	0	0	144	7.21
[E] CO	28	0	0	36	1.80
[F] H_2O	18	0	0	270	13.52
Total	kg/h	3000	52873	55873	
Mass IN	kg/h	55873	Mass OUT kg/h		55873
Closure	%	100			

SUMMARY

[1] Material balance calculations hinge on the general balance equation (GBE), which can be used in either the integral or differential form, depending on the type of problem:

Integral form of GBE: [For a defined system and a specified quantity]
Material ACC = Material IN – Material OUT + Material GEN – Material CON

Differential form of GBE: [For a defined system and a specified quantity]
Rate of material ACC = Rate of material IN – Rate of material OUT
+ Rate of material GEN – Rate of material CON

[2] Material balance calculations with the GBE are mostly a matter of bookkeeping, backed by knowledge of mathematics, physical properties, phase equilibria, and stoichiometry.
The main steps involved in doing material balances are as follows:

a. Interpreting the jargon and resolving ambiguities that sometimes obscure the problem.
b. Locating the process units and the known conditions and/or [in practical "on the job" situations] obtaining reliable data.
c. Translating a word problem into a flowsheet, with a corresponding stream table showing the known quantities of the material balance.
d. Defining the system envelope(s) that give the simplest solution.
e. Specifying the quantities of interest, along with a set of symbols that represent the quantities without ambiguity.
f. Writing a set of equations that relate the known quantities to the unknown quantities needed to complete the stream table.
g. Determining the degrees of freedom of the system, and avoiding the trap of inconsistent over-specification.
h. Arranging and sequencing the equations for a computer (e.g. spreadsheet) calculation.
i. Translating the output of the calculations into a practical result with the correct units and the appropriate number of significant figures.
j. Presenting the output in a transparent and unambiguous format, such as stream table together with a labelled flowsheet.

[3] Choosing the system envelope(s) and the quantity(s) are critical in starting a material balance, and can have a big effect on the complexity of subsequent calculations. The simplest cases use atom balances on overall systems, while more difficult problems may need atom and/or species balances on individual process units and/or combinations of units.

[4] Material balance calculations (in non-nuclear processes) may use either conserved or non-conserved quantities.
Conserved quantities (e.g. total mass, atoms, moles or mass of individual species without chemical reaction): GEN = CON = 0
Non-conserved quantities (e.g. moles or mass of individual species in a chemical reaction):
GEN and/or CON $\neq$ 0

[5] Each generic process unit (DIVIDE, MIX, SEPARATE, HEAT EXCHANGE, PUMP, REACT) is represented by a set of material balance equations plus subsidiary relations that uniquely define its function. A complete process flowsheet can be constructed by sequencing these generic process units with interconnecting process streams.

[6] Always check your material balance solutions for *closure* of the mass balance. Closing the mass balance means accounting for all mass entering, leaving and accumulating in the system, to ensure agreement with the principle of conservation of mass.

Closure of the mass balance is a necessary (but not sufficient) check on the correctness of material balance calculations.

For a continuous process at steady state the closure (%) is defined as:
Closure = e_M = 100 [Total mass flow rate OUT/ Total mass flow rate IN].
Exact balance calculations should give closure = 100%, although practical process design calculations may be satisfied by a closure of 99.9% or lower, depending on the application. A tight material balance (i.e. closure ≅ 100%) is usually needed when the process involves trace materials such as catalyst poisons and/or environmental contaminants that are problematic in low (e.g. ppm or ppb) concentrations.

[7] The *specification* of a material balance problem involves a comparison between the number of *stream variables* and the number of independent relations that connect the variables to known values and to each other.

The *degrees of freedom (D of F)* of the material balance is then defined as:
D of F = number of stream variables – number of (known values of stream variables + independent relations between stream variables)
= number of unknown quantities – number of independent equations

In material balance problems the *stream variables* are the flows (or amounts) of each species in each process stream. The number of independent chemical reactions is included with the stream variables for species (i.e. mass or mole) balances, but excluded for element (i.e. atom) balances.

The independent relations include the individual species or elemental balances plus subsidiary relations defining the function of each process unit.

If, D of F > 0
The material balance is *under-specified* and has no unique solution.
Typical of *design problems*, whose solution involves optimisation.
If, D of F = 0
The material balance is *fully-specified* and has one unique solution.
Typical of problems in beginners' engineering and science courses.

If, D of F < 0
The material balance is *over-specified* and is either redundant or incorrect.
When a balance is over-specified look at it carefully for errors.

[8] The solution of integral material balances and of differential balances at steady-state requires only algebra, whereas unsteady-state differential balances are solved using calculus. Algebraic balances may involve linear, non-linear and simultaneous equations and sometimes require numerical methods for their solution.

[9] Full material balances for multi-unit processes are set up by combining the individual unit material balances in sequence. The resulting multi-unit balances can be treated by either:

a. The method of *sequential modular solution,* where equations for each process unit are solved in sequence, OR

b. The method of *simultaneous solution,* where equations for the whole system are solved simultaneously.

The sequential modular method, which is used in most commercial process simulation software, is employed throughout this text.

[10] The inclusion of recycle streams in a process flowsheet complicates the material balance by coupling the equations for all process units in each recycle loop. Recycle material balances are done with the sequential modular method by adopting a *tear stream* and iterating the calculations until the balance is converged. *Convergence* of the material balance means that the mass balance is closed to an acceptable tolerance.

[11] In practical recycle systems there is usually an accumulation of undesirable material(s) in the recycle loops that leads to difficulties in long-term operation of the system. An undesired accumulating substance may be removed from the recycle system by purging a stream containing the substance (preferably in high concentration).

[12] A computer should be used for all except the most simple material balance calculations. The computer solution can use a spreadsheet, in-house code or commercial process simulation software.
The spreadsheet solution of material balances takes two common forms:

a. A *stream table* attached to a process flowsheet, with designated process units, streams and species, OR

b. Sets of stream variable values adjacent to each stream in the process flowsheet.

This text uses spreadsheets with form (a) to set-up, solve and present material balances (and also energy balances — see Chapter 5) .

FURTHER READING

[1] D. J. Jacob, *Atmospheric Chemistry,* Princeton University Press, Princeton, 1999.

[2] T. M. Duncan and J. A. Reimer, *Chemical Engineering Design and Analysis,* Cambridge University Press, 1998.

[3] R. M. Felder and R.W. Rousseau, *Elementary Principles of Chemical Processes,* John Wiley & Sons, New York, 2000.

[4] G. V. Reklaitis, *Introducion to Material and Energy Balances,* John Wiley & Sons, New York, 1983.

[5] D. M. Himmelblau, *Basic Principles and Calculations in Chemical Engineering,* Prentice Hall, Englewood Cliffs, 1989.

[6] P. M. Doran, *Bioprocess Engineering Principles,* Academic Press, San Diego, 1995.

[7] B. V. Liengme, *A Guide to Microsoft Excel for Scientists and Engineers,* Arnold, London, 1997.

[8] R. K. Sinnott, *Chemical Engineering Design,* Butterworth Heinmann, Oxford, 1999.

[9] L. T. Biegler, I. E.Grossmann and A. W. Westerberg, *Systematic Methods of Chemical Process Design,* Prentice Hall, Upper Saddle Rover, 1997.

[10] W. D. Seider, J. D. Seader and D. R. Lewin, *Process Design Principles,* John Wiley & Sons, New York, 1999.

[11] J. M. Douglas, *Conceptual Design of Chemical Processes,* McGraw-Hill, New York, 1988.

CHAPTER FIVE

ENERGY BALANCES

GENERAL ENERGY BALANCE

In this chapter energy balances are divorced from material balances and calculated with the assumption that the corresponding material balance is already known. Energy balance calculations then follow a similar pattern to the material balance calculations of Chapter 4.

Energy balances are based on the general balance equation of Chapter 1, reintroduced here as *Equation 5.01*. In a defined *system* and for a specified *quantity:*

Accumulation in system	=	**Input to system**	–	**Output from system**	+	**Generation in system**	–	**Consumption in system**	*Equation 5.01*

For energy balance problems the *system* is a space whose boundaries are defined to suit the problem at hand and the *quantity* in *Equation 5.01* is the amount of energy. The energy balance differs from the general material balance in that total energy (in non-nuclear processes) is a *conserved quantity,* so the generation and consumption terms in the total energy balance are both zero.

Equation 5.01, written in abbreviated form for the total energy, can be used as an *integral balance*:

ACC = IN – OUT *Equation 5.02*

where:

ACC	≡ (final amount of energy – initial amount of energy) in the system	kJ
IN	≡ energy into the system	kJ
OUT	≡ energy out of the system	kJ

or as a *differential balance*:

Rate ACC = Rate IN – Rate OUT *Equation 5.03*

where:

Rate ACC	≡ rate of accumulation of energy in the system	w.r.t. time (t)	kW
Rate IN	≡ rate of energy input to the system	w.r.t. time (t)	kW
Rate OUT	≡ rate of energy output from the system	w.r.t. time (t)	kW

The rate terms in *Equation 5.03* have units of energy/time, which is equal to power (e.g. kJ/s = kW). *Equations 5.02* and *5.03* are both equivalents of the *first law of thermodynamics*, which states that energy is conserved in any (non-nuclear) process.

Note that the "energy" in *Equations 5.02* and *5.03* embraces all forms of energy related to the system, including the heat and work transferred between the system and its surroundings. In the subsequent *Equations 5.04, 5.05, 5.06* and *5.07*, energy is considered to consist of two parts, the energy associated with the material of the system (E) and the energy *transferred* across the system boundary as heat and work (Q and W). Although E, Q and W are all measured in the same units (e.g. kJ) these terms differ in the important respect that E is a state function, while Q and W are both path functions. [1]

---

[1] A state function depends only on the state and is independent of the path to that state. A path function is one whose value depends on the path between two states.

Energy transferred to or from the system as radiation is usually lumped into the heat term, for example, energy transfer by infra-red radiation is the major source of "heat" in industrial furnaces. Photochemical processes are driven by other types of radiation.

To avoid ambiguity the quantities used in energy balance calculations must be clearly specified. For this purpose the nomenclature used in this text is given in *Table 5.01*.

Table 5.01. Nomenclature for energy balances (see also Chapter 2 and Table 4.01).

Symbol	Meaning	Typical Units
E	energy content of the material of the system, with respect to reference condition(s)[2]	kJ
h(j)	specific enthalpy of component "j", with respect to *compounds* at a reference condition	$kJ.kg^{-1}$, $kJ.kmol^{-1}$
h*(j)	specific enthalpy of component "j", with respect to *elements* at standard state	$kJ.kg^{-1}$, $kJ.kmol^{-1}$
H	enthalpy, with respect to a reference condition = U + PV	kJ
U	internal energy, with respect to a reference condition	kJ
E_k	kinetic energy = $(1E\text{-}3)0.5m\tilde{u}^2$	kJ
E_p	potential energy, with respect to a reference level = $(1E\text{-}3)mgL'$	kJ
PV	pressure–volume energy	kJ
Q	net heat <u>input</u> to the system	kJ
W	net work <u>output</u> from (i.e. work done by) the system	kJ
$\dot{H}(i)$	enthalpy flow of stream "i", with respect to *compounds* at a reference condition	kW
$\dot{H}^*(i)$	enthalpy flow of stream "i", with respect to *elements* at standard state	kW
$\dot{E}_k(i)$	kinetic energy flow of stream "i" = $(1E\text{-}3)0.5\bar{m}(i)\tilde{u}^2$	kW
$\dot{E}_p(i)$	potential energy flow of stream "i" = $(1E\text{-}3)\,\bar{m}(i)gL'$	kW
$\dot{Q}$	net heat rate of heat <u>input</u> to the system	kW
$\dot{W}$	net rate of work (i.e. power) <u>output</u> from the system	kW
P(i)	pressure of stream "i"	kPa(abs)
T(i)	temperature of stream "i"	K
T_{ref}	reference temperature for calculation of H and U	K
$\dot{V}(i)$	volumetric flow rate of stream "i"	$m^3.s^{-1}$
ũ	velocity	$m.s^{-1}$
g	gravitation constant	$m.s^{-2}$
L′	height above a reference level	m

Calculations with differential energy balances in which [Rate ACC ≠ 0] require the use of calculus and are treated in Chapter 7. All other energy balance problems can be solved by arithmetic and algebra, as shown in the examples following in Chapter 5.

[2] Some texts refer to E as the "total energy" of the system, but you should be aware that this term is ambiguous w.r.t. its meaning in the "total energy balance".

CLOSED SYSTEMS (BATCH PROCESSES)

A closed system is defined as one in which, over the time period of interest, there is zero transfer of *material* across the system boundary. The movement of *energy* across the boundaries of a closed system is permitted and occurs by the transfer of heat (Q) or work (W). The integral and differential energy balances for a closed system are respectively *Equations 5.02* and *5.03*. These equations are simplified to *Equations 5.04* and *5.05*.

Closed system – integral energy balance:

$$\mathbf{E_{final} - E_{initial} = Q - W} \qquad \textit{Equation 5.04}$$

Closed system – differential energy balance:

$$\mathbf{dE/dt} = \dot{Q} - \dot{W}. \qquad \textit{Equation 5.05}$$

where:

E	= $[U + E_k + E_p]$ content of the system	kJ
E_{final}	= final value of "E" for the system	kJ
$E_{initial}$	= initial value of "E" for the system	kJ
Q	= net heat input to system	kJ
$\dot{Q}$	= net rate of heat input to system	kW
t	= time	s
W	= net work output from the system	kJ
$\dot{W}$	= net rate of work (i.e. power) output from system	kW

The definition of energy "E" used here excludes exotic energy terms such as surface energy, potential energy in a magnetic field, etc. that are insignificant in most chemical process calculations.

Equations 5.04 and *5.05* may be familiar to you as forms of the *first law of thermodynamics* for a closed system.

Note that the work[3] (W) is defined here as positive for net work done BY the system (the opposite sign convention is used in some thermodynamics texts). This work usually takes the form of:

- mechanical "work" , such as expansion, compression, lifting, pushing or rotation of a shaft.
- electrical "work", for which: Power (kW) = (Voltage (V)) (current (kA)).

Examples 5.01, *5.02* and *5.03* illustrate some energy balances in closed systems.

---

[3] The term "work" is used loosely in many texts, to mean either work (i.e. (force)(distance)) or power (i.e. work/time).

EXAMPLE 5.01 Energy balance on a falling mass (closed system).

A 5.0 kg block of of iron at 298 K ($C_{v,m}$ ref 298 K = 0.50 kJ/(kg.K)) falls to earth from a height of 100 m under a constant gravity of 9.81 m/s^2, with air resistance assumed to be zero.

Problem: Calculate:

A. The velocity of the iron just before impact with the earth, assuming no change in temperature during the fall.

B. The temperature rise in the iron on impact with the earth, assuming zero heat and work transfer to air or earth.

Solution:

Define the system = block of iron (a closed system)

Specify the quantity = energy

Reference condition = element iron at T = 298 K, L´ = 0, ũ = 0

L´ = height above earth m

T = temperature K

ũ = velocity m/s

A. Integral energy balance on the closed system:

$E_{final} - E_{initial} = Q - W \qquad Q = W = 0$

Initial condition = block at L´ = 100 m

Final condition = block at L´ = 0, just before impact

$E_{initial} = m(C_{v,m}(T - 298) + h_{f,298K}) + mgL' + 0.5m\tilde{u}^2$

$= 5\text{ kg}(0.5\text{ kJ/(kg.K)}(298 - 298)\text{ K} + 0\text{ kJ/kg}) + (5\text{ kg})(9.81\text{ m/s}^2)(100\text{ m}) + (0.5)(5\text{ kg})(0\text{ m/s})^2$

$= 4.91E3\text{ J}$

$E_{final} = 5\text{ kg}(0.5\text{ kJ/(kg.K)}(298 - 298)\text{ K} + 0\text{ kJ/kg}) + (5\text{ kg})(9.81\text{ m/s}^2)(0\text{ m}) + (0.5)(5\text{ kg})(\tilde{u}\text{ m/s})^2$

$= 2.5\tilde{u}^2\text{ J}$

i.e. $2.5\tilde{u}^2 - 4.91E3 = 0 - 0$

$\tilde{u} = $ **<u>44 m/s</u>**

B. Integral energy balance on the closed system:

$E_{final} - E_{initial} = Q - W \qquad Q = W = 0$

Initial condition = block at L´ = 100 m

Final condition = block at L´ = 0, just after impact

$E_{initial} = m(C_{v,m}(T - 298) + h_{f,298K}) + mgL' + 0.5m\tilde{u}^2$

$= 5\text{ kg}(0.5\text{ kJ/(kg.K)}(298 - 298)\text{ K} + 0\text{ kJ/kg})(1E3\text{ J/kJ}) + (5\text{ kg})(9.81\text{ m/s}^2)(100\text{ m}) + (0.5)(5\text{ kg})(0\text{ m/s}^2) = 4.91E3\text{ J}$

$E_{final} = 5\text{ kg}(0.5\text{ kJ/(kg.K)}(T - 298)\text{ K} + 0\text{ kJ/kg})(1E3\text{ J/kJ}) + (5\text{ kg})(9.81\text{ m/s}^2)(0\text{ m}) + (0.5)(5\text{ kg})(0\text{ m/s}^2) = 2.5E3\,(T-298)\text{ J}$

2.5E3(T-298) – 4.91E3 = 0 – 0 Solve for T = 300 K Temperature rise on impact = **2.0 K**

NOTE: Mass cancels from the balances.
The three assumptions used in this problem would not apply in the real situation.

EXAMPLE 5.02 Energy balance for expansion of gas against a piston (closed system).

A vertical cylinder of 2.00 m internal diameter contains 1.00 kmol N_2 gas at 298 K, enclosed under a 5000 kg frictionless piston. The cylinder sits on the Earth with atmospheric pressure = 101.3 kPa(abs) and g = 9.81 m/s^2. N_2 gas $C_{v,m}$ ref 298 K = 21.2 kJ/(kmol.K).

Problem:
Calculate the amount of heat that must be transferred to the gas to raise the piston by 4.00 m.

Solution:

Define the system = gas inside the cylinder (a closed system)
Specify the quantity = energy
Reference condition = elements at standard state 298 K, L´ = 0, ũ = 0
Integral energy balance on the closed system: $E_{final} - E_{initial} = Q - W$
Initial condition = gas at 298 K, piston L´ = 0 m
Final condition = gas at T_f K, piston L´ = 4 m
Initial pressure in cylinder = P_i = 101.3 kPa(abs) + (1E-3 Pa/kPa)(5000 kg)(9.81 m/s^2)/(3.14)(1 m)^2 = 101.3 + 15.6
= 117 kPa(abs)
Initial volume of gas (Ideal gas, see *Equation 2.13*) = $V_i = nRT_i/P_i$ = (1 kmol)(8.314 kJ/(kmol.K)(298 K) / (117 kPa(abs)) = 21.2 m^3
Final pressure = P_f = initial pressure = 117 kPa(abs)
Final volume = V_f = 21.3 m^3 + (3.14)(1 m)^2 (4 m) = 21.3 + 12.6 = 33.9 m^3
Final temperature (ideal gas) = $T_f = P_fV_f/nR$ = (117 kPa(abs))(33.9 m^3) / (1 kmol)(8.314 kJ/(kmol.K)) = 477 K

$E_{initial}$ = $U + E_k + E_p$ = (1 kmol)(21.2 kJ/(kmol.K))(298 – 298)K + 0 kJ/kmol) + 0 + 0) = 0 kJ
E_{final} = $U + E_k + E_p$ = (1 kmol)(21.2 kJ/(kmol.K))(477 – 298)K + 0 kJ/kmol) + 0 + 0) = 3795 kJ
Q = ?
W = work done in expansion against the atmospheric pressure + weight of piston
= $P_{atm}(V_f - V_i) + mg(L'_f - L'_i)$
= (101.3 kPa)(33.9 – 21.2) m^3 + (1E-3 J/kJ)(5000 kg)(9.81 m/s^2)(4 – 0) m = 1287 + 196
= 1483 kJ
Or W = $P(V_2 - V_1)$ = 117 kPa(33.9 – 21.2) m^3 = 1483 kJ
(i.e. work done by expansion of gas against constant pressure)

Energy balance: 3795 – 0 = Q – 1483
Solve for **Q = 5278 kJ** (heat IN)

EXAMPLE 5.03 Energy balance for an electric battery (closed system).

A fully charged silver-zinc electric battery contains 0.01 kmol Ag_2O with 0.01 kmol Zn. The battery is initially at 298 K then discharged under adiabatic conditions at a constant voltage and current, respectively = 1.2 V and 50 A. By Faraday's law, in 1 hour of operation at 50 A, 9.33% of both reactants are consumed in the net cell discharge reaction:

$$Ag_2O + Zn \xrightarrow{2F} 2Ag + ZnO \qquad \textit{Reaction 1}$$

Data:

Component		Ag_2O	Zn	Ag	ZnO
$C_{v,m}$ ref 298 K	kJ/(kmol.K)	80	26	26	42
$h^o_{f,298K}$	kJ/kmol	-29E3	0	0	-348E3

Problem:
Calculate the battery (uniform) internal temperature 1 hour after the current is started.

Solution:

Define the system = battery (a closed system)
Specify the quantity = energy
Reference condition = elements at standard state, 298 K, $L' = 0, \tilde{u} = 0$
Integral energy balance on the closed system: $E_{final} - E_{initial} = Q - W$
Initial condition = fully charged battery, $t = 0$
Final condition = 9.33% discharged battery, $t = 3600$ s

$E_{initial} = \Sigma[n(j)(C_{v,m}(j)(298-298) + h^o_{298K}(j))] + 0 + 0 \qquad (E_k = E_p = 0)$

$= (0.01\text{ kmol})((80\text{ kJ/((kmol.K)})(298-298)\text{ K} + (\text{-29E3 kJ/kmol}))$
$+ (0.01\text{kmol})((26\text{ kJ/(kmol.K)})(298-298)\text{ K} + (0\text{ kJ/kmol})) = \text{-290 kJ}$

$E_{final} = \Sigma[n(j)(C_{v,m}(j)(T-298) + h^o_{298K}(j))] + 0 + 0 \qquad (E_k = E_p = 0)$

$= (1-0.0933)(0.01)((80)(T-298) + (\text{-29E3})) + (1-0.0933)(0.01)((26)(T-298) + (0))$
$+ (2)(0.0933)(0.01)((26))(T-298) + (0)) + (0.0933)(0.01)((42))(T-298) + (\text{-348E3}))$
$= 1.049(T-298) - 588\text{ kJ}$

Q = 0 (adiabatic operation)
W $= E_v I' t = (1.2\text{ V})(50\text{ A})(3600\text{ s}) = 216\text{E3 J} = 216\text{ kJ}$

Energy balance: $[1.049(T-298) - 588] - (-290) = 0 - 216$ Solve for: T = **<u>376 K</u>** at 1 hour

E_v = voltage V
I' = current A
t = time s

NOTE: When a battery is discharged, part of the initial chemical energy of the reactants is converted to electricity (work) and the remainder is converted to heat. A discharging battery gets very hot if it is insulated to prevent heat loss, particulary at high discharge rates or short circuit conditions.

OPEN SYSTEMS (CONTINUOUS PROCESSES)

An open system is defined as one in which, over the time period of interest, *material is transferred across the system boundary*. In this case energy can enter and/or leave the system in the form of the energy content of material, as well as by the transfer of heat and work. The integral and differential energy balances for an open system are now adapted to account for the energy content (E) of material that crosses the system boundary and the net pressure–volume energy ($PV_{output} - PV_{input}$), called the "flow work", needed to move material in and out of the system.

Open system – integral energy balance:

$$\mathbf{E_{final} - E_{initial} = E_{input} - E_{output} + Q - W - (PV_{output} - PV_{input})} \qquad \textit{Equation 5.06}$$

Open system – differential energy balance:

$$\mathbf{dE/dt = \dot{E}_{input} - \dot{E}_{output} + \dot{Q} - \dot{W} - (P\dot{V}_{output} - P\dot{V}_{input})} \qquad \textit{Equation 5.07}$$

where: Typical Units

E	$= [U + E_k + E_p]$ content of the system (i.e. inside the system boundary)	kJ
E_{input}	$= [U + E_k + E_p]$ sum for all material inputs to the system	kJ
E_{output}	$= [U + E_k + E_p]$ sum for all material outputs from the system	kJ
$\dot{E}_{input}$	$= [\dot{U} + \dot{E}_k + \dot{E}_p]$ sum for all material inputs to the system	kW
$\dot{E}_{output}$	$= [\dot{U} + \dot{E}_k + \dot{E}_p]$ sum for all material outputs from the system	kW
$E_{initial}$	= initial value of $[U + E_k + E_p]$ content of the system (i.e. inside the system boundary)	kJ
E_{final}	= final value of $[U + E_k + E_p]$ content of the system (i.e. inside the system boundary)	kJ
PV_{input}	= pressure–volume energy for material inputs	kJ
PV_{output}	= pressure–volume energy for material outputs	kJ
$P\dot{V}_{input}$	= pressure–volume power for material inputs	kW
$P\dot{V}_{output}$	= pressure–volume power for material outputs	kW
Q	= net heat input to system	kJ
$\dot{Q}$	= net rate of heat input to system	kW
t	= time	s
W	= net work output from the system (a.k.a. "shaft work"), excluding the flow work	kJ
$\dot{W}$	= net rate of work (a.k.a. "shaft power") output from system, excluding the flow power	kW

You should understand that the Q and W of the above equations are respectively energy inputs and energy outputs and do not correspond to generation and consumption terms, which are both zero in the total energy balance. Also, the word "energy" is often used in thermodynamics texts instead of the more accurate term "power" (i.e. energy/time) for the description of differential energy balances.

Example 5.04 illustrates an integral energy balance for an open system.
Examples 5.05 to *5.13*, as well as *Examples 6.01* to *6.05* and *7.04* to *7.07* illustrate a range of problems based on differential energy balances in open systems.

EXAMPLE 5.04 Energy balance for a personal rocket pack (open system).

A personal rocket pack powered by the thermo-chemical decomposition of hydrogen peroxide generates an exhaust of water vapour and oxygen with a nozzle velocity of 1050 m/s at 100 kPa(abs), 700 K.

$$2H_2O_2(l) \rightarrow 2H_2O(g) + O_2(g)$$ *Reaction 1*

A 75 kg RocketPerson is strapped into a 30 kg rocket pack (empty mass) initially filled with 25 kg of 90 wt% H_2O_2 in liquid water at 298 K.

Data:

Component		$H_2O_2(l)$	$H_2O(l)$	$H_2O(g)$	$O_2(g)$
$h^o_{f,298\ K}$	kJ/kmol	-188E3	-286E3	-242E3	0
$C_{p,m}$ ref 298 K	kJ/(kmol.K)	89	75	36	30

Problem:
Calculate the maximum height above ground that RocketPerson could reach in a vertical upward flight that uses all the peroxide (i.e. no peroxide remains for the descent). Assume: temperature of RocketPerson and rocket pack stay at 298 K. Zero air resistance and constant gravity.

Solution:
First do the material balance for the peroxide decomposition process from the reaction stoichiometry.

Component		$H_2O_2(l)$	$H_2O(l)$	$H_2O(g)$	$O_2(g)$	TOTAL MASS
Molar mass	kg/kmol	34.0	18.0	18.0	32.0	
Initial	kmol	0.662	0.139	0.000	0.000	25.0 kg
Final	kmol	0.000	0.000	0.801	0.331	25.0 kg

Define the system = RocketPerson with rocket pack (an open system)
Specify the quantity = Energy
Reference conditions: RocketPerson and rocket pack = compounds at 298 K
Fuel and exhaust = elements at 298 K

Let: L´ = maximum height reached by RocketPerson with the rocket pack.
Integral energy balance on the open system: ACC = (Final – Initial) = IN – OUT (GEN = CON = 0)

Initial energy[4] = Energy in (RocketPerson + rocket pack) + energy in fuel
$= (U + KE + PE)_{person+pack} + (U + KE + PE)_{fuel}$
$= (0+0+0) + ((0.662\ \text{kmol})(-188\text{E}3\ \text{kJ/kmol}) + (0.139\ \text{kmol})(-286\text{E}3\ \text{kJ/kmol}))$
$= -164.0\text{E}3\ \text{kJ}$

Final energy = Energy in (RocketPerson + rocket pack)
$= (U + KE + PE)_{person+pack}$
$= (0+0+(1\text{E-}3\ \text{kJ/J})(105\ \text{kg})(9.81\ \text{m/s}^2)(L'\ \text{m}))\ \text{kJ}$

Energy IN = 0 kJ

Energy OUT[5] $= (H + KE + PE)_{exhaust}$
$= (0.801\ \text{kmol})(36\ \text{kJ/(kmol.K)}(700-298)\ \text{K} - 242\text{E}3\ \text{kJ/kmol})$
$+ (0.331\ \text{kmol})(30\ \text{kJ/(kmol.K)}(700-298)\ \text{K} + 0\ \text{kJ/kmol})$
$+ (1\text{E-}3\ \text{kJ/J})((0.5)(25\ \text{kg})(1050\ \text{m/s}))^2$
$+ (1\text{E-}3\ \text{kJ/J})(0.5)(25\ \text{kg})(9.81\ \text{m/s}^2)(L'\ \text{m})\ \text{kJ}$

Energy balance: $1.030L' + 164.00\text{E}3 = 164.40\text{E}3 - 0.122L'$

Solve the energy balance to get: L´ = **345 m**

CONTINUOUS PROCESSES AT STEADY-STATE

Energy balances for continuous processes at steady-state use the differential energy balance *Equation 5.07*, with [Rate ACC = dE/dt = 0].
In this case the calculations are simplified by replacing the "U + PV" term by the enthalpy "H" to give:

$$\mathbf{dE/dt = 0 = [\dot{H} + \dot{E}_k + \dot{E}_p]_{input} - [\dot{H} + \dot{E}_k + \dot{E}_p]_{output} + \dot{Q} - \dot{W}}$$ *Equation 5.08*

where: Typical Units

$[\dot{H} + \dot{E}_k + \dot{E}_p]_{input}$	= sum for all streams entering the system	kW
$[\dot{H} + \dot{E}_k + \dot{E}_p]_{output}$	= sum for all streams leaving the system	kW
$\dot{Q}$	= net rate of heat transfer <u>into</u> the system	kW
$\dot{W}$	= net rate of work (a.k.a. shaft power) <u>output</u> from the system	kW

[4] Internal energy is assumed equal to enthalpy for the liquid fuel.

[5] Potential energy of the exhaust is effectively the PE of the total exhaust mass at elevation = 0.5L´.

Example 5.05 shows how *Equation 5.08* can be applied to calculate the power requirements of a hypothetical continuous process at steady-state, in which liquid water is pumped, elevated, heated, vaporised, expanded, heated and finally decomposed (i.e. reacted) to produce hydrogen and oxygen. By examination of *Example 5.05* you will see that the kinetic and potential energy terms are relatively small and that the energy balance is dominated by enthalpy and heat terms. This situation is common in energy balances on chemical processes, where the relative magnitude of the energy effects is usually:

Reaction > latent heat > sensible heat > gas compression or expansion > liquid pumping > kinetic and potential energy > exotics

Mechanical Energy Balance

For "mechanical" systems, in which thermal effects such as chemical reaction, phase transformation and heat transfer are not present (or negligible), the energy balance *Equations 5.04* to *5.08* reduce to a form called the "mechanical energy balance", the steady-state version of which is *Equation 5.08A*.

$$0 = [\dot{E}_k + \dot{E}_p]_{in} - [\dot{E}_k + \dot{E}_p]_{out} - \int_{Pin}^{Pout} \dot{V} dp - \dot{F}_e - \dot{W}$$ *Equation 5.08A*

where:

$\dot{E}_k, \dot{E}_p$	= kinetic, potential energy flow	kW
$\dot{F}_e$	= rate of energy degradation by friction in flowing fluid	kW
$\dot{W}$	= rate of work output from the system	kW
$\dot{V}$	= volumetric fluid flow	$m^3.s^{-1}$
P	= pressure	kPa

For the special case of incompressible fluid flow with zero friction loss and zero work done, *Equation 5.08A* reduces to *Equation 5.08B*, commonly known as the Bernoulli equation.

$$0 = 0.5[(\tilde{u}_{out})^2 - (\tilde{u}_{in})^2] + g[L'_{out} - L'_{in}] + [P_{out} - P_{in}]/\rho$$ *Equation 5.08B*

L′	= height above a reference level	m
g	= acceleration of gravity	ms^{-2}
P	= pressure	Pa
ũ	= fluid velocity	ms^{-1}
ρ	= fluid density	$kg.m^{-3}$

The mechanical energy balance is used for fluid flow and pressure changes in pipes and process units.

EXAMPLE 5.05 Energy balance for processing water (open system at steady-state).

Flowsheet [This is a hypothetical process]

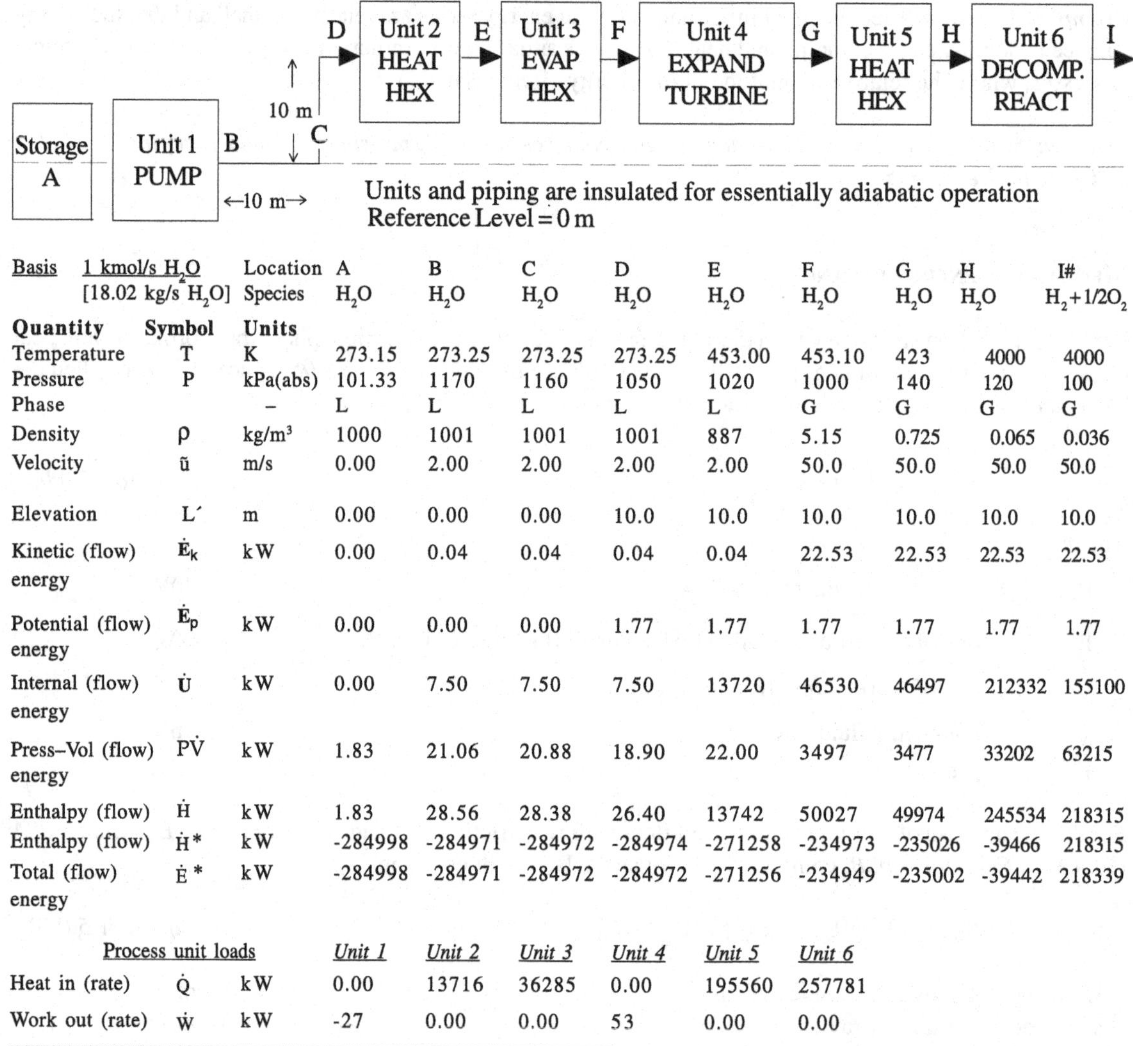

Basis 1 kmol/s H_2O [18.02 kg/s H_2O]		Location / Species	A H_2O	B H_2O	C H_2O	D H_2O	E H_2O	F H_2O	G H_2O	H H_2O	I# $H_2+1/2O_2$
Quantity	**Symbol**	**Units**									
Temperature	T	K	273.15	273.25	273.25	273.25	453.00	453.10	423	4000	4000
Pressure	P	kPa(abs)	101.33	1170	1160	1050	1020	1000	140	120	100
Phase		–	L	L	L	L	L	G	G	G	G
Density	ρ	kg/m³	1000	1001	1001	1001	887	5.15	0.725	0.065	0.036
Velocity	ũ	m/s	0.00	2.00	2.00	2.00	2.00	50.0	50.0	50.0	50.0
Elevation	L′	m	0.00	0.00	0.00	10.0	10.0	10.0	10.0	10.0	10.0
Kinetic (flow) energy	$\dot{E}_K$	kW	0.00	0.04	0.04	0.04	0.04	22.53	22.53	22.53	22.53
Potential (flow) energy	$\dot{E}_P$	kW	0.00	0.00	0.00	1.77	1.77	1.77	1.77	1.77	1.77
Internal (flow) energy	$\dot{U}$	kW	0.00	7.50	7.50	7.50	13720	46530	46497	212332	155100
Press–Vol (flow) energy	$P\dot{V}$	kW	1.83	21.06	20.88	18.90	22.00	3497	3477	33202	63215
Enthalpy (flow)	$\dot{H}$	kW	1.83	28.56	28.38	26.40	13742	50027	49974	245534	218315
Enthalpy (flow)	$\dot{H}$*	kW	-284998	-284971	-284972	-284974	-271258	-234973	-235026	-39466	218315
Total (flow) energy	$\dot{E}$*	kW	-284998	-284971	-284972	-284972	-271256	-234949	-235002	-39442	218339

Process unit loads			*Unit 1*	*Unit 2*	*Unit 3*	*Unit 4*	*Unit 5*	*Unit 6*
Heat in (rate)	$\dot{Q}$	kW	0.00	13716	36285	0.00	195560	257781
Work out (rate)	$\dot{W}$	kW	-27	0.00	0.00	53	0.00	0.00

The asterisk (*) indicates that the reference condition is the elements at their standard states (H_2 and O_2). No asterisk means the reference condition is the compounds. The reference pressure and temperature for both the element and compound reference conditions are 101.3 kPa(abs), 273.15 K. The effect of pressure from 0.61 kPa(abs) [steam table data] to 101.3 kPa(abs) [heat of formation data] on the reference condition is assumed negligible. Heat of formation of H_2O (liq) at 273 K = -285000 kJ/kmol. #assumes complete decomposition of H_2O at 4000 K. Values for points H and I are approximate, since thermodynamic data for 4000 K are not available.

ENTHALPY BALANCE

For chemical process energy balances, *Equation 5.08* is simplified by dropping the kinetic and potential energy terms, to give *Equation 5.09* and its equivalent *Equation 5.10*.

$$0 = \dot{H}_{input} - \dot{H}_{output} + \dot{Q} - \dot{W} \qquad \textit{Equation 5.09}$$

$$0 = \Sigma[\dot{H}(i)]in - \Sigma[\dot{H}(i)]out + \dot{Q} - \dot{W} \qquad \textit{Equation 5.10}$$

where:

$\dot{H}_{input}$	= sum of enthalpy flow carried by all input streams to the system	kW
$\dot{H}_{output}$	= sum of enthalpy flow carried by all output streams from the system	kW
$\dot{H}(i)$	= enthalpy flow of stream "i" w.r.t. a specified reference condition	kW
$\dot{Q}$	= thermal *utility load* on the system (+) ≡ heating, (–) ≡ cooling	kW
$\dot{W}$	= mechanical or electrical *utility load* on the system (+) ≡ power out; (–) ≡ power in	kW

The utility loads are the energy (a.k.a. power) requirements to operate the system, which are usually supplied by fuel, steam or cooling water and/or by electricity that is used to run machines such as compressors, pumps, heaters and refrigerators.

Equation 5.10 (called the "enthalpy balance" or when $\dot{W}$ is zero, "heat balance") is the main equation for calculating energy balances on continuous chemical processes at steady-state, where kinetic and potential energy effects are assumed insignificant. This equation is easy to use as long as you obey the rules for energy accounting regarding the reference state. This means that the enthalpy of reference state(s) must cancel out of the equation when the enthalpy OUT is subtracted from the enthalpy IN. The recommended procedure is thus to calculate all enthalpies in *Equation 5.10* with respect to the same reference state. This procedure is inconvenient in cases where enthalpy values are taken from different sources (e.g. steam tables and thermochemical tables), so it is permissible to use different reference states for terms in the enthalpy IN, provided the reference enthalpies cancel from corresponding terms in the enthalpy OUT (or if you know that the error introduced by un-cancelled reference states is acceptable for the objective of the energy balance).

The best reference state to use for energy balances on general chemical processes is the *elements* in their standard state at 298 K (see *Figure 2.05*). The procedure based on this reference state is sometimes called the *"heat of formation method"* of energy balance calculation. On this basis the enthalpy of each process stream is estimated by *Equation 5.11*, which assumes the stream is an ideal mixture and sums the components enthalpies calculated by *Equation 5.12*, i.e:

$$\dot{H}^*(i) = \Sigma[\dot{n}(i,j)\, h^*(j)] \qquad \text{[Ideal mixture]} \qquad \textit{Equation 5.11}$$

and:

$$h^*(j) \approx \int_{T_{ref}}^{T(i)} C_p(j)dT + h^o_{f,T_{ref}}(j) \qquad \text{[Respect the phase]} \qquad \textit{Equation 5.12}$$

where:

$C_p(j)$	=	heat capacity at constant pressure of the substance "j" in phase Π	$kJ.kmol^{-1}.K^{-1}$
$h^*(j)$	=	specific enthalpy of the substance "j" at temperature T(i) in phase Π w.r.t. *elements* at standard state (see *Figure 2.05*)	$kJ.kmol^{-1}$
$h^o_{f,Tref}(j)$	=	standard heat of formation of the substance "j" at T_{ref} in phase Π	$kJ.kmol^{-1}$
T(i)	=	temperature of stream "i"	K
T_{ref}	=	reference temperature = 298 K	K
$\dot{n}(i, j)$	=	flow of substance "j" in stream "i"	$kmol.s^{-1}$
$\dot{H}^*(i)$	=	total enthalpy flow of stream "i" w.r.t. *elements* at standard state	kW

In *Equation 5.12* the phase Π is the actual phase of the substance "j" in stream "i". If substance "j" exists in more than one phase in stream "i", then its flow is assigned to each phase in the appropriate split fractions.

For many practical calculations, *Equation 5.12* can be simplified by replacing the integral of heat capacity by the mean heat capacity defined in *Equation 2.60*, i.e.

$$\mathbf{h^*(j) \approx C_{p,m}(j)[T(i) - T_{ref}] + h^o_{f,Tref}} \qquad \text{[Respect the phase]} \qquad \textit{Equation 5.13}$$

where:

$C_{p,m}(j)$ = mean heat capacity of substance "j" in phase Π over the temperature range T_{ref} to the stream temperature T(i) $kJ.kmol^{-1}.K^{-1}$

For non-ideal mixtures a term is added to *Equation 5.11* to account for the heat of mixing.

$$\mathbf{\dot{H}^*(i) = \Sigma[\dot{n}(i,j)h^*(j)] + \dot{H}_{mix}} \qquad \text{[Non-ideal mixture]} \qquad \textit{Equation 5.14}$$

where:

$\dot{H}_{mix}$ = enthalpy of mixing (i.e. heat of mixing) for the flowing component kW

With the heat of formation method, using the enthalpy of *Equations 5.11* (or *5.14*) and *5.12*, the energy balance automatically takes care of any *heat of reaction* in a chemical process, so it is not necessary to add a special term for the heat of reaction. This energy balance still applies to process units where no reactions occur because the standard heats of formation in the enthalpy IN and enthalpy OUT are equal, and so cancel from the balance.

For energy balances on systems that involve only physical processes such as sensible heating, evaporation, compression, etc. (i.e. no chemical reactions) it is convenient and acceptable to calculate enthalpy relative to *compounds* at a reference condition. This enthalpy, defined in *Equations 5.15* and *5.16*, does not include heats of formation and therefore does not compensate the energy balance for heats of reaction.

$$\mathbf{\dot{H}(i) = \Sigma[\dot{n}(i,j)h(j)]} \qquad \text{[Ideal mixture]} \qquad \textit{Equation 5.15}$$

and: $\mathbf{h(j) \approx \int_{T_{ref}}^{T(j)} C_p(j)\, dT + h_{p,Tref}(j)}$ [Respect the phase]

plus $\mathbf{h(j) = C_{p,m}(j)[T(i) - T_{ref}] + h_{p,Tref}(j)}$ *Equation 5.16*

where:

$h(j)$	=	specific enthalpy of the substance "j" at temperature T in phase Π w.r.t. the *compound* at the reference state	$kJ.kmol^{-1}$
$C_p(j)$	=	heat capacity at constant pressure of the substance "j" in phase Π	$kJ.kmol^{-1}.K^{-1}$
$h_{p,Tref}(j)$	=	latent heat of any phase change(s), measured at T_{ref} of substance "j" from its reference state to phase Π	$kJ.kmol^{-1}$
$T(i)$	=	temperature of stream "i"	K
T_{ref}	=	reference temperature	K
$\dot{n}(i, j)$	=	flow of substance "j" in stream "i"	$kmol.s^{-1}$
$\dot{H}(i)$	=	enthalpy flow of stream "i" w.r.t. *compounds* at the reference state	kW

In *Equation 5.16* the phase Π is the actual phase of the substance "j" in stream "i". If substance "j" exists in more than one phase in stream "i", then its flow is assigned to each phase in the appropriate split fractions.

The enthalpy estimated by *Equations 5.15* and *5.16* is similar to that obtained from thermodynamic tables and charts such as the steam table and enthalpy-concentration diagrams, which also do not include heats of formation. This enthalpy is normally used for energy balances on "mechanical" systems such as compressors, refrigeration cycles and steam engines, as well as on process operations like heat transfer, mixing and separation (e.g. distillation, extraction, etc.) where reactions do not occur.

If the enthalpy from *Equations 5.15* and *5.16* is used for an energy balance on a chemical reactor then an *extra term* must be added to the balance equation to account for the *heat(s) of reaction*. This *"heat of reaction method"* is used in many texts (see *Refs. 1–6*) but it is not emphasised in this text due to its relative complexity when dealing with reactive systems with incomplete conversion and multiple reactions. *Table 5.02* compares the "heat of formation" and "heat of reaction" methods of setting up energy balances.

Table 5.02. Comparison of energy balance reference conditions.

Reference condition	Non-reactive (physical) processes	Reactive (chemical) processes	Steady-state energy balance
Elements at 298 K [Respect the phases]	CORRECT. Heats of formation cancel	CORRECT. Heats of formation capture heat(s) of reaction	$0 = \Sigma[\dot{H}^*(i)in] - \Sigma[\dot{H}^*(i)out] + \dot{Q} - \dot{W}$ *"Heat of formation"* method
Compounds at T_{ref} [Respect the phases]	CORRECT. Heats of formation not needed	NOT SUFFICIENT. Must add heat(s) of reaction term to balance	$0 = \Sigma[\dot{H}(i)in] - \Sigma[\dot{H}(i)out] + \dot{Q} - \dot{W} - \Sigma[\dot{\varepsilon}(\ell)h_{rx}(\ell)]$ *"Heat of reaction"* method

Examples 5.06 and *5.09* to *5.13* show the application of *Equations 5.11* and *5.12* to energy balances on systems both with and without chemical reactions.

Example 5.07 A and B illustrates the application of *Equations 5.15* and *5.16* to energy balances on systems without chemical reaction. *Example 5.08 A and B* shows how steam tables and enthalpy-concentration diagrams can be used to solve energy balances in systems without chemical reaction.

HEAT AND WORK

In a continuous chemical process the *heat* ($\dot{Q}$) and *work* ($\dot{W}$) terms[6] of the energy balance are called the *utility loads* on the process. The *utility loads* are the rates of energy transfer (a.k.a. power) across the system envelope required to operate the process. Utilities are usually supplied by fuel, steam and cooling water, as well as by electricity to drive machines such as compressors, pumps, heaters and refrigerators.

Heat is transferred to the system across *heat transfer surfaces.* A *heat transfer surface* may be the shell of a process unit, through which heat transfers from/to the surroundings, or may be contained in a dedicated heat exchanger, where heat is transferred from/to a utility or another process stream.

Heat transfer through the shells of insulated units from/to the surroundings is usually a minor part of the energy balance for industrial chemical processes, accounting for up to about 5% of the total energy flow. However in small-scale systems without insulation (e.g. laboratory apparatus) heat transfer from/to the surroundings can be important, due to the relatively high ratio of surface area to volume for small process units and associated piping.

Heat transfer via dedicated indirect heat exchangers is typically a major feature of the energy balance for industrial chemical processes. The rate of heat transfer can be calculated by *Equation 5.17* (see *Ref. 9*).

$$\dot{Q} = U_{hex} A_{hex} \Delta T \qquad \textit{Equation 5.17}$$

where:

$\dot{Q}$	=	rate of heat transfer into the system (i.e. the *thermal load* or *"thermal duty"* of the exchanger)	kW
U_{hex}	=	heat transfer coefficient, which depends in part on the thermal conductivity of the process streams and the heat transfer wall	$kW.m^{-2}.K^{-1}$
A_{hex}	=	area of the heat transfer surface	m^2
ΔT	=	temperature difference between the hot and cold streams (logarithmic mean value[7])	K

In this convention ΔT is (+) when the "system" temperature is below that of the heat transfer medium, and negative (–) in the reverse condition.

---

[6] Remember that heat and work are path functions, whose values depend on the path taken between the inlet and outlet process conditions.

[7] Logarithmic mean $\Delta T = [\Delta T_1 - \Delta T_2]/\ln[\Delta T_1/\Delta T_2]$ where subscripts 1 and 2 refer to the respective ends of the heat exchanger.

Work (a.k.a. power) is usually in the form of mechanical work or electrical work. This work is sometimes called "shaft work" (W_s) to separate it from the "flow work" (PV) done on the material entering/exiting the system and/or the work done by changing the system volume. Mechanical work is associated with gas compressors and liquid pumps, as well as with turbines that transfer energy by gas expansion or by a pressure drop in liquid flow. The shaft power in compression and expansion of ideal gases can be calculated by *Equations 5.18* or *5.19*. Power in pumping liquids is given by *Equation 5.20*, which is simpler than the gas equations because liquids are nearly incompressible.

$$\dot{W} = -(P_1\dot{V}_1)(r/(r-1))[(P_2/P_1)^{((r-1)/r)}-1]$$ [Adiabatic power, ideal gas, and *Equation 5.18*

$$T_2 = T_1(P_2/P_1)^{((r-1)/r)}$$ Adiabatic temperature]

$$\dot{W} = -(P_1\dot{V}_1)\ln(P_2/P_1)$$ [Isothermal power, ideal gas] *Equation 5.19*

$$\dot{W} = -\dot{V}_1(P_2-P_1)$$ [Liquid (any conditions)] *Equation 5.20*

where:

$\dot{W}$	= rate of mechanical energy transfer (i.e. power) out of the system	kW
P_1	= inlet stream pressure	kPa(abs)
P_2	= outlet stream pressure	kPa(abs)
$\dot{V}_1$	= inlet stream volumetric flow rate	$m^3.s^{-1}$
r	= C_p/C_v = ratio of gas heat capacities (e.g. r ≅ 1.4 for air)	–
T_1	= inlet stream temperature	K
T_2	= outlet stream temperature	K

Note that the work (power) from *Equations 5.18*, *5.19* and *5.20* is the rate transferred *across the system envelope*. This is the quantity used in the energy balance on the system, but the work (power) in the surroundings must be adjusted by the efficiency of the transfer machine. For example, a gas compressor that transfers 30 kW to the system with an efficiency of 60% requires a power input of (30/0.6) = 50 kW, while a turbine that transfers 50 kW from the system with an efficiency of 60% will give a useful power output of (50 * 0.6) = 30 kW.

Electrical work can take several forms, such as power to a resistance heater or from a photovoltaic cell. Electrical work is also a major part of the energy balance for electrochemical systems such as batteries, fuel cells and electro-synthesis reactors. In each case the electric work (power) can be calculated from *Equation 5.21*.

$\dot{W} = E_v I'$ [Electric power] *Equation 5.21*

where:

$\dot{W}$ = rate of electrical energy transfer (i.e. power) out of the system kW
E_v = voltage drop across the system Volt
I' = current kA

For a photovoltaic cell, battery or fuel cell $\dot{W}$ is (+), whereas for an electric heater or electro-synthesis reactor $\dot{W}$ is (–).

SPECIFICATION OF ENERGY BALANCE PROBLEMS

Material flows in a chemical process are specified by the set of equations and variables summarised in *Table 4.02*. The energy balance adds one independent equation to this set and introduces four more variables, the pressure (P) and temperature (T) needed to find the phase(s) and calculate the enthalpy of each stream, the heat ($\dot{Q}$) and the work ($\dot{W}$). Altogether, the material flow, pressure, temperature and phase(s) of each stream, with the heat transfer and work done, specifies every process unit and fully defines the performance of multi-unit processes. The *stream table* for a full M&E balance is thus expanded to include the phase(s), pressure, temperature, volume and enthalpy of each stream, plus the values of $\dot{Q}$ and $\dot{W}$ assigned to each unit in the flowsheet.

Table 5.03 summarises the number of variables and equations associated with the energy balance on individual process units and on a combined multi-unit process, where all material flows are already known. The degrees of freedom (D of F) of the energy balance is that of *Equation 4.10*.

The comprehensive treatment of energy balance problems is difficult because it requires simultaneous solution of *Equation 5.07* with the equations of state, the entropy balance, momentum balance, heat transfer rate, etc. which is beyond the scope of this text. Fortunately, for most chemical process calculations you can make simplifying assumptions to fully specify the problem and solve the energy balance. The assumptions depend on the situation and can be one or more of the following:

A. Specific enthalpy is independent of pressure and depends only on temperature and phase, as shown in *Equations 5.12* and *5.16*. This approximation is good for solids and liquids, and for gases away from their critical state. The specific enthalpy from thermodynamic tables (e.g. the steam table) incorporates the effect of pressure but still depends mainly on temperature and phase, except near the critical state.

B. Heat transfer to a process unit is zero, i.e. an *"adiabatic process"* ($\dot{Q} = 0$). This approximation is good for most industrial process units, which are usually insulated to suppress heat transfer with the surroundings, unless they are explicitly designed for that heat transfer.

C. Zero work is done by a process unit ($\dot{W} = 0$). This approximation is usually good for all process units except compressors, expansion turbines, pumps, electric heaters and electrochemical reactors. In some cases it is necessary to include work in energy balances on mixers and bio or thermo-chemical reactors when they use a lot of mechanical energy (a.k.a. power) for mixing.

D. Temperature is unchanged through the unit, i.e. an *"isothermal process"* (T = constant). This approximation is good for process units such as dividers, some mixers and separators ($\dot{Q} = 0$, $\dot{W} = 0$) and is usually adequate for pumps with a net liquid flow. Other process units such as compressors and reactors can be forced to operate isothermally by the appropriate heat transfer (i.e. $\dot{Q} \neq 0$). Heat exchangers are not isothermal, except in their condensing and vaporising sections.

E. The pressures and the temperatures of all streams leaving a process unit are equal. This approximation is good for dividers and for some separators such as evaporators, centrifuges and clarifiers, which operate with a single stage, but not for multi-stage units such as gas absorbers and fractional distillation columns.

Table 5.03. Specification of multi-unit process energy balance.

Process unit	Divide	Mix	Separate	Hex	Pump	React	Overall
Stream variables (P, T)	2I	2I	2I	2I	2I	2I	2I
Unit variables ($\dot{Q}$, $\dot{W}$)	2	2	2	2	2	2	2
Energy balance equations	1	1	1	1	1	1	1
Entropy balance equations	1	1	1	1	1	1	1
Momentum balance equations	1	1	1	1	1	1	1
State equations	I	I	I	I	I	I	I
Other relations	Heat transfer rate: $Q = U_{hex}A_{hex}\Delta T$ Mechanical energy balance. Pressure drop.						
Phase equilibria	Phase rule, Raoult's law, Henry's law, distribution coefficient, etc.						
Simplifying assumptions [Depend on situation, see notes above] Enthalpy is independent of pressure. $\dot{Q} = 0$ *(adiabatic process)* $\dot{W} = 0$ Zero work done T = constant *(isothermal process)* P and T the same for all streams leaving a process unit. Pressure drop zero, or estimate by rule of thumb.							
Specified quantities	Stream pressures (P) Heat transfer ($\dot{Q}$) to process units			Stream temperatures (T) Work done ($\dot{W}$) by process units			
I = number of streams All material flows are assumed known from an independent material balance. Phase splits in each stream are not assumed known. Degrees of freedom (D of F) of energy balance = number of unknown quantities – number of independent equations							

F. Pressure drop through a process unit is zero, or is estimated by a rule of thumb.[8] Since most energy balances are insensitive to pressure these assumptions are usually good for all process units except compressors, expansion turbines, pumps, and some separators and reactors that function with a high pressure drop. For the examples in this chapter all pressure drops are estimated by the rules of thumb used in process design (see *Refs. 9–10*).

The specified stream pressures and temperatures are mostly fixed by design considerations for efficient reaction and separation. Heat transfer and work are then usually determined to meet these process requirements — ultimately through an optimisation procedure with an objective such as maximum return on investment.

ENERGY BALANCE CALCULATIONS BY COMPUTER (SPREADSHEET)

Energy balance calculations in this chapter start from the assumption that the material balance has been fixed and summarised in a stream table, as shown in Chapter 4. The energy balance is treated here as a fully-specified problem (i.e. D of F = 0) that hinges on calculation of the stream enthalpies. The solved energy balance is added to the material balance to form a complete M&E balance. The M&E balance is then presented in an expanded stream table that shows the material balance plus the phase(s), pressure, temperature, volume and enthalpy of each process stream, together with the heat and work transfers to each process unit. The basic format of the stream table is shown in *Figure 5.01*. The completed stream table should be presented within view of the process flowsheet (e.g. on the same page) so the process can be quickly understood and examined for consistency on both a conceptual and quantitative level.

Component	M	Stream			
	kg/kmol	1	2	3	4
		Flow (kmol/h)			
A					
B					
C					
Total	kg/h				
Phase	-				
Pressure	kPa(abs)				
Temperature	K				
Volume	m^3/h				
Enthalpy	kJ/h				
		Unit 1	Unit 2	Unit 3	Overall
Q	kJ/h				
W	kJ/h				
Mass balance check					
Mass IN	kg/h				
Mass OUT	kg/h				
Closure	%				
Energy balance check					
Energy IN	kJ/h				
Energy OUT	kJ/h				
Closure	%				

Figure 5.01. Generic material and energy balance stream table.

As a check on the calculations it is recommended to include in the stream table values for the closure of both the mass and the energy balances on each process unit and on the overall process. For a more detailed analysis it is sometimes useful to track down errors with individual element balances (a.k.a. atom balances) across each process unit and on the overall process.

The mass balance closure (e_M = mass out / mass in) is defined in *Equation 4.09*. The energy balance closure is defined in *Equation 5.22*.

[8] A rule of thumb is an approximation that holds in most circumstances. In process design, rules of thumb are based on trade-offs between capital and operating costs that aim to minimise the total cost and/or maximise return on investment.

$$e_E = 100[\Sigma\dot{H}^*(i)out + \dot{W}]/[\Sigma\dot{H}^*(i)in + \dot{Q}]$$ *Equation 5.22*

where:

e_E	=	energy balance closure (steady-state)	%
$\Sigma\dot{H}^*(i)out$	=	sum of enthalpy rates carried by all output streams to the system	kW
$\Sigma\dot{H}^*(i)in$	=	sum of enthalpy rates carried by all input streams from the system	kW
$\dot{Q}$	=	net rate of heat transfer into the system	kW
$\dot{W}$	=	net rate of work output by the system	kW

Spreadsheet calculations for the energy balance follow the pattern of those for the material balance outlined in Chapter 4. The major issue is finding the stream enthalpies. To estimate enthalpies by *Equation 5.11* or *5.15* the data needed to calculate the *specific enthalpy* of each component (i.e. C_p, h_p, h_f) must be tabulated on the spreadsheet, either as part of the stream table or in a separate "properties" section. If the specific enthalpies are taken from thermodynamic tables or charts you need a "look-up" function that refers to thermodynamic data in the spreadsheet, or better, a mathematical correlation that calculates specific enthalpy from the stream conditions.

Energy balance calculations in the spreadsheet usually fall into one of two categories:

1. Given the conditions (flow, phase, P, T) for each stream, find $\dot{Q}$ and/or $\dot{W}$.
   In this case you first calculate the stream enthalpies to complete the "enthalpy" row of the stream table. Next, calculate any accessible and significant mechanical or electrical work (power) transfer by one of the *Equations 5.18* to *5.21* given above. Finally, solve *Equation 5.10* for the unknown $\dot{Q}$ or $\dot{W}$.
2. Given $\dot{Q}$ and $\dot{W}$ plus some stream conditions, find the other stream condition(s).
   In this case you first calculate all the known stream enthalpies to partially complete the "enthalpies" row of the stream table. Next, calculate significant but unspecified mechanical or electrical work (power) transfer. Finally, solve *Equation 5.10* for the unknown condition(s). For example, if the unknown condition is the outlet temperature from a utility HEX heating a single component fluid *without changing its phase,*[9] then *Equation 5.10*, with *Equations 5.11* and *5.12*, can be solved explicitly for that temperature to give:

   $$T(out) = T_{ref} + [(\dot{H}^*(in) + \dot{Q} - \dot{W})/\bar{n}(in) - h^o_{f,ref}]/C_{p,m}$$ *Equation 5.23*

where:

T (out) = outlet stream temperature K

---

[9] If part or all of the process fluid does change phase in the HEX then this solution must be modified because the value of $C_{p,m}$ is not defined across a phase change, plus the enthalpy change will include a latent heat term.

T_{ref}	=	reference temperature for enthalpy calculations	K
$\dot{H}^*(in)$	=	enthalpy flow of inlet stream	kW
$\dot{Q}$	=	rate of heat transfer (i.e. the utility load)	kW
$\dot{W}$	=	rate of work done by process unit	kW
$\bar{n}(in)$	=	inlet flow rate of process stream	$kmol.s^{-1}$
C_{pm}	=	mean heat capacity of process stream across the HEX	$kJ.kmol^{-1}.K^{-1}$
$h^o_{f,Tref}$	=	heat of formation of the process fluid component at T_{ref} in phase Π	$kJ.kmol^{-1}$

Note that phase Π here is the phase of the inlet stream, and that the heat of formation includes latent heat for any phase changes from the standard state to the inlet stream condition.

When using a spreadsheet it is not necessary for you to write an explicit solution for the unknown quantity in the energy balance. If the balance is fully-specified the unknown value can be found by setting up the stream table with known conditions and using the "Goal Seek" or "Solver" to find the value of the unknown condition that closes the energy balance at steady-state (i.e. gives $e_E = 100\%$).

ENERGY BALANCES WITH THE HEAT OF COMBUSTION

An important class of energy balance problems involves reactions with materials such as fuels and biological substrates (i.e. biomass) that are characterised only by an empirical formula and a "heat of combustion" or a "heating value". Examples of these materials are wood, coal, fuel oil, waste mixtures, bio-sludge, bio-oil, enzymes, yeast, etc. There are two ways to handle such materials in the energy balance, the first (A) has general application; the second (B) applies only to combustion. These calculations correspond respectively to the "heat of formation" and the "heat of reaction" methods of *Table 5.02*.

A. Calculate the effective heat of formation of the material from its heat of combustion and use this heat of formation in the general energy balance.

To find the effective heat of formation, first write the stoichiometry of the combustion reaction from the empirical formula of the reactant, e.g. for a material with empirical formula $C_aH_bS_cN_d$:

$$C_aH_bS_cN_d + (a + b/4 + c)O_2(g) \rightarrow (a)CO_2(g) + (b/2)H_2O(l) + (c)SO_2(g) + (d/2)N_2(g)$$ *Reaction 5.01* [10]

(The material balance is most easily calculated by atom balances on C, H, S, N and O)

---

[10] This stoichiometry ignores the formation of nitrogen oxides (NO, NO_2, N_2O), that normally accounts for a small fraction of the nitrogen fed to combustion processes.

Next, solve *Equation 2.53* for the heat of formation of the reactants:

$$\Sigma(\nu(j)h_f^\circ(j)) \text{ reactants} = -h_{rxn} + \Sigma(\nu(j)h_f^\circ(j)) \text{ products}$$

Substitute:

$\nu(j) = 1$ for the reactant (i.e. the "fuel" of the combustion reaction). Then:
$\Sigma(\nu(j)h_f^\circ(j)) \text{ reactants} = h_f^\circ(\text{fuel})$

$h^o_{f,O_2(g)} = h^o_{f,N_2(g)} = 0$ and $h_{rxn} = h^o_c$

then:

$$h^o_{f,C_aH_bS_cN_d} = -h^o_c + [(a)\,h^o_{f,CO_2(g)} + (b/2)\,h^o_{f,H_2O(1)} + (c)\,h^o_{f,SO_2(g)} + (d/2)\,h^o_{f,N_2(g)}]$$

where:

$h^o_{f,C_aH_bS_cN_d}$ = standard heat of formation of the reactant (a.k.a. the fuel) kJ.kmol^{-1}

h^o_c = standard *gross* heat of combustion of the reactant (a.k.a. the fuel) kJ.kmol^{-1}

$h^o_{f,CO_2(g)}$, $h^o_{f,H_2O(1)}$, $h^o_{f,SO_2(g)}$, $h^o_{f,N_2(g)}$ = standard heats of formation of the reaction products [note phases] kJ.kmol^{-1}

a, b, c, d = stoichiometric coefficients in the combustion *Reaction 5.01* –

Note that h^o_c above is the *gross* heat of combustion.
If you use the *net* heat of combustion here the $h^o_{f,H_2O(1)}$ must be replaced by $h^o_{f,H_2O(g)}$.

B. In typical "combustion" processes the fuel undergoes complete combustion in air or oxygen to give CO_2, H_2O, SO_2, N_2, etc. by a reaction such as *5.01*. Method A can be used for energy balances with complete or incomplete combustion, but for complete combustion it is sometimes easier to use the "heat of reaction" method to solve combustion problems. For the "heat of reaction" method the heat of combustion is added as an extra term to the energy balance, in which the stream enthalpies *do not include heats of formation. Equation 5.10* then becomes:

$$0 = \Sigma[\dot{H}(i)]in - \Sigma[\dot{H}(i)]out + \dot{Q} - \dot{W} - \dot{n}_{fuel}h^o_c \qquad \textit{Equation 5.24}$$

where:

$\Sigma[\dot{H}(i)]in$ = sum of enthalpy rates for all inlet streams, w.r.t. compounds at reference state kW

$\Sigma[\dot{H}(i)]in$ = sum of enthalpy rates of all outlet streams, w.r.t. compounds at reference state kW

$\dot{Q}$	=	net rate of heat transfer <u>into</u> the system	kW
$\dot{W}$	=	net rate of work transfer <u>out</u> of the system	kW
$\dot{n}_{fuel}$	=	rate of fuel flow into the system	$kmol.s^{-1}$
h_c^o	=	standard heat of combustion of fuel	$kJ.kmol^{-1}$

Equation 5.24 can cause difficulty because the reference state for the enthalpy terms [$\dot{H}(i)$] may not be same as the standard state for the heat of combustion (usually 101.3 kPa(abs), 298 K). Strictly the references states should be corrected to cancel from the balance. However the heat of combustion is typically so large that the relatively small difference in the usual reference states (e.g. 273 K vs. 298 K) has little effect on the energy balance.

In almost all practical combustion problems, the product water leaves the system as a gas, in which case the heat of combustion in *Equation 5.24* must be the *net* heat of combustion. The net heat of combustion can be found from the gross heat of combustion by *Equation 5.25*:

$$\mathbf{h^o_{c,\,net} = h^o_{c,\,gross} + (b/2)\ h_{v,\,H_20}} \qquad \textit{Equation 5.25}$$

where:

$h^o_{c,net}$	=	standard net heat of combustion of the fuel (= - lower heating value)	$kJ.kmol^{-1}$
$h^o_{c,\,gross}$	=	standard gross heat of combustion of the fuel (= - higher heating value)	$kJ.kmol^{-1}$
b/2	=	moles of H_2O produced per mole of fuel (cf. *Reaction 5.01*)	–
$h_{v,\,H_2O}$	=	heat of vaporisation of water (e.g. 44E3 kJ/kmol at 298K)	$kJ.kmol^{-1}$

Beware that combustion is an *exothermic* reaction, so h^o_c should always be negative (–), whereas the "heating value" is usually given as positive (+).

SINGLE PROCESS UNITS

Chapter 3 describes six generic process units used to build chemical process flowsheets and Chapter 4 *(Figure 4.01 and Example 4.04)* shows the steady-state material balance for each generic unit. The corresponding steady-state energy balances on each of these generic process units are illustrated in *Example 5.06 A, B, C, D, E* and *F*.

Example 5.06 uses only the *heat of formation method* for energy balances, in which stream enthalpies include the heats of formation of the components, as in *Equations 5.11* and *5.12*. For the five generic process units without chemical reaction (i.e. DIVIDE, MIX, SEPARATE, HEAT EXCHANGE and PUMP) the heats of formation are redundant and cancel from the energy balances. However the heats of formation are essential in the energy balance on the REACTOR, to account for the heat liberated or absorbed by chemical reaction(s).

The heat of formation method is preferred for general chemical process energy balance calculations because it automatically deals with both non-reactive and reactive processes. Nonetheless, for purely non-reactive processes it is sometimes convenient to calculate stream enthalpies without the heats of formation, as in *Equations 5.15* and *5.16*. This situation arises particularly with non-reactive processes when enthalpies are obtained from thermodynamic diagrams or tables (e.g. the steam table) whose reference conditions are not the elements at standard state. *Example 5.07 A* and *B* shows such energy balance calculations on two non-reactive systems (SEPARATOR and PUMP) using *Equations 5.15* and *5.16* to estimate the stream enthalpies. *Example 5.08 A and B* illustrates non-reactive energy balance calculations on a MIXER and a HEAT EXCHANGER using thermodynamic diagrams and tables as the source of enthalpy values.

Example 5.09 shows a more "advanced" problem involving a chemical reactor whose temperature is controlled by indirect heat transfer from condensing steam. In this case, since the steam does not mix with the reactants, it is convenient and acceptable to take the *element* reference state for the reactor side and the *compound* reference state for the steam side of the REACT/HEX unit. *Example 5.09* (as well as *Example 5.08B*) demonstrates that enthalpy values can be taken from sources with different reference conditions, provided the individual reference conditions cancel from the energy balance.

MULTIPLE PROCESS UNITS

Energy balance calculations on multiple process units (i.e. complete process flowsheets) follow the same procedure as material balances on multiple process units described in Chapter 4. As for material balances the system should first be checked for its specification (degrees of freedom), guided by the rules of *Table 5.03*. The coupled energy balances of a fully-specified system may then be solved by either the sequential modular or the simultaneous solution method. In the sequential modular method the solution is iterated to convergence through tear streams to deal with recycle loops.

Example 5.10 demonstrates an energy balance for a simple process flowsheet with a mixer-reactor-separator + recycle pump, based on the material balance of *Example 4.10*.

Note that in *Example 5.10* the unknown values of T(2), T(6), $\dot{Q}(2)$ and $\dot{Q}(3)$ are obtained by solving the energy balances on the respective process units. These results can come from either:

A. Explicit algebraic solution.

or

B. Using the spreadsheet "Goal Seek" or "Solver" tool with the unknown value as a variable to set closure: $e_E = 100\%$ (see *Equation 5.22*) in the appropriate energy balance.

If necessary, more difficult cases may be dealt with by programming a spreadsheet macro with a numerical method for solving non-linear equations (e.g. bisection, Newton's method, etc.).

Examples 5.11, 5.12 and *5.13* show the spreadsheet solutions to a set of material and energy balances on respectively, a biochemical, an electrochemical and a thermo-chemical process, each with three process units coupled through a recycle stream. In each of *Examples 5.11, 5.12* and *5.13* the material balance is independent of the energy balance, so the separate material and energy balances are solved in sequence. As in *Example 5.10* the unknown values in the energy balance are found using the "Goal Seek" or "Solver" tool to set closure = 100% in *Equation 5.22*. This procedure is exemplified in *Example 5.13* by the solution for T(2) = 2148 K, which is effectively the *adiabatic flame temperature* in the fuel oil combustion process.

EXAMPLE 5.06 Energy balances for generic process units (open system at steady-state).

Note that these problems are simplified by using mean heat capacities instead of integrating the C_p polynomials (see *Equations 2.59, 2.60* and *2.62*).

A. Divider

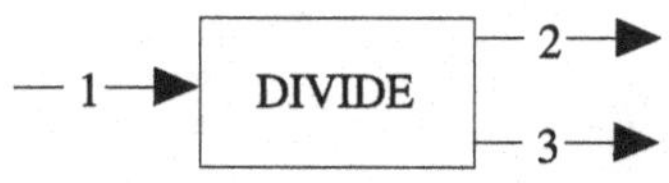

No phase change
No reaction
For material balance, see *Example 4.04*

Stream Table		Divider		Problem
Species	M	Stream		
		1	2	3
	kg/kmol	kmol/h		
A (liq)	20	4	1	3
B (liq)	30	8	2	6
C (liq)	40	12	3	9
Total	kg/h	800	200	600
Phase		L	L	L
Temp.	K	400	?	?
Press.	kPa(abs)	150	140	140
Enthalpy	kJ/h	-1.05E+06	?	?

Properties	Species	A (liq)	A (gas)	B (liq)	B (gas)	C (liq)	C (gas)
$C_{p,m}$(ref.298K)	kJ/kmol.K	50	35	70	45	90	55
$h_{f,298K}$	kJ/kmol	-2.00E+05	-1.80E+05	-1.00E+05	-7.00E+04	3.00E+04	4.00E+04
$h_{v,298K}$	kJ/kmol	-	2.00E+04	-	3.00E+04	-	1.00E+04

Solution: Energy balance. Reference condition = *elements* at standard state, 298K.

$$0 = \Sigma[\dot{H}^*(in)] - \Sigma[\dot{H}^*(out)] + \dot{Q} - \dot{W}$$

For a divider $\dot{Q} = \dot{W} = 0$ and $T(2) = T(3)$ i.e.

$$0 = \dot{H}^*(1) - \dot{H}^*(2) - \dot{H}^*(3) + 0 - 0 \quad [1]$$

where:

$$\dot{H}^*(i) = \Sigma[\dot{n}(i,j)h^*(j)] \text{ and } h^*(j) = C_{p,m}(j)(T(i) - T_{ref}) + h^o_{f,Tref}(j) \quad \text{[Respect the phase]}$$

$$\dot{H}^*(1) = (4\text{ kmol/h})((50\text{ kJ/kmol.K})(400\text{ K} - 298\text{ K}) + (\text{-2E5 kJ/kmol})) + (8\text{ kmol/h})((70\text{ kJ/kmol.K})(400\text{ K} - 298\text{ K}) + (\text{-1E5 kJ/kmol})) + (12\text{ kmol/h})((90\text{ kJ/kmol.K})(400\text{ K} - 298\text{ K}) + (\text{3E4 kJ/kmol})) = \text{-1.05E+06 kJ/h}$$

$$\dot{H}^*(2) = (1\text{ kmol/h})((50\text{ kJ/kmol.K})(T(2)\text{ K} - 298\text{ K}) + (\text{-2E5 kJ/kmol})) + (2\text{ kmol/h})((70\text{ kJ/kmol.K})(T(2)\text{ K} - 298\text{K}) + (\text{-1E5 kJ/kmol})) + (3\text{ kmol/h})((90\text{ kJ/kmol.K})(T(2)\text{ K} - 298\text{ K}) + (\text{3E4 kJ/kmol})) = f(T(2)) \text{ kJ/h}$$

$$\dot{H}^*(3) = (3\text{ kmol/h})((50\text{ kJ/kmol.K})(T(3)\text{ K} - 298\text{ K}) + (\text{-2E5 kJ/kmol})) + (6\text{ kmol/h})((70\text{ kJ/kmol.K})(T(3)\text{ K} - 298\text{K}) + (\text{-1E5 kJ/kmol})) + (9\text{ kmol/h})((90\text{ kJ/kmol.K})(T(3)\text{ K} - 298\text{ K}) + (\text{3E4 kJ/kmol})) = f(T(3)) \text{ kJ/h}$$

Solve equation [1] for: $T(2) = T(3) = 400$ K

Stream Table	***Divider***			***Solution***
Species	M		Stream	
		1	*2*	*3*
	kg/kmol		kmol/h	
A (liq)	20	4	1	3
B (liq)	30	8	2	6
C (liq)	40	12	3	9
Total	kg/h	800	200	600
Phase		L	L	L
Temp.	K	400	400	400
Press.	kPa(abs)	150	140	140
Enthalpy*	kJ/h	-1.05E+06	-2.63E+05	-7.89E+05
Energy balance check				
Energy IN	kJ/h	-1.05E+06	Closure %	
Energy OUT	kJ/h	-1.05E+06	100.0	

* *Enthalpy values include heats of formation*

B. Mixer

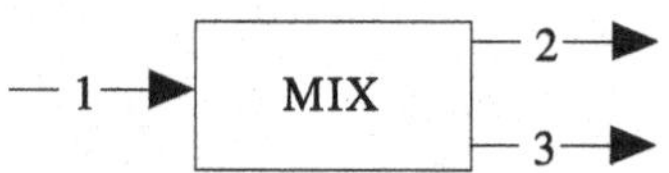

No phase change
No reaction
For material balance, see *Example 4.04*

Stream	Mixer			
Species	M	Stream		
		1	2	3
	kg/kmol	kmol/h		
A (liq)	20	4	3	7
B (liq)	30	8	5	13
C (liq)	40	12	7	19
Total	kg/h	800	490	1290
Phase		L	L	L
Temp.	K	400	300	?
Press.	kPa(abs)	150	150	140
Enthalpy	kJ/h	-1.05E+06	-8.88E+05	?

Properties	Species	A (liq)	A (gas)	B (liq)	B (gas)	C (liq)	C (gas)
$C_{p,m}$(ref.298K)	kJ/kmol.K	50	35	70	45	90	55
$h_{f,298K}$	kJ/kmol	-2.00E+05	-1.80E+05	-1.00E+05	-7.00E+04	3.00E+04	4.00E+04
$h_{v,298K}$	kJ/kmol	-	2.00E+04	-	3.00E+04	-	1.00E+04

Solution:

Energy balance.
Reference condition = *elements* at standard state, 298K.

$$0 = [\dot{H}^*(in)] - [\dot{H}^*(out)] + \dot{Q} - \dot{W} \qquad \text{For a mixer} \qquad \dot{Q} = \dot{W} = 0 \qquad \text{i.e.}$$

$$0 = \dot{H}^*(1) + \dot{H}^*(2) - \dot{H}^*(3) + 0 - 0 \qquad [1]$$

where:

$$\dot{H}^*(i) = \Sigma[\dot{n}(i,j)h^*(j)] \text{ and } h^*(j) = C_{p,m}(j)(T(i) - T_{ref}) + h^o_{f,Tref}(j) \qquad \text{[Respect the phase]}$$

$$\begin{aligned}\dot{H}^*(1) = &(4\,\text{kmol/h})((50\,\text{kJ/kmol.K})(400\,\text{K} - 298\,\text{K}) + (\text{-2E5 kJ/kmol}))\\ &+ (8\,\text{kmol/h})((70\,\text{kJ/kmol.K})(400\,\text{K} - 298\,\text{K}) + (\text{-1E5 kJ/kmol}))\\ &+ (12\,\text{kmol/h})((90\,\text{kJ/kmol.K})(400\,\text{K} - 298\,\text{K}) + (\text{3E4 kJ/kmol})) \quad = -1.05\text{E}+06 \quad \text{kJ/h}\end{aligned}$$

$$\begin{aligned}\dot{H}^*(2) = &(3\,\text{kmol/h})((50\,\text{kJ/kmol.K})(300\,\text{K} - 298\,\text{K}) + (\text{-2E5 kJ/kmol}))\\ &+ (5\,\text{kmol/h})((70\,\text{kJ/kmol.K})(300\,\text{K} - 298\,\text{K}) + (\text{-1E5 kJ/kmol}))\\ &+ (7\,\text{kmol/h})((90\,\text{kJ/kmol.K})(300\,\text{K} - 298\,\text{K}) + (\text{3E4 kJ/kmol})) \quad = -8.88\text{E}+05 \quad \text{kJ/h}\end{aligned}$$

$$\begin{aligned}\dot{H}^*(3) = &(7\,\text{kmol/h})((50\,\text{kJ/kmol.K})(T(3)\,\text{K} - 298\,\text{K})\\ &+ (\text{-2E5 kJ/kmol}))\\ &+ (13\,\text{kmol/h})((70\,\text{kJ/kmol.K})(T(3)\,\text{K} - 298\,\text{K})\\ &+ (\text{-1E5 kJ/kmol}))\\ &+ (19\,\text{kmol/h})((90\,\text{kJ/kmol.K})(T(3)\,\text{K} - 298\,\text{K})\\ &+ (\text{3E4 kJ/kmol}))\\ = &\ f(T(3))\ \text{kJ/h}\end{aligned}$$

Solve equation [1] for: T(3) = **362 K**

Stream	Mixer			
Species	M	Stream		
		1	2	3
	kg/kmol	kmol/h		
A (liq)	20	4	3	7
B (liq)	30	8	5	13
C (liq)	40	12	7	19
Total	kg/h	800	490	1290
Phase		L	L	L
Temp.	K	400	300	362
Press.	kPa(abs)	150	150	140
Enthalpy*	kJ/h	-1.05E+06	-8.88E+05	-1.94E+06
Energy balance check				
Energy IN	kJ/h	-1.94E+06	Closure %	
Energy OUT	kJ/h	-1.94E+06	100.0	

* *Enthalpy values include heats of formation*

C. Separator

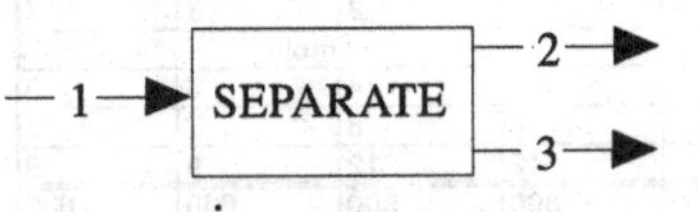

$\dot{Q} = 5.00E+05$ kJ/h $\quad T(2) = T(3)$

No reaction

For material balance, see *Example 4.04*

Stream Table		Separator		Problem
Species	M	Stream		
		1	2	3
	kg/kmol	kmol/h		
A	20	4	3	1
B	30	8	4	4
C	40	12	4	8
Total	kg/h	800	340	460
Phase		L	G	L
Temp.	K	300	?	?
Press.	kPa(abs)	600	580	580
Enthalpy	kJ/h	-1.24E+06	?	?

Properties	Species	A (liq)	A (gas)	B (liq)	B (gas)	C (liq)	C (gas)
$C_{p,m}$(ref.298K)	kJ/kmol.K	50	35	70	45	90	55
$h_{f,298K}$	kJ/kmol	-2.0E+05	-1.8E+05	-1.0E+05	-7.0E+04	3.00E+04	4.00E+04
$h_{v,298K}$	kJ/kmol	-	2.00E+04	-	3.00E+04	-	1.00E+04

Solution: Energy balance. Reference condition = *elements* at standard state, 298K.

$$0 = [\dot{H}^*(in)] - [\dot{H}^*(out)] + \dot{Q} - \dot{W} \quad \text{For a separator} \quad \dot{W} = 0 \quad \text{i.e.}$$

$$0 = \dot{H}^*(1) - \dot{H}^*(2) - \dot{H}^*(3) + \dot{Q} - 0 \qquad [1]$$

where:

$$\dot{H}^*(i) = \sum[\dot{n}(i,j)h^*(j)] \quad \text{and} \quad h^*(j) = C_{p,m}(j)(T(i) - T_{ref}) + h^o_{f,Tref}(j) \qquad \text{[Respect the phase]}$$

$$\begin{aligned}\dot{H}^*(1) = {} & (4\,\text{kmol/h})((50\,\text{kJ/kmol.K})(300\,\text{K} - 298\,\text{K}) + (\text{-2E5 kJ/kmol}))\\ & + (8\,\text{kmol/h})((70\,\text{kJ/kmol.K})(300\,\text{K} - 298\,\text{K}) + (\text{-1E5 kJ/kmol}))\\ & + (12\,\text{kmol/h})((90\,\text{kJ/kmol.K})(300\,\text{K} - 298\,\text{K}) + (\text{3E4 kJ/kmol})) \quad = \text{-1.24E+06} \quad \text{kJ/h}\end{aligned}$$

$$\begin{aligned}\dot{H}^*(2) = {} & (3\,\text{kmol/h})((35\,\text{kJ/kmol.K})(T(2)\,\text{K} - 298\,\text{K}) + (\text{-1.8E5 kJ/kmol}))\\ & + (4\,\text{kmol/h})((45\,\text{kJ/kmol.K})(T(2)\,\text{K} - 298\text{K}) + (\text{-7E4 kJ/kmol}))\\ & + (4\,\text{kmol/h})((55\,\text{kJ/kmol.K})(T(2)\,\text{K} - 298\,\text{K}) + (\text{4E4 kJ/kmol})) \quad = f(T(2)) \quad \text{kJ/h}\end{aligned}$$

$$\begin{aligned}\dot{H}^*(3) = {} & (1\,\text{kmol/h})((50\,\text{kJ/kmol.K})(T(3)\,\text{K} - 298\,\text{K}) + (\text{-2E5 kJ/kmol}))\\ & + (4\,\text{kmol/h})((70\,\text{kJ/kmol.K})(T(3)\,\text{K} - 298\text{K})\\ & + (\text{-1E5 kJ/kmol}))\\ & + (8\,\text{kmol/h})((90\,\text{kJ/kmol.K})(T(3)\,\text{K} - 298\,\text{K})\\ & + (\text{3E4 kJ/kmol}))\\ = {} & f(T(3))\,\text{kJ/h}\end{aligned}$$

Solve equation [1] for: T(2) = T(3) = **481K**

Stream Table		Separator		Solution
Species	M	Stream		
		1	2	3
	kg/kmol	kmol/h		
A	20	4	3	1
B	30	8	4	4
C	40	12	4	8
Total	kg/h	800	340	460
Phase		L	G	L
Temp.	K	300	481	481
Press.	kPa(abs)	600	580	580
Enthalpy*	kJ/h	-1.24E+06	-5.68E+05	-1.68E+05
Energy balance check				
Energy IN	kJ/h	-7.36E+05	Closure %	
Energy OUT	kJ/h	-7.36E+05	100.0	

* *Enthalpy values include heats of formation.*

D. Heat exchanger (indirect contact)

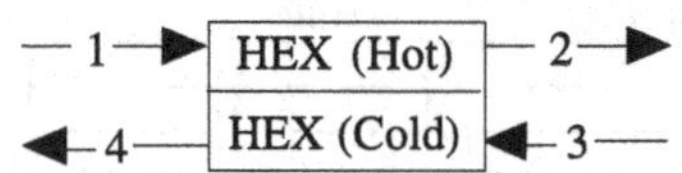

Indirect contact
No reaction
For material balance, see *Example 4.04*

Stream Table		Heat Exchanger			Problem
Species	M	Stream			
		1	2	3	4
	kg/kmol	kmol/h			
A	20	4	4	3	3
B	30	8	8	7	7
C	40	12	12	9	9
Total	kg/h	800	800	630	630
Phase		G	L	L	G
Temp.	K	500	420	300	?
Press.	kPa(abs)	600	570	200	170
Enthalpy	kJ/h				

Properties	Species	A (liq)	A (gas)	B (liq)	B (gas)	C (liq)	C (gas)
$C_{p,m}$ (ref.298K)	kJ/kmol.K	50	35	70	45	90	55
$h_{f,298K}$	kJ/kmol	-2.0E+05	-1.8E+05	-1.0E+05	-7.0E+04	3.00E+04	4.00E+04
$h_{v,298K}$	kJ/kmol	-	2.00E+04	-	3.00E+04	-	1.00E+04

Solution:

Energy balance.
Reference condition = *elements* at standard state, 298K.

$$0 = \Sigma[\dot{H}^*(in)] - \Sigma[\dot{H}^*(out)] + \dot{Q} - \dot{W}$$ For a heat exchanger $\dot{Q} = \dot{W} = 0$ i.e.

$$0 = \dot{H}^*(1) - \dot{H}^*(2) + \dot{H}^*(3) - \dot{H}^*(4) + 0 - 0$$ [overall HEX balance] [1]

where: $\dot{H}^*(i) = \Sigma[\dot{n}(i,j)h^*(j)]$ and $h^*(j) = C_{p,m}(j)(T(i) - T_{ref}) + h^o_{f,Tref}(j)$ [Respect the phase]

$$\dot{H}^*(1) = (4\text{ kmol/h})((35\text{ kJ/kmol.K})(500\text{ K} - 298\text{ K}) + (-1.8E5\text{ kJ/kmol})) + (8\text{ kmol/h})((45\text{ kJ/kmol.K})(500\text{ K} - 298\text{ K}) + (-7E4\text{ kJ/kmol})) + (12\text{ kmol/h})((55\text{ kJ/kmol.K})(500\text{ K} - 298\text{ K}) + (4E4\text{ kJ/kmol})) = -5.66E+05\text{ kJ/h}$$

$$\dot{H}^*(2) = (4\text{ kmol/h})((50\text{ kJ/kmol.K})(400\text{ K} - 298\text{ K}) + (-2E5\text{ kJ/kmol})) + (8\text{ kmol/h})((70\text{ kJ/kmol.K})(400\text{ K} - 298\text{ K}) + (-1E5\text{ kJ/kmol})) + (12\text{ kmol/h})((90\text{ kJ/kmol.K})(400\text{ K} - 298\text{ K}) + (3E4\text{ kJ/kmol})) = -1.08E+06\text{ kJ/h}$$

Stream Table		Heat Exchanger			Solution
Species	M	Stream			
		1	2	3	4
	kg/kmol	kmol/h			
A	20	4	4	3	3
B	30	8	8	7	7
C	40	12	12	9	9
Total	kg/h	800	800	630	630
Phase		G	L	L	G
Temp.	K	500	420	300	473
Press.	kPa(abs)	600	570	200	170
Enthalpy*	kJ/h	-5.66E+05	-1.08E+06	-1.03E+06	-5.10E+05
Energy balance check.					
Energy IN	kJ/h	-1.59E+06	Closure %		
Energy OUT	kJ/h	-1.59E+06	100.0		

* *Enthalpy values include heats of formation*

$$\dot{H}^*(3) = (3\text{ kmol/h})((50\text{ kJ/kmol.K})(350\text{ K} - 298\text{ K}) + (-2E5\text{ kJ/kmol})) + (7\text{ kmol/h})((70\text{ kJ/kmol.K})(350\text{ K} - 298\text{ K}) + (-1E5\text{ kJ/kmol})) + (9\text{ kmol/h})((90\text{ kJ/kmol.K})(350\text{ K} - 298\text{ K}) + (3E4\text{ kJ/kmol})) = -1.03E+06\text{ kJ/h}$$

$$\dot{H}^*(4) = (3\text{ kmol/h})((35\text{ kJ/kmol.K})(T(4)\text{ K} - 298\text{ K}) + (-1.8E5\text{ kJ/kmol})) + (7\text{ kmol/h})((45\text{ kJ/kmol.K})(T(4)\text{ K} - 298\text{ K}) + (-7E4\text{ kJ/kmol})) + (9\text{ kmol/h})((90\text{ kJ/kmol.K})(T(4)\text{ K} - 298\text{ K}) + (4E4\text{ kJ/kmol})) = f(T(4))\text{ kJ/h}$$

Solve equation [1] for T(4) = 473K

Thermal duty of HEX = $\dot{H}^*(1) - \dot{H}^*(2) = \dot{H}^*(4) - \dot{H}^*(3)$
= 5.18E+05 kJ/h = **144 kW**

E. Pump

Stream Table		Pump	Problem
Species	M	Stream	
		1	2
	kg/kmol	kmol/h	
A	20	4	4
B	30	8	8
C	40	12	12
Total	kg/h	800	800
Phase		L	L
Temp.	K	300	?
Press.	kPa(abs)	150	2150
Volume	m^3/h	0.89	0.89
Enthalpy	kJ/h	-1.24E+06	?

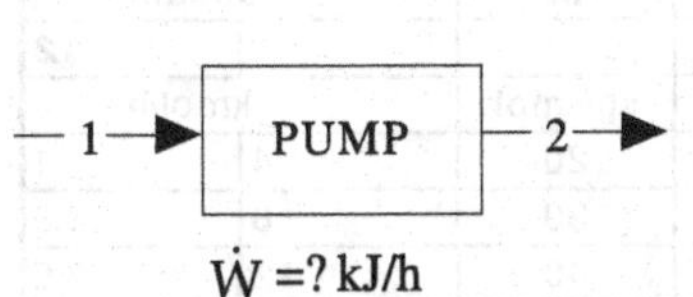

$\dot{W}$ =? kJ/h

For material balance, see *Example 4.04*

Properties	Species	A (liq)	A (gas)	B (liq)	B (gas)	C (liq)	C (gas)
$C_{p,m}$ (ref.298K)	kJ/kmol.K	50	35	70	45	90	55
$h_{f,298K}$	kJ/kmol	-2.0E+05	-1.8E+05	-1.0E+05	-7.0E+04	3.00E+04	4.00E+04
$h_{v,298K}$	kJ/kmol	-	2.00E+04	-	3.00E+04	-	1.00E+04
Density	kg/m^3	900	-	900	-	900	-

Solution:

Energy balance.

Reference condition = *elements* at standard state, 298K.

$$0 = \Sigma[\dot{H}^*(in)] - \Sigma[\dot{H}^*(out)] + \dot{Q} - \dot{W} \qquad \text{For a pump} \qquad \dot{Q} = 0$$

$$0 = \dot{H}^*(1) - \dot{H}^*(2) + 0 - \dot{W} \qquad [1]$$

where: $\dot{H}^*(i) = \Sigma[\dot{n}(i,j)h^*(j)]$ and $h^*(j) = C_{p,m}(j)(T(i) - T_{ref}) + h^o_{f,Tref}(j)$ [Respect the phase]

$$\begin{aligned}\dot{H}^*(1) &= (4\ \text{kmol/h})((50\ \text{kJ/kmol.K})(300\ \text{K} - 298\ \text{K}) + (\text{-2E5 kJ/kmol}))\\ &\quad + (8\ \text{kmol/h})((70\ \text{kJ/kmol.K})(300\ \text{K} - 298\ \text{K}) + (\text{-1E5 kJ/kmol}))\\ &\quad + (12\ \text{kmol/h})((90\ \text{kJ/kmol.K})(300\ \text{K} - 298\ \text{K}) + (\text{3E4 kJ/kmol}))\\ &= \text{-1.24E+06} \quad \text{kJ/h}\end{aligned}$$

Stream Table		Pump	Solution
Species	M	Stream	
		1	2
	kg/kmol	kmol/h	
A	20	4	4
B	30	8	8
C	40	12	12
Total	kg/h	800	800
Phase		L	L
Temp.	K	300	***301***
Press.	kPa(abs)	150	2150
Volume	m^3/h	0.89	0.89
Enthalpy*	kJ/h	-1.24E+06	-1.23E+06
Energy balance check			
Energy IN	kJ/h	-1.24E+06	Closure %
Energy OUT	kJ/h	-1.24E+06	100.0

$$\begin{aligned}\dot{H}^*(2) &= (4\ \text{kmol/h})((50\ \text{kJ/kmol.K})(T(2)\ \text{K} - 298\ \text{K}) + (\text{-2E5 kJ/kmol}))\\ &\quad + (8\ \text{kmol/h})((70\ \text{kJ/kmol.K})(T(2)\ \text{K} - 298\ \text{K}) + (\text{-1E5 kJ/kmol}))\\ &\quad + (12\ \text{kmol/h})((90\ \text{kJ/kmol.K})(T(2)\ \text{K} - 298\ \text{K}) + (\text{3E4 kJ/kmol}))\\ &= f(T(2)) \quad \text{kJ/h}\end{aligned}$$

$$\begin{aligned}\dot{W} &= \dot{V}(1)[P(1) - P(2)] = (0.89\ m^3/h)(150\ \text{kPa} - 2150\ \text{kPa})\\ &= \text{-1780 kJ/h}\end{aligned}$$

Solve equation [1] for T(2) = **301K**

* *Enthalpy values include heats of formation*

F. Reactor

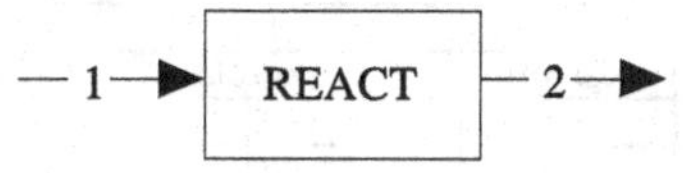

$\dot{Q} = ?$ kJ/h

Reaction: $2A + 3B \rightarrow C + 2D$

Conversion of A = 75%

For material balance, see *Example 4.04*

Stream Table		Reactor	Problem
Species	M	Stream	
		1	*2*
	kg/kmol	kmol/h	
A	20	4	1
B	30	8	3.5
C	40	12	13.5
D	45	3	6
Total	kg/h	935	935
Phase		L	G
Temp.	K	400	500
Press.	kPa(abs)	500	450
Enthalpy	kJ/h	?	?

Solution:

Energy balance.

Properties	Species	A (liq)	A (gas)	B (liq)	B (gas)	C (liq)	C (gas)	D (liq)	D (gas)
$C_{p,m}$(ref.298K)	kJ/kmol.K	50	35	70	45	90	55	105	65
$h_{f,298K}$	kJ/kmol	-2.0E+05	-1.8E+05	-1.0E+05	-7.0E+04	3.00E+04	4.00E+04	-1.00E+05	-6.00E+04
$h_{v,298K}$	kJ/kmol	-	2.00E+04	-	3.00E+04	-	1.00E+04	-	4.00E+04

Reference condition = *elements* at standard state, 298K.

$$0 = \Sigma[\dot{H}^*(in)] - \Sigma[\dot{H}^*(out)] + \dot{Q} - \dot{W} \quad \text{For a reactor} \quad \dot{W} = 0 \quad \text{i.e.}$$

$$0 = \dot{H}^*(1) - \dot{H}^*(2) + \dot{Q} - 0 \qquad [1]$$

where: $\dot{H}^*(i) = \Sigma[\dot{n}(i,j)h^*(j)]$ and $h^*(j) = C_{p,m}(j)(T(i) - T_{ref}) + h^o_{f,Tref}(j)$ [Respect the phase]

$$\begin{aligned}\dot{H}^*(1) &= (4\text{ kmol/h})((50\text{ kJ/kmol.K})(400\text{ K} - 298\text{ K}) + (\text{-2E5 kJ/kmol}))\\ &+ (8\text{ kmol/h})((70\text{ kJ/kmol.K})(400\text{ K} - 298\text{ K}) + (\text{-1E5 kJ/kmol}))\\ &+ (12\text{ kmol/h})((90\text{ kJ/kmol.K})(400\text{ K} - 298\text{ K}) + (\text{3E4 kJ/kmol}))\\ &+ (3\text{ kmol/h})((105\text{ kJ/kmol.K})(400\text{ K} - 298\text{ K}) + (\text{-1E5 kJ/kmol})) = \text{-1.32E+06 kJ/h}\end{aligned}$$

Stream Table		Reactor	Solution
Species	M	Stream	
		1	*2*
	kg/kmol	kmol/h	
A	20	4	1
B	30	8	3.5
C	40	12	13.5
D	45	3	6
Total	kg/h	935	935
Phase		L	G
Temp.	K	400	500
Press.	kPa(abs)	500	450
Enthalpy*	kJ/h	***-1.32E+06***	***2.27E+04***
Energy balance check			
Energy IN	kJ/h	2.27E+04	Closure %
Energy OUT	kJ/h	2.27E+04	100%

$$\begin{aligned}\dot{H}^*(2) &= (1\text{ kmol/h})((35\text{ kJ/kmol.K})(500\text{ K} - 298\text{ K}) + (\text{-1.8E5 kJ/kmol}))\\ &+ (3.5\text{ kmol/h})((45\text{ kJ/kmol.K})(500\text{ K} - 298\text{ K}) + (\text{-7E4 kJ/kmol}))\\ &+ (13.5\text{ kmol/h})((55\text{ kJ/kmol.K})(500\text{ K} - 298\text{ K}) + (\text{4E4 kJ/kmol}))\\ &+ (6\text{ kmol/h})((65\text{ kJ/kmol.K})(500\text{ K} - 298\text{ K}) + (\text{-6E4 kJ/kmol}))\\ &= \text{2.27E+04 kJ/h}\end{aligned}$$

Solve equation [1] for

$Q = \dot{H}^*(2) - \dot{H}^*(1)$ = **<u>1.34E+06 kJ/h = 373 kW</u>**

* *Enthalpy values include heats of formation*

EXAMPLE 5.07 Energy balances without chemical reaction (open system at steady-state).

A. Compressor

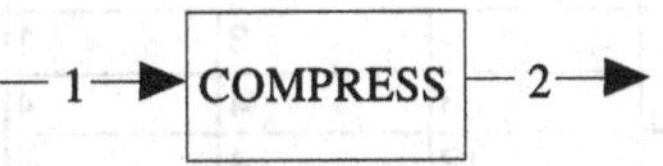

$\dot{W}$ = ? kJ/h

Adiabatic compression. Ideal gas.
No reaction.

Stream Table		Compressor Problem	
Species	M	Stream	
		1	2
	kg/kmol	kmol/h	
[A] N_2	28	4	4
[B] O_2	32	8	8
[C] CO_2	44	12	12
Total	kg/h	896	896
Phase		G	G
Temp.	K	300	?
Press	kPa(abs)	100	1100
Volume	m^3/h	599	?
Enthalpy	kJ/h	?	?

Properties	Species	A (gas)	B(gas)	C(gas)
$C_{p,m}$(ref.298K)	kJ/kmol.K	29.1	29.4	37.2

Solution: Energy Balance. Reference condition = *compounds* at standard state, 298K.

$$0 = \Sigma[\dot{H}(in)] - \Sigma[\dot{H}(out)] + \dot{Q} - \dot{W} \quad \text{For an adiabatic compressor } \dot{Q} = 0 \text{ i.e.}$$

$$0 = \dot{H}(1) - \dot{H}(2) + \dot{Q} - \dot{W} \qquad [1]$$

where: $\dot{H}(i) = \Sigma[\dot{n}(i, j)h(j)]$

$h(j) = C_{p,m}(j)(T(i) - T_{ref}) + h_{p,Tref}(j)$ [Respect the phase]

$$\dot{H}(1) = (4\,\text{kmol/h})((29.1\,\text{kJ/kmol.K})(300\,\text{K} - 298\,\text{K}) + 0) + (8\,\text{kmol/h})((29.4\,\text{kJ/kmol.K})(300\,\text{K} - 298\,\text{K}) + 0) + (12\,\text{kmol/h})((37.2\,\text{kJ/kmol.K})(300\,\text{K} - 298\,\text{K}) + 0) = 1596\ \text{kJ/h}$$

$$\dot{H}(2) = (4\,\text{kmol/h})((29.1\,\text{kJ/kmol.K})(T(2)\,\text{K} - 298\,\text{K}) + 0) + (8\,\text{kmol/h})((29.4\,\text{kJ/kmol.K})(T(2)\,\text{K} - 298\,\text{K}) + 0) + (12\,\text{kmol/h})((37.2\,\text{kJ/kmol.K})(T(2)\,\text{K} - 298\,\text{K}) + 0) = f(T(2))\ \text{kJ/h}$$

$$C_{p,m} \text{ gas mixture} = \Sigma\,[y(j)C_{p,m}(j)] = (4/24)(29.1) + (8/24)(29.4) + (12/24)(37.2) = 33.3\ \text{kJ/kmol.K}$$

$$r = C_p/C_v = C_p/(C_p - R) = 33.3/(33.3 - 8.31) = 1.333$$

$$\dot{W} = [(P_1\dot{V}_1(r/(r-1))][(P_2/P_1)^{((r-1)/r)} - 1] \quad \text{(see \textit{Equation 5.18})}$$

$$= -(100\,\text{kPa})(599\,\text{m}^3\text{/h})(1.333/(1.333-1))[(1100\,\text{kPa}/100\,\text{kPa})^{((1.333-1)/1.333)} - 1]$$

$$= -1.97\text{E+05 kJ/h} \qquad = -54.7\ \text{kW}$$

Solve equation [1] for T(2) = **546 K**

or alternatively, by *Equation 5.18*:

$$T(2) = T(1)\,[P(2)/P(1)]^{((r-1)/r)} = (300)\,(1100/100)^{((1.333-1)/1.333)}$$

$$= \mathbf{546\ K}$$

Stream Table		Compressor Solution	
Species	M	Stream	
		1	2
	kg/kmol	kmol/h	
[A] N_2	28	4	4
[B] O_2	32	8	8
[C] CO_2	44	12	12
Total	kg/h	896	896
Phase		G	G
Temp.	K	300	***546***
Press	kPa(abs)	100	1100
Volume	m^3/h	599	***99***
Enthalpy#	kJ/h	***1.60E+03***	***1.98E+05***
Energy balance check			
Energy IN	kJ/h	1.98E+05	Closure %
Energy OUT	kJ/h	1.98E+05	100

\# *Enthalpy values exclude heats of formation*

B. Separator

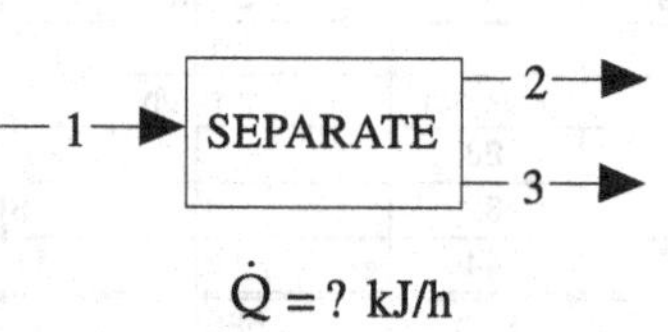

$\dot{Q} = ?$ kJ/h

Stream Table		Separator		Problem
Species	M	Stream		
		1	2	3
	kg/kmol	kmol/h		
A	20	4	3	1
B	30	8	4	4
C	40	12	4	8
Total	kg/h	800	340	460
Phase		L	G	L
Temp.	K	300	500	500
Press.	kPa(abs)	400	200	200
Enthalpy	kJ/h	?	?	?

Solution:

Energy Balance.

Properties	Species	A (liq)	A (gas)	B (liq)	B (gas)	C (liq)	C (gas)
$C_{p,m}$ wrt.298K	kJ/kmol.K	50	35	70	45	90	55
$h_{v,298K}$	kJ/kmol	-	2.00E+04	-	3.00E+04	-	1.00E+04

Reference condition = *compounds* at standard state, 298K.

$$0 = \Sigma[\dot{H}(\text{in})] - \Sigma[\dot{H}(\text{out})] + \dot{Q} - \dot{W} \qquad \text{For a separator:} \quad \dot{W} = 0$$

$$0 = \dot{H}(1) - \dot{H}(2) - \dot{H}(3) + \dot{Q} - \dot{W} \qquad [1]$$

where:

$$\dot{H}(i) = \Sigma[\dot{n}(i,j)h(j)]$$

$$h(j) = C_{p,m}(j)(T(i) - T_{ref}) + h_{p,Tref}(j) \qquad \text{[Respect the phase]}$$

$$\dot{H}(i) = (4\,\text{kmol/h})((50\,\text{kJ/kmol.K})(300\,\text{K} - 298\,\text{K}) + 0) + (8\,\text{kmol/h})((70\,\text{kJ/kmol.K})(300\,\text{K} - 298\,\text{K}) + 0)$$
$$+ (12\,\text{kmol/h})((90\,\text{kJ/kmol.K})(300\,\text{K} - 298\,\text{K}) + 0) \quad = 3.68\text{E+03} \quad \text{kJ/h}$$

$$\dot{H}(2) = (3\,\text{kmol/h})((35\,\text{kJ/kmol.K})(500\,\text{K} - 298\,\text{K}) + 2\text{E4})$$
$$+ (4\,\text{kmol/h})((45\,\text{kJ/kmol.K})(500\,\text{K} - 298\,\text{K}) + 3\text{E4})$$
$$+ (4\,\text{kmol/h})((55\,\text{kJ/kmol.K})(500\,\text{K} - 298\,\text{K}) + 1\text{E4})$$
$$= 1.02\text{E+05 kJ/h}$$

$$\dot{H}(3) = (1\,\text{kmol/h})((50\,\text{kJ/kmol.K})(500\,\text{K} - 298\,\text{K}) + 0)$$
$$+ (4\,\text{kmol/h})((70\,\text{kJ/kmol.K})(500\,\text{K} - 298\,\text{K}) + 0)$$
$$+ (8\,\text{kmol/h})((90\,\text{kJ/kmol.K})(500\,\text{K} - 298\,\text{K}) + 0)$$
$$= 2.12\text{E+05 kJ/h}$$

Solve equation [1] for

$$\dot{Q} = \dot{H}(2) + \dot{H}(3) - \dot{H}(1) = \mathbf{3.10E+05\ kJ/h}$$
$$= \mathbf{86.2\ kW}$$

Stream Table		Separator		Solution
Species	M	Stream		
		1	2	3
	kg/kmol	kmol/h		
A	20	4	3	1
B	30	8	4	4
C	40	12	4	8
Total	kg/h	800	340	460
Phase		L	G	L
Temp.	K	300	500	500
Press.	kPa(abs)	400	200	200
Enthalpy#	kJ/h	***3.68E+03***	***1.02E+05***	***2.12E+05***
Energy balance check				
Energy IN	kJ/h	3.14E+05	Closure %	
Energy OUT	kJ/h	3.14E+05	100.0	

 # Enthalpy values exclude heats of formation

EXAMPLE 5.08 Energy balances without chemical reaction (open system at steady-state).
Using enthalpy values from thermodynamic diagrams and tables

A. Mixer

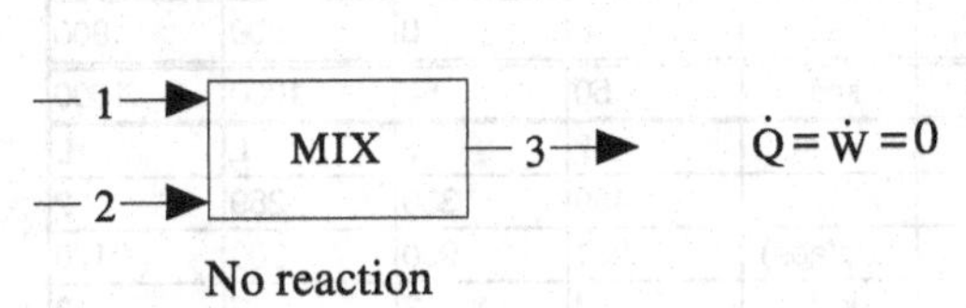

Stream Table	Mixer			Problem
Species	M	Stream		
		1	2	3
		kg/h		
[A]H_2O	18	50	10	60
[B]H_2SO_4	98	0	90	90
Total	kg/h	50	100	150
Phase		L	L	L
Temp	K	311	298	?
Press	kPa(abs)	150	150	140
Enthalpy	kJ/h	?	?	?

Solution:
Energy Balance.
Reference condition = H_2SO_4 and $H_2O(l)$ at 273 K, 101.3 kPa(abs).

$$0 = \Sigma[\dot{H}(in)] - \Sigma[\dot{H}(out)] + \dot{Q} - \dot{W} \quad [1]$$

$$0 = \dot{H}(1) + \dot{H}(2) - \dot{H}(3) + 0 - 0$$

where:

$$\dot{H}(i) = \overline{m}(i)h(i)$$

Values of h(i) are taken from the enthalpy-concentration chart for $H_2SO_4 - H_2O$ *Figure 2.06*

Stream 1: H_2O liquid at 311K, 150 kPa(abs)

$$\dot{H}(1) = (50\ kg/h)(149\ kJ/kg) = 7450 \quad kJ/h$$

Stream 2: 90 wt% $H_2SO_4 - H_2O$ liquid at 289 K, 150 kPa(abs)

$$\dot{H}(2) = (100\ kg/h)(-186\ kJ/kg) = -18600 \quad kJ/h$$

Stream Table	Mixer			Solution
Species	M	Stream		
		1	2	3
		kg/h		
[A]H_2O	18	50	10	60
[B]H_2SO_4	98	0	90	90
Total	kg/h	50	100	150
Phase		L	L	L
Temp	K	310	298	***344***
Press	kPa(abs)	150	150	140
Enthalpy#	kJ/h	***7450***	***-18600***	***-11150***
Energy balance check				
Energy IN	kJ/h	-11150	Closure %	
Energy OUT	kJ/h	-11150	100.0	

Stream 3: 60 wt% $H_2SO_4 - H_2O$ at T(3), 140 kPa(abs)

$$\dot{H}(3) = (150\ kg/h)(h(3)\ kJ/kg) = f(T(3))\ kJ/h$$

Solve equation [1] for
h(3) = - 74 kJ/kg liquid

By examination of the enthalpy-concentration chart, *Figure 2.06:*
T(3) = **<u>344 K</u>**

 # Enthalpy values <u>exclude</u> heats of formation

B. Heat exchanger (indirect contact)

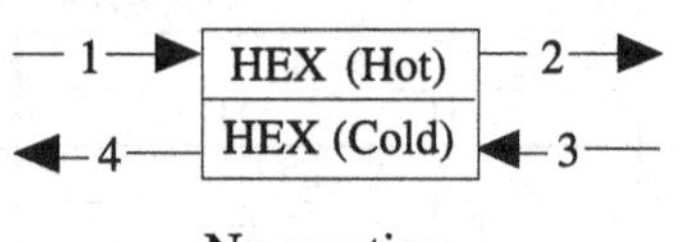

No reaction

Stream Table		Heat Exchanger			Problem
Species	M	Stream			
		1	2	3	4
	kg/kmol	kg/h			
[A]H_2O	18	50	50	100	100
[B]H_2SO_4	98	0	0	900	900
Total	kg/h	50	50	1000	1000
Phase		?	?	L	L
Temp.	K	450	350	289	?
Press.	kPa(abs)	931	900	200	170
Enthalpy	kJ/h	?	?	?	?

Solution:

Energy balance. Reference conditions:

Streams 1 and 2 *compounds* at triple point of water = 273.16 K, 0.61 kPa(abs)

Streams 3 and 4 *compounds* at 273 K, 101.3 kPa(abs)

$$0 = \Sigma[\dot{H}(in)] - \Sigma[\dot{H}(out)] + \dot{Q} - \dot{W}$$ For a heat exchanger $\dot{Q} = \dot{W} = 0$ i.e.

$$0 = \dot{H}(1) - \dot{H}(2) + \dot{H}(3) - \dot{H}(4) + 0 - 0$$ [overall HEX balance] [1]

where: $\dot{H}(i) = \overline{m}(i)h(i)$ h(i) = specific enthalpy

For streams 1 and 2 values of h(i) are taken from the steam table (*Table 2.20*). For streams 3 and 4 values of h(i) are taken from the $H_2SO_4 - H_2O$ enthalpy-concentration chart (*Figure 2.06*). Note that different reference conditions are acceptable here because streams 1–2 and 3–4 are kept separate and the individual reference conditions cancel from equation [1].

Stream 1: From the steam table: at 450 K the vapour pressure of H_2O = 931.5 kPa(abs)

Since P(1) < p*H_2O at 450 K, 931 kPa(abs), stream 1 is a ***GAS.***

$$\dot{H}(1) = (50\ kg/h)(2775\ kJ/kg) = 138750\ kJ/h$$

Stream Table		Heat Exchanger			Solution
Species	M	Stream			
		1	2	3	4
	kg/kmol	kg/h			
[A]H_2O	18	50	50	100	100
[B]H_2SO_4	98	0	0	900	900
Total	kg/h	50	50	1000	1000
Phase		***G***	***L***	L	L
Temp.	K	450	350	289	***355***
Press.	kPa(abs)	931	900	200	170
Enthalpy[#]	kJ/h	***1.39E+05***	***1.62E+04***	***-1.86E+05***	***-6.34E+04***
Energy balance check					
Energy IN	kJ/h	-47250	Closure %		
Energy OUT	kJ/h	-47250	100.0		

[#] *Enthalpy values exclude heats of formation*

Stream 2: From the steam table:

At 350 K the vapour pressure of H_2O = 41.7 kPa(abs)

Since P(2) > p*H_2O at 350 K, 900 kPa(abs), stream 2 is a ***LIQUID.***

$$\dot{H}(2) = (50\ kg/h)(323\ kJ/kg) = 16150\ kJ/h$$

Stream 3: 90 wt% H_2SO_4 at 289 K, 200 kPa(abs)

$$\dot{H}(3) = (1000\ kg/h)(-186\ kJ/kg) = -186000\ kJ/h$$

Stream 4: 90 wt% H_2SO_4 at T(4) K, 170 kPa(abs)

$$\dot{H}(4) = (1000\ kg/h)(h(4)) = f(T(4))\ kJ/h$$

Solve equation [1] for h(4) = - 63 kJ/kg

By examination of the $H_2SO_4 - H_2O$ enthalpy concentration chart (*Figure 2.06*), **T(4) = 355 K**

EXAMPLE 5.09 Energy balance with chemical reaction (open system at steady-state).

Using enthalpy values based on both the *element* and the *compound* reference states.

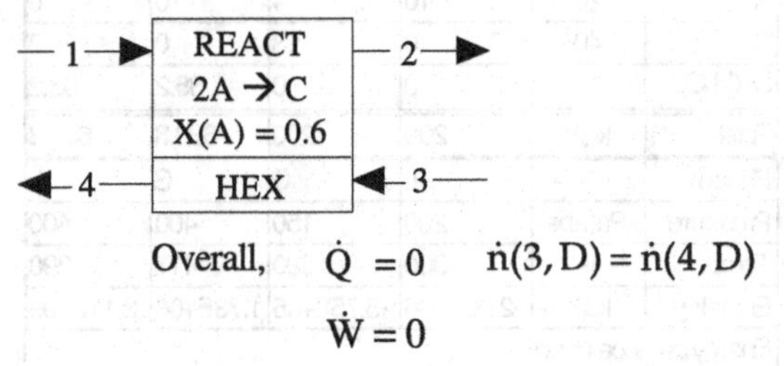

Stream Table		Reactor with Steam Heating			Problem
Species	M	Stream			
		1	2	3	4
	kg/kmol	kmol/h			
A	20	10	4	0	0
C	40	0	3	0	0
D (H_2O)	18	0	0	?	?
Total	kg/h	200	200	?	?
Phase		L	G	G	L
Pressure	kPa(abs)	200	150	400	400
Temp.	K	300	380	417	390
Enthalpy	kJ/h	-2.00E+06	-5.75E+05	?	?

Solution: Energy Balance. Reference conditions:

Streams 1 and 2 (h*) *Elements* at standard state, 298 K
Streams 3 and 4 (h) *Compound* liquid water at triple-point, 273 K

$$0 = \Sigma[\dot{H}(in)] - \Sigma[\dot{H}(out)] + \dot{Q} - \dot{W}$$

$$0 = \dot{H}^*(1) + \dot{H}(3) - \dot{H}^*(2) - \dot{H}(4) + 0 - 0$$

Overall balance.
Note that the two separate reference states both cancel out of this energy balance.

Reference state cancels out   Reference state cancels out

where: Reference State

$\dot{H}^*(i) = \Sigma[\dot{n}(i,j)h^*(j)]$ and $h^*(j) = C_{p,m}(j)(T(i) - T_{ref}) + h^o_{f,Tref}(j)$ *Elements* [Respect the phase]

$\dot{H}(i) = \Sigma[\dot{n}(i,j)h(j)]$ and $h(j) = C_{p,m}(j)(T(i) - T_{ref}) + h_{p,Tref}(j)$ *Compound* [Respect the phase]

$\dot{H}^*(1) = (10\ \text{kmol/h})((50\ \text{kJ/kmol.K})(300\ \text{K} - 298\ \text{K}) + (\text{-2.0E5}))$
$+ (0\ \text{kmol/h})((90\ \text{kJ/kmol.K})(300\ \text{K} - 298\ \text{K}) + (\text{3.0E4}))$ = -2.00E+06 kJ/h

$\dot{H}^*(2) = (4\ \text{kmol/h})((35\ \text{kJ/kmol.K})(300\ \text{K} - 298\ \text{K}) + (\text{-1.8E5}))$
$+ (3\ \text{kmol/h})((55\ \text{kJ/kmol.K})(300\ \text{K} - 298\ \text{K}) + (\text{4.0E4}))$ = -5.75E+05 kJ/h

$\dot{H}(3) = (18\ \text{kg/kmol})(\dot{n}(3,E)\ \text{kmol/h})(2739^{\#}\ \text{kJ/kg})$ = 4.93E+04 $\dot{n}(3,D)$ kJ/h

$\dot{H}(4) = (18\ \text{kg/kmol})(\dot{n}(4,E)\ \text{kmol/h})(490.4^{\#}\ \text{kJ/kg})$ = 8.83E+03 $\dot{n}(4,D)$ kJ/h

 # = specific enthalpy from the steam table (<u>excludes</u> heat of formation)

Solve equation [1] for

$\dot{n}(3,D) = \dot{n}(4,D) = \underline{\mathbf{35.2\ kmol/h}}$

COMMENT:
The energy balance finds the flow of condensing steam required to supply the heat of the endothermic reaction, with incomplete conversion, plus to vaporise the reaction product mixture.

Steam heating utility load:

$\dot{H}^*(2) - \dot{H}^*(1) = \dot{H}(3) - \dot{H}(4) = \underline{\mathbf{396\ \ kW}}$

Stream Table		***Reactor with Steam Heating Solution***			
Species	M	Stream			
		1	***2***	***3***	***4***
	kg/kmol	kmol/h			
A	20	10	4	0	0
C	40	0	3	0	0
D (H_2O)	18	0	0	35.2	35.2
Total	kg/h	200	200	633.3	633.3
Phase		L	G	G	L
Pressure	kPa(abs)	200	150	400	400
Temp.	K	300	380	417	390
Enthalpy	kJ/h	-2.00E+06	-5.75E+05	1.73E+06	3.11E+05
Energy balance check					
Energy IN	kJ/h	-2.64E+05			
Energy OUT	kJ/h	-2.64E+05 Closure = 100%			

EXAMPLE 5.10 Energy balance on a multi-unit recycle process (open system at steady-state).

Problem: Complete the stream table energy balance.

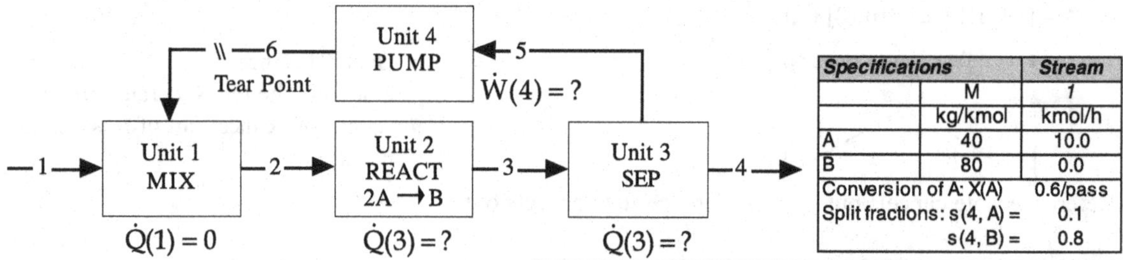

Specifications		***Stream***
	M	*1*
	kg/kmol	kmol/h
A	40	10.0
B	80	0.0
Conversion of A: X(A)		0.6/pass
Split fractions: s(4, A) =		0.1
s(4, B) =		0.8

Properties	**Species**	**A (liq)**	**A (gas)**	**B (liq)**	**B (gas)**
$C_{p,m}$ (ref.298K)	kJ/kmol.K	50	35	70	45
$h_{f,298K}$	kJ/kmol	-2.0E+05	-1.8E+05	-1.0E+05	-7.0E+04
$h_{v,298K}$	kJ/kmol	-	2.00E+04	-	3.00E+04
Density	kg/m^3	900	-	900	-

Solution:

Energy balance. Reference condition = *elements* at standard state, 298K.

$$0 = \Sigma[\dot{H}^*(in)] - \Sigma[\dot{H}^*(out)] + \dot{Q} - \dot{W}$$

where:

$$\dot{H}^*(i) = \Sigma\,[n(i,j)\,h^*(j)]$$

$$h^*(j) = C_{p,m}(j)(T(i) - T_{ref}) + h^o_{f,Tref}(j)$$

[Respect the phase]

Stream Table		***Mixer-reactor-separator + recycle***					***Problem***
Species	M	Stream					
		1	***2***	***3***	***4***	***5***	***6***
	kg/kmol	kmol/h					
A	40	10.00	15.63	**6.25**	**0.63**	**5.63**	**5.63**
B	80	0.00	1.17	**5.86**	**4.69**	**1.17**	**1.17**
Total	*kg/h*	*400*	*719*	*719*	*400*	*319*	*319*
Phase		L	L	G	L	L	L
Temp.	K	300	*?*	500	400	400	*?*
Press.	kPa(abs)	200	180	150	130	130	200
Volume	m^3/h	*?*	*?*	*?*	*?*	*?*	*?*
Enthalpy	kJ/h	*?*	*?*	*?*	*?*	*?*	*?*

$$\dot{H}^*(1) = (10.00\ \text{kmol/h})[(50\ \text{kJ/kmol.K})(300\ \text{K} - 298\ \text{K}) + (\text{-}2.0\text{E}5\ \text{kJ/kmol})]$$
$$+ (0.00\ \text{kmol/h})[(70\ \text{kJkmol.K})(300\ \text{K} - 298\ \text{K}) + (\text{-}1.0\text{E}5\ \text{kJ/kmol})] = \text{-}2.00\text{E+}06 \quad \text{kJ/h}$$

$$\dot{H}^*(2) = (15.59\ \text{kmol/h})[(50\ \text{kJ/kmol.K})(T(2)\ \text{K} - 298\ \text{K}) + (\text{-}2.0\text{E}5\ \text{kJ/kmol})]$$
$$+ (1.16\ \text{kmol/h})[(70\ \text{kJkmol.K})(T(2)\ \text{K} - 298\ \text{K}) + (\text{-}1.0\text{E}5\ \text{kJ/kmol})] = f(T(2)) \quad \text{kJ/h}$$

$$\dot{H}^*(3) = (6.24\ \text{kmol/h})[(35\ \text{kJ/kmol.K})(500\ \text{K} - 298\ \text{K}) + (\text{-}1.8\text{E}5\ \text{kJ/kmol})]$$
$$+ (5.84\ \text{kmol/h})[(45\ \text{kJkmol.K})(500\ \text{K} - 298\ \text{K}) + (\text{-}7.0\text{E}4\ \text{kJ/kmol})] = \text{-}1.44\text{E+}06 \quad \text{kJ/h}$$

$$\dot{H}^*(4) = (0.62\ \text{kmol/h})[(50\ \text{kJ/kmol.K})(400\ \text{K} - 298\ \text{K}) + (\text{-}2.0\text{E}5\ \text{kJ/kmol})]$$
$$+ (4.67\ \text{kmol/h})[(70\ \text{kJkmol.K})(400\ \text{K} - 298\ \text{K}) + (\text{-}1.0\text{E}5\ \text{kJ/kmol})] = \text{-}5.57\text{E+}05 \quad \text{kJ/h}$$

$$\dot{H}^*(5) = (5.61\ \text{kmol/h})[(50\ \text{kJ/kmol.K})(400\ \text{K} - 298\ \text{K}) + (\text{-}2.0\text{E}5\ \text{kJ/kmol})]$$
$$+ (1.17\ \text{kmol/h})[(70\ \text{kJkmol.K})(400\ \text{K} - 298\ \text{K}) + (\text{-}1.0\text{E}5\ \text{kJ/kmol}) = \text{-}1.21\text{E+}06 \quad \text{kJ/h}$$

$$\dot{H}^*(6) = (6.24\ \text{kmol/h})[(50\ \text{kJ/kmol.K})(T(6)\ \text{K} - 298\ \text{K}) + (\text{-}2.0\text{E}5\ \text{kJ/kmol})]$$
$$+ (5.84\ \text{kmol/h})[(70\ \text{kJkmol.K})(T(6)\ \text{K} - 298\ \text{K}) + (\text{-}1.0\text{E}5\ \text{kJ/kmol})] = f(T(6)) \quad \text{kJ/h}$$

Iterative sequential modular solution of recycle material and energy balance. For unit "k":

$$0 = \Sigma[\dot{H}^*(i)]\text{in} - \Sigma[\dot{H}^*(i)]\text{out} + \dot{Q}(k) - \dot{W}(k)$$

<u>UNIT 1 (Mixer)</u>: $\dot{Q}(1) = 0$ $\dot{W}(1) = 0$ $\quad 0 = \dot{H}^*(1) + \dot{H}^*(6) - \dot{H}^*(2) + 0 - 0$ [1]

<u>UNIT 2 (Reactor)</u>: $\dot{Q}(2) = ?$ $\dot{W}(2) = 0$ $\quad 0 = \dot{H}^*(2) + \dot{H}^*(3) + \dot{Q}(2) - 0$ [2]

<u>UNIT 3 (Separator)</u>: $\dot{Q}(3) = ?$ $\dot{W}(3) = 0$ $\quad 0 = \dot{H}^*(3) - \dot{H}^*(4) - \dot{H}^*(5) + \dot{Q}(3) - 0$ [3]

<u>UNIT 4 (Pump)</u>: $\dot{Q}(4) = 0$ $\dot{W}(4) = ?$ $\quad 0 = \dot{H}^*(5) - \dot{H}^*(6) + 0 - \dot{W}(4)$ [4]

For each liquid stream: $\dot{V}(i) = \overline{m}(i)/\rho(i)$ e.g. $\dot{V}(1) = (400\ \text{kg/h})/(900\ \text{kg/m}^3) = 0.44$ m^3/h

For each gas stream (ideal gas): $\dot{V}(i) = \overline{n}(i)RT(i)/P(i)$

e.g. $\dot{V}(3) = (12.08\ \text{kmol/h})(8.31\ \text{kJ/kmol.K})(500\ \text{K})/(150\ \text{kPa}) = 336$ m^3/h

Stream Table		***Mixer-reactor-separator + recycle***					***Solution***
Species	M	Stream					
		1	***2***	***3***	***4***	***5***	***6***
	kg/kmol	kmol/h					
A	40	10.00	15.63	**6.25**	**0.63**	**5.63**	**5.63**
B	80	0.00	1.17	**5.86**	**4.69**	**1.17**	**1.17**
Total	*kg/h*	*400*	*719*	*719*	*400*	*319*	*319*
Phase		L	L	G	L	L	L
Temp.	K	300	***342***	500	400	400	***400.01***
Press.	kPa(abs)	200	180	150	130	130	200
Volume	m^3/h	***0.44***	***0.80***	***336***	***0.44***	***0.35***	***0.35***
Enthalpy*	kJ/h	***-2.00E+06***	***-3.20E+06***	***-1.44E+06***	***-5.57E+05***	***-1.21E+06***	***-1.21E+06***
Energy balance check			Unit 1	Unit 2	Unit 3	Unit 4	Overall
Energy IN	kJ/h		-3.20E+06	-1.44E+06	-1.76E+06	-1.21E+06	-5.57E+05
Energy OUT	kJ/h		-3.20E+06	-1.44E+06	-1.76E+06	-1.21E+06	-5.57E+05
Closure %			100.0	100.0	100.0	100.0	100.0

** Enthalpy values include heats of formation*

$\dot{W}(4) = \text{-}(0.44\ \text{m}^3/\text{h})(200\ \text{kPa} - 130\ \text{kPa})$

$= \text{-}31$ kJ/h

Solve equation [4] for T(6) = <u>**400.01**</u> K

Solve equation [1] for T(2) = <u>**342**</u> K

Solve equation [2] for $\dot{Q}(2)$

= <u>**1.77E+06**</u> kJ/h

= 4.91E+02 kW heating

Solve equation [3] for $\dot{Q}(3)$

= <u>**-3.25E+05**</u> kJ/h

= -9.02E+01 kW cooling

EXAMPLE 5.11 Material and energy balance on a biochemical process (ethanol from glucose).

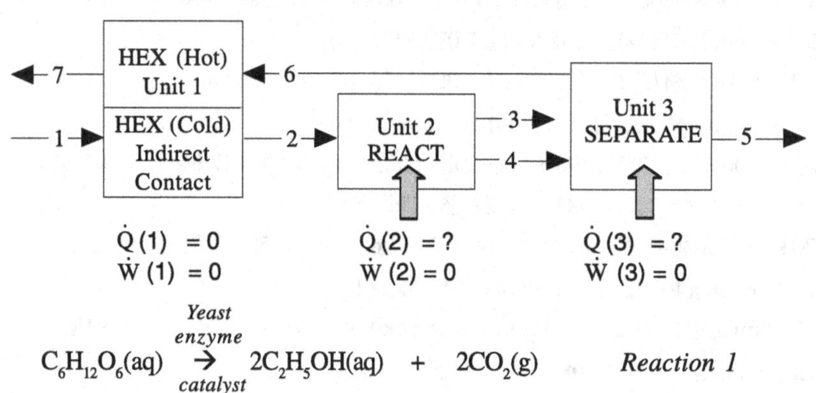

The figure shows a simplified flowsheet of a biochemical process for the production of ethanol by fermentation of glucose. In this process a solution of glucose in water *(Stream 1)* is preheated prior to delivery to a reactor *(Unit 2)* where glucose is converted to ethanol by *Reaction 1*. Carbon dioxide gas is released from the reactor *(Stream 3)* and the reaction liquid product mixture *(Stream 4)* is separated by evaporation *(Unit 3)* to a waste liquor *(Stream 5)* and an ethanol rich vapour *(Stream 6)*, which is used to preheat the process feed *(Stream 1)* in a heat exchanger *(Unit 1)*.

The process specifications are as follows:

Stream 1:	5 wt% liquid solution of glucose in water	40E3 kg/h		140 kPa(abs), 288 K
Stream 2:	5 wt% liquid solution of glucose in water	40E3 kg/h		110 kPa(abs), T(2) K
Stream 3:	CO_2 gas with ethanol and water vapour		Flow unspecified	110 kPa(abs), 313 K
Stream 5:	Liquid solution of glucose + ethanol in water	Zero CO_2	Flow unspecified	110 kPa(abs), 363 K
Stream 6:	Vapour mixture of ethanol + water	96 wt% ethanol	Flow unspecified	140 kPa(abs), 363 K
Stream 7:	Liquid mixture of ethanol + water	96 wt% ethanol	Flow unspecified	110 kPa(abs), 298 K
Unit 2:	Conversion of glucose = 90%			
Unit 3:	Separation efficiency of ethanol from Stream 4 into Stream 6 = 95%			

Properties	Species	$C_6H_{12}O_6$	$C_2H_5OH(l)$	$C_2H_5OH(g)$	$CO_2(g)$	$H_2O(l)$	$H_2O(g)$
$C_{p,m}$ (ref.298K)	kJ/kmol.K	100	66	***40***	39	75	34
$h_{f,298K}$	kJ/kmol	-7.0E+05	-2.78E+05	-2.35E+05	-3.94E+05	-2.86E+05	-2.42E+05
Density	kg/m^3	1500	789	gas law	gas law	1000	gas law

Vapour pressure of water: $p^* = \exp[16.5362 - 3985.44/(T - 38.9974)]$ kPa, T = K

Vapour pressure of ethanol: $p^* = \exp[16.1952 - 3423.53/(T - 55.7152)]$ kPa, T = K

Assume that ethanol-water forms an ideal liquid mixture that follows Raoult's law and has negligible heat of mixing.

Problem:

A. Make a degrees of freedom analysis of the problem.

B. Use a spreadsheet to calculate the stream table for the steady-state material and energy balance.

Solution:

Material balance

In this example the material balance is independent of the energy balance. The material balance problem is fully defined and is solved by the methods of Chapter 4, using mole balances on each species, stream compositions and reaction stoichiometry to give the following stream table. The compositions of streams 3 and 4 are obtained by iterating on the vapour–liquid equilibria.

Vapour pressure at 313 K; Ethanol = 18.0 kPa, Water = 7.3 kPa

Stream Table		*Bio-synthesis of Ethanol*			*Material Balance*			*Solution*
Species	M				Stream			
		1	2	3	4	5	6	7
	kg/kmol				kmol/h			
[A] $C_6H_{12}O_6$	180	11.11	11.11	0.00	1.11	1.11	0.00	0.00
[B] C_2H_5OH	46	0.00	0.00	0.03	19.97	1.00	18.97	18.97
[C] CO_2	44	0.00	0.00	20.00	0.00	0.00	0.00	0.00
[D] H_2O	18	2111.11	2111.11	1.41	2109.70	2107.70	2.00	2.00
Total	kg/h	*40000*	*40000*	*907*	*39093*	*38185*	*909*	*909*
Mass balance check		*Unit 1*	*Unit 2*	*Unit 3*	*Overall*			
Mass IN		40909	40000	39093	40000			
Mass OUT		40909	40000	39093	40000			
Closure %		100.0	100.0	100.0	100.0			

Energy balance

A. Examine the specification of each unit and of the overall process (see *Table 5.03*).

	Unit 1	**Unit 2**	**Unit 3**	**OVERALL**
Number of unknowns (stream + unit variables)	2I + 2 = 10	2I + 2 = 8	2I + 2 = 8	2I + 2 = 10
Number of independent equations (see below)	10	8	8	10
D of F	0	0	0	0

Overall unknowns = 2I + 2 = (2)(4) + 2 = 10

Independent equations = 1 energy balance + 1 work + 8 specified T and P = 10 FULLY-SPECIFIED

B. Write the equations.

Reference condition = *elements* at standard state, 298K.

$$\dot{H}^*(i) = \Sigma[\dot{n}(i,j)(C_{p,m}(j)(T(i) - T_{ref}) + h^o_{f,Tref}(j))]$$ [Respect the phase]

UNIT 1 (Heat exchanger)

$$0 = \dot{H}^*(1) + \dot{H}^*(6) - \dot{H}^*(2) - \dot{H}^*(7) + \dot{Q}(1) - \dot{W}(1) \quad [1]$$

Stream conditions:

Stream 1: $P(1) = 140\,kPa$, $T(1) = 288\,K$ [2][3]
Stream 2: $P(2) = 110\,kPa$, [4]
Stream 6: $P(6) = 140\,kPa$, $T(6) = 363\,K$ [5][6]
Stream 7: $P(7) = 110\,kPa$, $T(7) = 298\,K$ [7][8]

Unit loads: $\dot{Q}(1) = 0$ $\dot{W}(1) = 0$ [9][10]

UNIT 2 (Reactor)

$$0 = \dot{H}^*(2) - \dot{H}^*(3) - \dot{H}^*(4) + \dot{Q}(2) - \dot{W}(2) \quad [11]$$

Stream conditions:

Stream 3: $P(3) = 110\,kPa$, $T(3) = 313\,K$ [12][13]
Stream 4: $P(4) = 110\,kPa$, $T(4) = 313\,K$ [14][15]

Unit loads: $\dot{W}(2) = 0$ [16]

UNIT 3 (Separator)

$$0 = \dot{H}^*(4) - \dot{H}^*(5) - \dot{H}^*(6) + \dot{Q}(3) - \dot{W}(3) \quad [17]$$

Stream conditions:

Stream 5: $P(5) = 110\,kPa$, $T(5) = 363\,K$ [18][19]

Unit loads: $\dot{W}(3) = 0$ [20]

Stream Table		***Bio-synthesis of Ethanol***			***Material and Energy Balance***			***Solution***
Species	M	Stream						
		1	*2*	*3*	*4*	*5*	*6*	*7*
	kg/kmol	kmol/h						
[A] $C_6H_{12}O_6$	180	11.11	11.11	0.00	1.11	1.11	0.00	0.00
[B] C_2H_5OH	46	0.00	0.00	0.03	19.97	1.00	18.97	18.97
[C] CO_2	44	0.00	0.00	20.00	0.00	0.00	0.00	0.00
[D] H_2O	18	2111.11	2111.11	1.41	2109.70	2107.70	2.00	2.00
Total	kg/h	*40000*	*40000*	*907*	*39093*	*38185*	*909*	*909*
Phase		L	L	G	L	L	G	L
Temp.	K	288	***294***	313	313	363	363	298
Press.	kPa(abs)	140	110	110	110	110	140	110
Volume	m^3/h	39.3	39.3	507	39.3	38.1	452	1.1
Enthalpy*	kJ/h	-6.13E+08	-6.12E+08	-8.22E+06	-6.07E+08	-5.94E+08	-4.89E+06	-5.84E+06
Mass balance check		*Unit 1*	*Unit 2*	*Unit 3*	*Overall*			
Mass IN	kg/h	40909	40000	39093	40000			
Mass OUT	kg/h	40909	40000	39093	40000			
Closure %		100.0	100.0	100.0	100.0			
Energy balance check								
Energy IN	kJ/h	-6.18E+08	-6.16E+08	-5.98E+08	-6.08E+08			
Energy OUT	kJ/h	-6.18E+08	-6.16E+08	-5.98E+08	-6.08E+08			
Closure %		100.0	100.0	100.0	100.0			

** Enthalpies include heats of formation*

EXAMPLE 5.12 Material and energy balance on an electrochemical process (hydrogen/air fuel cell).

This figure shows a simplified flowsheet for a hydrogen/air fuel cell used to power a bus. In this process atmospheric air *(Stream 1 – Stream 2)* is taken by an on-board compressor *(Unit 1)* and delivered to the fuel cell *(Unit 2)*, where part of the O_2 is consumed by the fuel cell *Reaction 1*. Exhaust air from the fuel cell *(Stream 3)* containing the reaction product water is passed through a separator *(Unit 3)* where part of the water is recovered then recycled *(Stream 5)* to cool the fuel cell and to keep the fuel cell membrane properly humidified. Hydrogen from on-board compressed gas tanks *(Stream 6)* is delivered to the fuel cell where it undergoes partial conversion by *Reaction 1*, to generate electric power [$\dot{W}(2)$]. In this example both the exhaust air *(Stream 4)* and the unconverted hydrogen *(Stream 7)* are rejected to atmosphere, but in practice the energy in these streams would be recovered to help run the fuel cell.

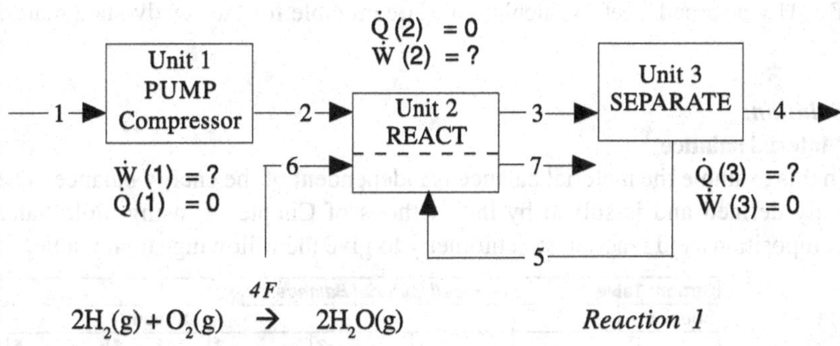

$$2H_2(g) + O_2(g) \xrightarrow{4F} 2H_2O(g) \qquad \textit{Reaction 1}$$

NOTE: The fuel cell is an electrochemical reactor in which the air and the hydrogen reactant streams are kept apart by a water permeable proton exchange membrane.

Stream 1:	Air at 30% relative humidity	Flow unspecified	100 kPa(abs), 290 K
Stream 2:	Air	Flow and T(2) unspecified	350 kPa(abs), T(2)
Stream 3:	Exhaust "air" + H_2O	Flow and phase unspecified	320 kPa(abs), 370 K
Stream 4:	Exhaust "air" + H_2O	Flow and phase unspecified	310 kPa(abs), 350 K
Stream 5:	Recycle liquid H_2O	Flow to match the electro-osmotic flux	310 kPa(abs), 350 K
Stream 6:	20 kg/h H_2		310 kPa(abs), 290 K
Stream 7:	Exhaust H_2 saturated with H_2O vapour	Flow unspecified	280 kPa(abs), 370 K

Air stoichiometry O_2 feed = 3 times the stoichiometric rate for *Reaction 1*
H_2 conversion = 90%
Fuel cell operating voltage= 0.6 Volt/cell
Electro-osmotic flux = 2 moles H_2O per mole H^+ crossing the membrance

Properties	Species	$O_2(g)$	$N_2(g)$	$H_2(g)$	$H_2O(l)$	$H_2O(g)$
$C_{p,m}$ (ref.298K)	kJ/kmol.K	30	29	29	75	34
$h_{f,298K}$	kJ/kmol	0.00E+00	0.00E+00	0.00E+00	-2.86E+05	-2.42E+05
Density	kg/m^3	Ideal gas	Ideal gas	Ideal gas	1000	Ideal gas

Problem:

A. Make a degrees of freedom analysis of the problem.

B. Use a spreadsheet to calculate the stream table for the steady-state material and energy balance.

Solution:

Material balance

In this example the material balance is independent of the energy balance. The material balance problem is fully defined and is solved by the methods of Chapter 4, using mole balances on each species, stream compositions and reaction stoichiometry to give the following stream table.

Stream Table		*Fuel-cell Material Balance*						*Solution*
Species	M	Stream						
		1	2	3	4	5	6	7
	kg/kmol	kmol/h						
[A] O_2	32	13.50	13.50	9.00	9.00	0.00	0.00	0.00
[B] N_2	28	50.79	50.79	50.79	50.79	0.00	0.00	0.00
[C] H_2	2	0.00	0.00	0.00	0.00	0.00	10.00	1.00
[D] H_2O	18	1.25	1.25	45.78	9.78	36.00	0.00	0.47
Total	kg/h	*1876*	*1876*	*2534*	*1886*	*648*	*20*	*10*

Energy Balance

A. Examine the specification of each unit and of the overall process.

	Unit 1	Unit 2	Unit 3	OVERALL
Number of unknowns (stream + unit variables)	2I + 2 = 6	2I + 2 = 12	2I + 2 = 8	2I + 2 = 10
Number of independent equations (see below)	6	11	8	10
D of F	0	1	0	0

Overall Energy Balance

Unknowns = 2I + 2 = (2)(4) + 2 = 10

Independent equations = 1 energy balance + 1 work + 8 specified T and P = 10 FULLY-SPECIFIED

B. Write the equations. *Reference condition = elements* at standard state, 298K.

$$\dot{H}^*(i) = \Sigma[\dot{n}(i,j)(C_{p,m}(j)(T(i) - T_{ref}) + h^o_{f,Tref}(j))]$$ [Respect the phase]

UNIT 1 (Pump)

$$0 = \dot{H}^*(1) - \dot{H}^*(2) + \dot{Q}(1) - \dot{W}(1)$$ [1]

Stream conditions:

Stream 1: $P(1) = 100\,kPa$, $T(1) = 290\,K$ [2][3]

Stream 2: $P(2) = 350\,kPa$, [4]

Compressor: $\dot{Q}(1) = 0$ $\dot{W}(1) = -P(1)\ \dot{V}(1)(r/(r-1))\ [((P(2)/P(1))^{((r-1)/r)} - 1]$ = **-2.36E+05 kJ/h** [5][6]

UNIT 2 (Reactor)

$$0 = \dot{H}^*(2) + \dot{H}^*(5) + \dot{H}^*(6) - \dot{H}^*(3) - \dot{H}^*(7) + \dot{Q}(2) - \dot{W}(2) \qquad [7]$$

Stream conditions:

Stream 3:	$P(3) = 320\,kPa$,	$T(3) = 370\,K$	[8][9]
Stream 5:	$P(5) = 310\,kPa$,	$T(5) = 350\,K$	[10][11]
Stream 6:	$P(6) = 310\,kPa$,	$T(6) = 290\,K$	[12][13]
Stream 7:	$P(7) = 280\,kPa$,	$T(7) = 370\,K$	[14][15]

Faraday's law: $I' = 2F(\dot{n}(6,C) - \dot{n}(7,C))/3600 = (2)(96480)(10-1)/3600 = 482\,kA$

Electric power: $\dot{W}(2) = E_v I' = (0.6\ Volt)(482\ kA) = 289\ kW =$ **<u>1.04E+06 kJ/h</u>** [16]

UNIT 3 (Separator)

$$0 = \dot{H}^*(3) - \dot{H}^*(4) - \dot{H}^*(5) + \dot{Q}(3) - \dot{W}(3) \qquad [17]$$

Stream conditions:

Stream 4: $P(4) = 310\,kPa$, $T(4) = 350\,K$ [18][19]

Separator: $\dot{W}(3) = 0$ [20]

NOTE: Calculation of $\dot{H}^*(3)$ and $\dot{H}^*(4)$ requires the distribution of H_2O between the liquid and gas phase. This is obtained from the Antoine equation assuming that when $(y(D))P > p^*$, the gas phase of *Stream 3* and *Stream 4* is saturated with H_2O vapour.

Vapour pressure of H_2O: $p(D)^* = \exp[16.5362 - 3985.44/(T - 38.9974)]$ kPa [21]

Stream	$p^*(D)$(kPa)
3	89.6
4	41.3
7	89.6

Solve for :

$Q(2) =$ **<u>-5.94E+05</u>** kJ/h (cooling)

$Q(3) =$ **<u>-6.76E+05</u>** kJ/h (cooling)

Stream Table		***Fuel-cell Material and Energy Balance***					***Solution***	
Species	M	Stream						
		1	***2***	***3***	***4***	***5***	***6***	***7***
		kmol/h						
[A] O_2(g)	32	13.50	13.50	9.00	9.00	0.00	0.00	0.00
[B] N_2(g)	28	50.79	50.79	50.79	50.79	0.00	0.00	0.00
[C] H2(g)	2	0.00	0.00	0.00	0.00	0.00	10.00	1.00
[D] H_2O(g)	18	1.25	1.25	23.26	9.19	0.00	0.00	0.47
H_2O(l)	18	0.00	0.00	22.51	0.58	36.00	0.00	0.00
Total	kg/h	*1876*	*1876*	*2534*	*1886*	*648*	*20*	*10*
Phase		G	G	G+L	G+L	L	G	G
Temp.	K	290	***413***	*370*	350	350	290	*370*
Press.	kPa(abs)	100	350	320	310	310	310	280
Volume	m^3/h	1579	643	1014	508	0.648	78	16
Enthalpy*	kJ/h	-3.17E+05	-8.04E+04	-1.18E+07	-2.28E+06	-1.02E+07	-2.32E+03	-1.11E+05
Mass balance check			*Unit 1*	*Unit 2*	*Unit 3*	*Overall*		
Mass IN	kg/h		1876	2544	2534	1896		
Mass OUT	kg/h		1876	2544	2534	1896		
Closure %			100.0	100.0	100.0	100.0		
Energy balance check								
Energy IN	kJ/h		-8.04E+04	-1.08E+07	-1.24E+07	-1.59E+06		
Energy OUT	kJ/h		-8.04E+04	-1.08E+07	-1.24E+07	-1.59E+06		
Closure %			100.0	100.0	100.0	100.0		

* *Enthalpies <u>include</u> heats of formation*

EXAMPLE 5.13 Material and energy balance on a thermochemical process (oil fired boiler).

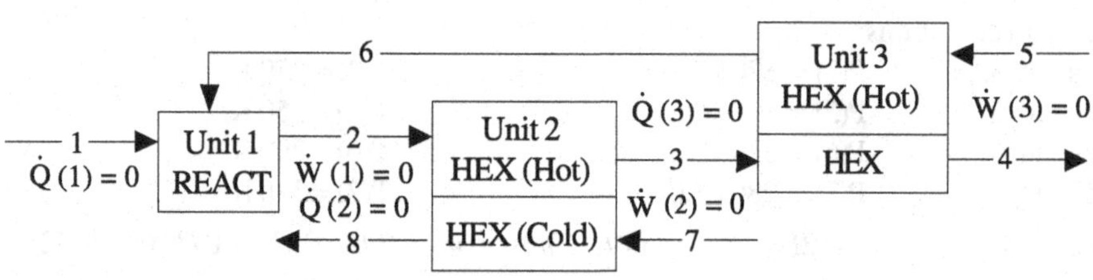

This figure shows a simplified flowsheet for part of a power generation cycle in an oil burning power station. In this process a fuel oil *(Stream 1)* is burned with excess preheated air *(Stream 6)* in a furnace *(Unit 1)*, where the oil undergoes complete combustion to CO_2, H_2O and SO_2 *(Stream 2)*. The hot combustion product gas then passes through a heat exchanger called a "boiler" *(Unit 2)* where heat is transferred to water inside steel tubes. Water enters the tubes as a liquid *(Stream 7)* and leaves as steam *(Stream 8)* which is subsequently used to drive turbines that generate electricity. The cooled combustion gas *(Stream 3)* passes through a second heat exchanger *(Unit 3)* where part of the remaining heat is transferred to the incoming air *(Stream 5)* to raise the thermal efficiency of the system. The exhaust gas *(Stream 4)* is then rejected through a stack into the atmosphere.
The process specifications are as follows:

Stream 1:	Fuel oil with empirical formula $CH_2S_{0.1}$	50,000 kg/h	130 kPa(abs), 298 K
Stream 2:	Combustion product gas mixture	Flow unspecified	110 kPa(abs), T(2) K
Stream 3:	Combustion product gas mixture	Flow unspecified	100 kPa(abs), 700 K
Stream 4:	Combustion product gas mixture	Flow unspecified	100 kPa(abs), 500 K
Stream 5:	Air at 60% relative humidity	20% excess for oil combustion	140 kPa(abs), 300 K
Stream 6:	Preheated air	Flow unspecified	130 kPa(abs), T(6) K
Stream 7:	Water	489,381 kg/h	2650 kPa(abs), 300 K
Stream 8:	Water	489,381 kg/h	2637 kPa(abs), T(8) K
Fuel oil gross heat of combustion		$h^{\circ}_{c,298K}$ = -3.00E+04	kJ/kg

Properties	Species	Fuel oil	$O_2(g)$	$N_2(g)$	$H_2O(l)$	$H_2O(g)$	$CO_2(g)$	$SO_2(g)$
$C_{p,m}$ (ref.298K)	kJ/kmol.K	-	31	30	75	35	42	43
$h_{f,298K}$	kJ/kmol	-	0.0E+00	0.0E+00	-2.86E+05	-2.42E+05	-3.94E+05	-2.97E+05
Density	kg/m^3	900	Ideal gas	Ideal gas	1000	Ideal gas	Ideal gas	Ideal gas

Problem:

A. Make a degrees of freedom analysis of the problem.

B. Use a spreadsheet to calculate the stream table for the steady-state material and energy balance.

Solution: **Material balance**

In this example the material balance is independent of the energy balance. The material balance problem is fully defined and is solved by the methods of Chapter 4, using atom balances on each element and the known stream compositions. The material balance solution is given in the top seven rows of the M&E balance stream table presented below.

Energy balance

A. Examine the specification of each unit and of the overall process.

	Unit 1	Unit 2	Unit 3	OVERALL
Number of unknowns (stream + unit variables)	2I + 2 = 8	2I + 2 = 10	2I + 2 = 10	2I + 2 = 12
Number of independent equations (see below)	7	9	10	12
D of F	1	1	0	0

Total number of stream + unit variables = 2I + 2K = (2)(8) + 2(3) = 22

Requires 22 independent equations

Overall Energy Balance: Unknowns = 2I + 2 = (2)(5) + 2 = 12

Independent equations = 1 energy balance + 1 heat + 1 work + 9 specified T and P = 12, FULLY-SPECIFIED.

B. Reference conditions:

Stream 1 to 6 ref. = *elements* at standard state 298 K

$$\dot{H}^*(i) = \sum[\dot{n}(i,j)(C_{p,m}(j)(T(i) - T_{ref}) + h^o_{f,Tref}(j))] \qquad \text{[Respect the phase]}$$

Stream 7 and 8 ref. = *compound* water (liquid) at its triple-point, i.e. 273.16 K, 0.61 kPa(abs)

$$\dot{H}(i) = \sum[\dot{n}(i,j)h(j)] \qquad \text{Specific enthalpy h(j) obtained from steam table}$$

Note that the different reference conditions are no problem here since they cancel from the energy balance.

UNIT 1 (Adiabatic reactor)

$$0 = \dot{H}^*(1) + \dot{H}^*(6) - \dot{H}^*(2) + \dot{Q}(1) - \dot{W}(1) \qquad [1]$$

Stream conditions.

Stream 1: P(1) = 130 kPa, T(1) = 300 K [2][3]

Stream 2: P(2) = 110 kPa, [4]

Stream 6: P(6) = 130 kPa, [5]

$\dot{Q}(1) = 0$ $\dot{W}(1) = 0$ [6][7]

UNIT 2 (Heat exchanger)

$$0 = \dot{H}^*(2) + \dot{H}(7) - \dot{H}^*(3) - \dot{H}(8) + \dot{Q}(2) - \dot{W}(2) \qquad [8]$$

Stream conditions:

Stream 3: P(3) = 100 kPa, T(3) = 700 K [9][10]

Stream 7: $P(7) = 2650$ kPa, $T(7) = 300$ K [11][12]

Stream 8: $P(8) = 2637$ kPa [13]

$\dot{Q}(2) = 0$ $\dot{W}(2) = 0$ [14][15]

<u>UNIT 3 (Heat exchanger)</u>

$$0 = \dot{H}^*(3) + \dot{H}^*(5) - \dot{H}^*(4) - \dot{H}^*(6) + \dot{Q}(3) - \dot{W}(3) \quad [16]$$

Stream conditions:

Stream 4: $P(4) = 100$ kPa, $T(4) = 500$ K [17][18]

Stream 5: $P(5) = 140$ kPa, $T(5) = 300$ K [19][20]

$\dot{Q}(3) = 0$ $\dot{W}(3) = 0$ [21][22]

Note that T(2) is effectively the *"adiabatic flame temperature"*, calculated on the simplifying assumptions of constant heat capacities and stable reaction products. These assumptions inflate T(2) by about 200 K.

Fuel oil (see *Equation 2.53* and *Reaction 5.01*)

$$h^o_{f,298K} = -(-30E3)(17.2) + [(1)(-3.94E5) + (2/2)(-2.86E5) + (0.1)(-2.97E5)] = -1.50E+05 \quad \text{kJ/kmol}$$

T(8) is found by solving the energy balance for the specific enthalpy of steam in Stream 8 = 2803 kJ/kg, and matching it in the steam table at P(8).

Stream Table		***Furnace (thermo-reactor) and Boiler***				***Material and Energy Balance***		***Solution***	
Species	M	Stream							
		1	***2***	***3***	***4***	***5***	***6***	***7***	***8***
	kg/kmol	kmol/h							
[A] $CH_2S_{0.1}$	17.2	2907	0	0	0	0	0	0	0
[B] O_2	32	0	930	930	930	5581	5581	0	0
[C] N_2	28	0	20997	20997	20997	20997	20997	0	0
[D] H_2O	18	0	3315	3315	3315	408	408	27188	27188
[E] CO_2	44	0	2907	2907	2907	0	0	0	0
[F] SO_2	64	0	291	291	291	0	0	0	0
Total	kg/h	50000	823860	823860	823860	773860	773860	489381	489381
Phase		L	G	G	G	G	G	L	G
Temp.	K	298	*2148*	700	500	300	*523*	300	*500*
Press	kPa(abs)	130	110	100	100	140	130	2650	2637
Volume	m3/h	56	4615714	1654345	1181675	480550	901429	489	42839
Enthalpy*	kJ/h	-4.36E+08	-3.51E+08	-1.67E+09	-1.85E+09	-9.72E+07	8.47E+07	5.47E+07	1.37E+09

Mass Balance Check		*Unit 1*	*Unit 2*	*Unit 3*	*Overall*
Mass IN	kg/h	823860	1313241	1597721	1313241
Mass OUT	kg/h	823860	1313241	1597721	1313241
Closure %		100.0	100.0	100.0	100.0
Energy Balance Check					
Energy IN	kJ/h	-3.51E+08	-2.97E+08	-1.77E+09	-4.79E+08
Energy OUT	kJ/h	-3.51E+08	-2.97E+08	-1.77E+09	-4.79E+08
Closure %		100.0	100.0	100.0	100.0

* *Enthalpies <u>include</u> heats of formation, except Streams 7 and 8*

SUMMARY

[1] In the general M&E balance problem the energy balance may be coupled to or uncoupled from the material balance. This chapter treats the energy balance as uncoupled from the material balance and assumes that the material balance has been solved before the energy balance is considered.

[2] Energy balances are derived from the general balance equation (GBE) in which the specified quantity is "energy" in all of its forms. Since (in non-nuclear processes) energy[11] is a conserved quantity, the integral and differential forms of the GBE are simplified by dropping the generation and consumption terms, as follows:

Integral form of GBE: Energy ACC = Energy IN – Energy OUT
Differential form of GBE: Rate of energy ACC = Rate of energy IN – Rate of energy OUT

[3] As a rule of thumb, in chemical process calculations the relative order of magnitude of terms in the energy balance is:

Reaction (heat of reaction) > phase change (latent heat) > temperature change (sensible heat) > gas compression/expansion > liquid pumping > kinetic and potential energy > exotics

Consequently energy balances on chemical processes can often be simplified by neglecting the exotic energy terms and some or all mechanical energy effects. The latter simplification is not valid in thermal power generation cycles, where about 40% of the chemical energy (heat of combustion) is converted to mechanical energy via an expansion turbine. Also in electric batteries, fuel cells and electro-synthesis processes the chemical energy terms are balanced by large electrical energy effects.

[4] The energy balance on a closed system (batch process) translates to a commonly identified case of *the first law of thermodynamics*, for which the integral form is:

$$E_{final} - E_{initial} = Q - W$$ (see *Equation 5.04*)

where: Typical units

E	$= [U + E_k + E_p]$ content of the system	kJ
E_{final}	= final value of "E" for the system	kJ
$E_{initial}$	= initial value of "E" for the system	kJ
Q	= net heat <u>input</u> to system	kJ
W	= net work <u>output</u> from the system	kJ

[5] The energy balance on an open system (continuous process) includes the energy content of *material that crosses the system boundary*, so the integral form of this energy balance is:

$$E_{final} - E_{initial} = E_{input} - E_{output} + Q - W - (PV_{output} - PV_{input})$$ (see *Equation 5.06*)

---

11 a.k.a. the total energy, considered with *Equations 5.02* and *5.03*.

where: Typical units

			Typical units
E_{final}	=	final value of $[U + E_k + E_p]$ content of the system	kJ
$E_{initial}$	=	initial value of $[U + E_k + E_p]$ content of the system	kJ
E_{input}	=	$[U+E_k+E_p]$ sum for all material inputs to the system	kJ
E_{output}	=	$[U+E_k+E_p]$ sum for all material outputs from the system	kJ
$(PV_{output} - PV_{input})$	=	net (flow) work to move material in/out of system	kJ
Q	=	net heat input to system	kJ
W	=	net work output from the system (a.k.a. shaft work)	kJ

[6] The energy balance on an open system for a chemical process at steady-state is usually simplified by replacing U + PV with H and dropping the kinetic and potential energy terms, to give the differential form:

$$dE/dt = 0 = \dot{H}_{input} - \dot{H}_{output} + \dot{Q} - \dot{W} \qquad \text{(see } \textit{Equation 5.09}\text{)}$$

where:

			Typical units
$E \cong U$	=	internal energy of the system	kJ
$\dot{H}_{input}$	=	sum of enthalpy flow carried by all material input streams to the system	kW
$\dot{H}_{output}$	=	sum of enthalpy flow carried by all material output streams from the system	kW
$\dot{Q}$	=	thermal *utility load* on the system (+) heating, (–) cooling	kW
$\dot{W}$	=	mechanical or electrical *utility load* on the system (+) power out, (–) power in	kW
t	=	time	s

[7] The best reference state to use for energy balances on general chemical processes is the *elements* in their standard state at 298K. The procedure based on this reference state is called the *"heat of formation method"* of energy balance calculation because the energy of each stream, as defined in *Equations 5.11* and *5.12*, includes the heat of formation of the stream components. The heat of formation method automatically accounts for the energy effects of chemical reactions, so it does not require addition of an extra term to the energy balance to deal with the heat of reaction.

[8] An alternative reference state for energy balances is the relevant *compounds* at some reference condition. The compound reference state is common in thermodynamic diagrams and tables such as enthalpy-concentration charts, psychrometric charts and the steam table.

Energy balances with non-reactive systems (i.e. physical processes) can use either the element or the compound references state without adjustment. However when the compound reference state is used with reactive systems it is necessary to add an extra term to the energy balance to deal with the heat of reaction. The procedure based on the compound reference state is called the *"heat of reaction method"* of energy balance calculation because the energy of each stream, as defined in *Equations 5.15* and *5.16*, does not include the heat of formation of the stream components, so this method requires that the heat of reaction be added as an explicit term in the energy balance.

[9] The heat (Q) and work (W) terms in the energy balance correspond to transfer of energy without the transfer of material across the system envelope. The Q and W terms usually represent respectively heat transfer across a heat exchange surface and mechanical work done by a rotating shaft or piston or electrical work done by an electrochemical, photoelectric or resistive system.

[10] The energy balance gives a single equation for each process unit that requires the specification of five variables: the phase (Π), pressure (P) and temperature (T) of each process stream, plus the heat (Q) and work (W) crossing the system envelope. Fully specifying the energy balance may require simplifying assumptions such as: adiabatic or isothermal conditions, zero work done, enthalpy independent of pressure, etc. as in *Table 5.03*. When such simplifications are not available, a more comprehensive analysis is called for, using relations such as: the entropy balance, momentum balance, heat transfer rate and pressure drop factors (not covered in this text), possibly combined with rules of thumb common to process design (see *Refs. 9–10*). The example energy balance problems in this text are made fully-specified by a combination of simplifying assumptions and rules of thumb.

[11] Energy balance problems that involve a *heat of combustion* can be treated by either:

- the *heat of formation method*, first calculating the heat of formation from the heat of combustion.
- the *heat of reaction method,* where the heat of combustion is addedas the heat of reaction term in combustion processes.

[12] As for material balances, energy balances can be written for each generic process unit (DIVIDE, MIX, SEPARATE, HEAT EXCHANGE, PUMP and REACT) and sequenced to model a complete process flowsheet. For a fully-specified continuous multi-unit process operating at steady-state the set of energy balance equations is conveniently solved in a spreadsheet by the iterative sequential modular method. The "Goal Seek" and "Solver" are handy spreadsheet tools that can be used to solve energy balances by making: *closure* = e_E = 100% at steady-state.

[13] The solution to the steady-state energy balance is presented in an expanded M&E balance stream table (see *Figure 5.01*) that shows the material balance with the phase(s), pressure, temperature, plus the volume and enthalpy flow of each process stream, the rates of heat and work transfer to each process unit and the *closure* of both the mass and the energy balances on each process unit and on the overall process.

FURTHER READING

[1] R. M. Felder and R.W. Rousseau, *Elementary Principles of Chemical Processes*, John Wiley & Sons, New York, 2000.

[2] G. V. Reklaitis, *Introduction to Material and Energy Balances*, John Wiley & Sons, New York, 1983.

[3] D. M. Himmelblau, *Basic Principles and Calculations in Chemical Engineering*, Prentice Hall, Englewood Cliffs, 1989.

[4] P. M. Doran, *Bioprocess Engineering Principles*, Academic Press, San Diego, 1995.

[5] O. A. Hougen, K. M. Watson and R. A. Ragatz, *Chemical Process Principles*, John Wiley & Sons, New York, 1956.

[6] T. M. Duncan and J. A. Reimer, *Chemical Engineering Design and Analysis*, Cambridge University Press, 1998.

[7] S. I. Sandler, *Chemical and Engineering Thermodynamics*, John Wiley & Sons, New York, 1999.

[8] J. Winnick, *Chemical Engineering Thermodynamics*, John Wiley & Sons, New York, 1997.

[9] S. M. Walas, *Chemical Process Equipment — Selection and Design*, Butterworth-Heinemann, Boston, 1990.

[10] R. K. Sinnott, *Chemical Engineering Design*, Butterworth-Heinemann, Oxford, 1999.

[11] L. T. Biegler, I. E. Grossmann and A.W. Westerberg, *Systematic Methods of Chemical Process Design*, Prentice Hall, Saddle River, 1997.

[12] W. D. Seider, J. D. Seader and D. R. Lewin, *Process Design Principles*, John Wiley & Sons, New York, 1999.

CHAPTER SIX

SIMULTANEOUS MATERIAL AND ENERGY BALANCES

SIMULTANEOUS BALANCES

In Chapters 4 and 5 the material balance and the energy balance on a system were introduced as separate problems that are solved in sequence:

(1) Material balance → (2) Energy balance

However, for some systems the material balance and the energy balance are coupled in such a way that it is not possible to solve them separately. In these cases the material balance and the energy balance must be solved simultaneously to obtain the distribution of material and energy in the process.

You can encounter simultaneous material and energy balance problems in many practical situations. In chemical process flowsheets, for example, such problems may arise in mixers, separators, reactors and direct contact heat exchangers. A characteristic of these process units is that the stream flows interact with the pressure and/or temperature in the unit. For example, the outlet component flows from a reactor depends on the reactor temperature, since temperature affects the reaction rate and determines the reactant conversion. Similarly, the outlet stream flows from a gas/liquid separator are affected by both the pressure and the temperature in the separator, which set the gas phase partial pressures for the phase "split". [1]

SPECIFICATION OF SIMULTANEOUS MATERIAL AND ENERGY BALANCES

When you approach a problem involving simultaneous material and energy balances the first thing you usually notice is that the material balance is *under-specified* (i.e. D of F > 0). The under-specification of a material balance can have two causes:

A. The material balance is part of a fully-specified M&E balance but must be solved simultaneously with the energy balance.

B. The material balance is part of a *design problem,* in which the M&E balance is truly under-specified.

The specification (i.e. D of F) of a M&E balance problem can be checked by combining the criteria of *Tables 4.02* and *5.03*. With this check you will often find that a material balance with one D of F can be resolved when an energy balance is added to the set of equations for the system, so the complete M&E balance becomes fully-specified. If the M&E balance is truly under-specified then its solution requires design and optimisation procedures outside the scope of this text (see *Refs. 4–6*).

Examples 6.01 to *6.10* and associated comments illustrate some typical problems involving simultaneous material and energy balances. These examples show both closed (batch) and open (continuous) systems and are set up as fully-specified M&E balances.

---

[1] The *phase split* is the distribution of components between the phases in a process stream or from a process unit.

CLOSED SYSTEMS (BATCH PROCESSES)

EXAMPLE 6.01 Simultaneous M&E balance for vapour/liquid equilibrium (closed system).

A rigid, closed and evacuated vessel initially holds a sealed container filled with 5 kg of liquid water at 100 kPa(abs), 300 K. The container is subsequently opened and 4380 kJ of heat is transferred into the vessel's contents, whose total volume is fixed at 1.00 m³. No work is transferred and the mass of vessel and container are assumed negligible.

Problem: Find the final (equilibrium) mass of *liquid* water in the vessel (i.e. find the phase split).

Solution: The final mass of *liquid* water will depend on the (unknown) final pressure and temperature in the vessel.

Define the *system* = closed vessel and contents. Let

m_l = final mass of water liquid kg
m_v = final mass of water vapour kg

Material balance on water *liquid* in the closed system:

ACC = IN – OUT + GEN – CON (IN = OUT = 0 for material in a closed system)
$m_l - 5 = 0 - 0 + 0 - m_v$ (mass balance) [1]

Note that the material balance is under-specified, i.e. 1 equation, 2 unknowns (m_l and m_v).

Energy balance on the closed system. Reference condition = *liquid water* at its triple-point.
Using the steam table, specific internal energy of liquid water at 100 kPa(abs), 300 K = 112.5 kJ/kg

ACC = IN – OUT + GEN – CON $Q = 4830$ kJ, $W = 0$ (GEN = CON = 0 for energy)
$(m_l u_l + m_v u_g) - (5 \text{ kg})(112.5 \text{ kJ/kg}) = 4830 \text{ kJ} - 0 + 0 - 0$ (energy balance) [2]

Also, since the total volume of the mixture is fixed at 1 m³: $m_l v_l + m_v v_g = 1$ [3]

where:

u_l = specific internal energy of water liquid at final conditions kJ.kg^{-1}
u_g = specific internal energy of water vapour at final conditions kJ.kg^{-1}
v_l = specific volume of water liquid at final conditions m^3.kg^{-1}
v_g = specific volume of water vapour at final conditions m^3.kg^{-1}

The final pressure and temperature are related by the state equations for water, which are embodied in the steam table (*Table 2.20*). Equations [1 to 3] are solved simultaneously by trial and error with the help of the steam table, to give:

At P = 246 kPa(abs), T = 400 K u_l = 532.6 kJ/kg u_g = 2536.2 kJ/kg
v_l = 0.001067 m³/kg v_g = 0.7308 m³/kg m_l = **<u>3.64 kg</u>**

EXAMPLE 6.02 Simultaneous M&E balance for a batch chemical reactor (closed system).

The reversible liquid phase thermo-chemical reaction and its temperature dependent equilibrium constant are defined by:

$A(l) \leftrightarrow B(l)$ $\quad\quad K_{eq} = [B]/[A] = \exp(200/T)$

where:

[A]	=	concentration of A	$kmol.m^{-3}$
[B]	=	concentration of B	$kmol.m^{-3}$
K_{eq}	=	equilibrium constant	–
T	=	reaction equilibrium temperature	K

Data	COMPONENT		A	B
	$h^o_{f,298K}$	kJ/kmol	-10E3	-20E3
	C_{vm}, ref. 298 K	kJ/kmol.K	80	80

A batch reactor of fixed volume contains initially only pure liquid A at 298 K. The reaction is then initiated and allowed to go essentially to equilibrium (i.e. $X \approx X_{eq}$) under adiabatic conditions, without a change of phase or transfer of work.

Problem: Calculate the final conversion of A. Assume the reaction vessel has zero heat capacity.

Solution: The final conversion of A will depend on the (unknown) final reaction temperature.

Define the *system* = closed reactor plus contents. Let:

X = final (equilibrium) conversion of A = (initial moles A – final moles A) / initial moles A

T_f = final (equilibrium) temperature K

Material balance on A in the closed system

ACC = IN – OUT + GEN – CON [IN = OUT = 0 for material in a closed system]

$$- [A]_i V_R X = 0 - 0 + 0 - K_{eq} [A]_f V_R = - K_{eq} [A]_i V_R (1 - X)$$

$$X = K_{eq} (1 - X) = (\exp(200/T_f))(1 - X) \quad [1]$$

where:

$[A]_i$	=	initial concentration of A	$kmol.m^{-3}$
$[A]_f$	=	final (equilibrium) concentration of A	$kmol.m^{-3}$
V_R	=	reactor volume (= fixed volume of reaction mixture)	m^3

Note that the material balance is under-specified, i.e. 1 equation, 2 unknowns (X and T_f).

Energy balance on the closed system. *Reference condition = elements at standard state, 298 K*

ACC = IN – OUT + GEN – CON

$$U_f - U_i = 0 - 0 + 0 - 0 = 0 \quad \text{[Energy is conserved. Q = 0, W = 0]} \quad [2]$$

U_i	=	initial internal energy of system	kJ
U_f	=	final internal energy of system	kJ

$$U_i = [A]_i V_R (C_{vm}(A)(T_i - 298) + h^o_{f,298K}(A)) = 1\text{ kmol }(80\text{ kJ/(kmol.K)}(298-298)\text{ K} + (-10E3\text{ kJ/kmol}))$$
$$= -10E3\text{ kJ}$$

$$U_f = [A]_f V_R (C_{vm}(A)(T_f - 298) + h^o_{f,298K}(A) + [B]_f V_R (C_{vm}(B)(T_f - 298) + h^o_{f,298K}(B))$$
$$= (1-X)(80\text{ kJ/(kmol.K)}(T_f - 298)\text{ K} + (-10E3\text{ kJ/kmol})) + X(80\text{ kJ/(kmol.K)}(T_f - 298)\text{ K} + (-20E3\text{ kJ/kmol}))$$

Note that the reactor volume does not affect the equilibrium conversion of A and drops from the material and energy balance equations.

Equations [1 and 2] make the system fully-specified and are solved simultaneously (by spreadsheet "Solver") to give: $T_f = 334$ K $K_{eq} = 1.82$ $X_{eq} = \underline{0.65}$

OPEN SYSTEMS (CONTINUOUS PROCESSES)

SINGLE PROCESS UNITS

Example 6.03 A, B, C, D, E and F illustrates simultaneous M&E balance problems for the generic process units: MIX, SEPARATE, HEAT EXCHANGE (indirect and direct), PUMP and REACT in open systems operating at steady-state. In single unit problems of this sort a common feature is an interdependent combination of an unknown flow (or flows) with an unknown stream temperature, as typified in *Example 6.03B*. In less common cases where enthalpy is a function of pressure an unknown stream pressure may also be combined with an unknown flow. Such features are usually easy to see in a "problem statement" stream table that lays out all known and unknown values of the case at hand. Simultaneous M&E balances do not arise in the generic process unit DIVIDE because by definition the composition and the temperature are equal for all streams in and out of this unit.

MULTIPLE PROCESS UNITS

Simultaneous material and energy balances are more difficult to identify and to solve when they occur in multi-unit processes, especially in processes with recycle streams. In some cases such problems can be solved by the *overall balance* approach in which the overall material balance is combined with the overall energy balance to give the desired result. A simple case of the overall balance approach is shown in *Example 6.04*.

When a complete stream table is needed (e.g. for process design) simultaneous M&E balances may have to be solved for some or all of the process units in the flowsheet. *Example 6.05* shows a recycle M&E balance similar to that of *Example 5.10*, with the complication that the conversion in the reactor depends on the reaction temperature, which comes in turn from energy balances on the pump, mixer and reactor in the recycle loop. Problems such as this can be solved by an extension of the ***iterative sequential modular method*** shown in *Chapters 4* and *5*, with provision to deal with the coupled material and energy equations that arise in some process units. The spreadsheet solution of this type of problem may use manual iteration and/or macros to close the simultaneous balances on individual process units.

Example 6.05 demonstrates effects that are characteristic of chemical processes. It also provides an insight to the relevance of complex systems with respect to the "ingenuity gap". The ingenuity gap is the divide between the *need for* and the *supply of* ideas to solve the problems facing society in the 21st century. This gap relates to ecological, economic and technological systems whose complexity exceeds the ability of humans to predict and/or control their behavior (see *Ref. 7*). Such complexity is due to the connectedness between many sub-systems, with non-linearity and interaction of multiple variables, driven by feedback loops among the system parts.

Example 6.05 is a simple illustration of the concepts of coupling, non-linearity, interaction[2] and recycling in a multi-unit, multi-variable system. You can see that even at this low level of complexity it is difficult to have an intuitive understanding of the system's behavior because *"everything depends on everything else"*. The simultaneous M&E balance provides a predictive model of the system that helps to narrow the ingenuity gap. Simultaneous M&E balances are used in engineering design, with sophisticated chemical process simulation software and complex mathematical algorithms linked to large data banks and associated correlations. On a larger scale, simultaneous differential M&E balances are at the heart of modelling the global climate change on planet Earth (see *Example 7.07*). In this case the high level of complexity leads to the possibility of system instabilities with "threshold effects"[3] that threaten to devastate our civilisation!

Finally, *Example 6.06* shows a simultaneous M&E balance on a hypothetical "environmentally balanced" process in which solar radiation and natural rainfall are integrated with hydrogen fuel cells to power the world's automobiles. This example shows how simultaneous M&E balances may be used to predict some of the consequences of a "sustainable" energy policy on the global environment.

---

[2] An interaction exists between two variables A and B when the effect of variable A on the objective function depends on the value of variable B.

[3] A threshold effect is a sudden change in behaviour of a system that occurs when process variable(s) cross a critical value.

EXAMPLE 6.03 Simultaneous M&E balances for generic process units (open system at steady-state).

This example shows how the known output temperature is used to find an unknown input flow.

A. Mixer

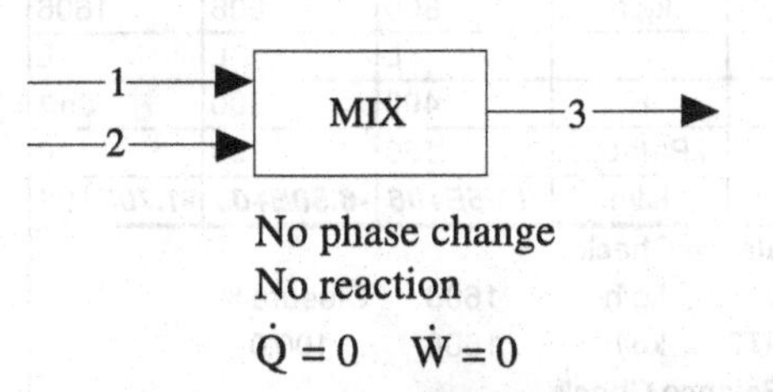

Stream Table				***Mixer Problem***
Species	M	Stream		
		1	*2*	*3*
	kg/kmol	kmol/h		
A (liq)	20	4	3	?
B (liq)	30	8	5	?
C (liq)	40	12	?	?
Total	kg/h	800	?	?
Phase		L	L	L
Temp.	K	400	300	350
Press.	kPa(abs)	150	150	140
Enthalpy	kJ/h	-1.05E+06	?	?

Properties	**Species**	**A (liq)**	**A(gas)**	**B(liq)**	**B(gas)**	**C(liq)**	**C(gas)**
$C_{p,m}$(ref.298K)	kJ/kmol.K	50	35	70	45	90	55
$h_{f,298K}$	kJ/kmol	-2.0E+05	-1.8E+05	-1.0E+05	-7.0E+04	3.00E+04	4.00E+04
$h_{v,298K}$	kJ/kmol	-	2.00E+04	-	3.00E+04	-	1.00E+04

***Solution*: Material balance**

Mole balance on A $\quad 0 = 4 + 3 - \dot{n}(3, A)$ $\quad$ kmol/h $\quad$ [1]

Mole balance on B $\quad 0 = 8 + 5 - \dot{n}(3, B)$ $\quad$ kmol/h $\quad$ [2]

Mole balance on C $\quad 0 = 12 + \dot{n}(2, C) - \dot{n}(3, C)$ $\quad$ kmol/h $\quad$ [3]

Three independent equations. Four unknowns. The material balance is UNDER-SPECIFIED.

Energy balance Reference condition = *elements* at standard state, 298K

$$0 = \Sigma[\dot{H}^*(in)] - \Sigma[\dot{H}^*(out)] + \dot{Q} - \dot{W}$$

For a mixer $\dot{Q} = \dot{W} = 0$ i.e.

$$0 = \dot{H}^*(1) + \dot{H}^*(2) - \dot{H}^*(3) + 0 - 0 \qquad [4]$$

where:

$$\dot{H}^*(i) = \Sigma\,[\dot{n}(i,j)(C_{p,m}(j)(T(i) - T_{ref}) + h^o_{f,Tref}(j))] \quad \text{[Respect the phase]}$$

$$\begin{aligned}\dot{H}^*(1) &= (4\ \text{kmol/h})((50\ \text{kJ/kmol.K})(400\ \text{K} - 298\ \text{K}) + (\text{-2E5 kJ/kmol}))\\ &\quad + (8\ \text{kmol/h})((70\ \text{kJ/kmol.K})(400\ \text{K} - 298\ \text{K}) + (\text{-1E5 kJ/kmol}))\\ &\quad + (12\ \text{kmol/h})((90\ \text{kJ/kmol.K})(400\ \text{K} - 298\ \text{K}) + (\text{3E4 kJ/kmol}))\\ &= -1.05E+06\ \text{kJ/h}\end{aligned}$$

$$\dot{H}*(2) = (3\ \text{kmol/h})((50\ \text{kJ/kmol.K})(300\ \text{K} - 298\ \text{K}) + (\text{-2E5 kJ/kmol})) + (5\ \text{kmol/h})((70\ \text{kJ/kmol.K})(300\ \text{K} - 298\ \text{K}) + (\text{-1E5 kJ/kmol})) + (\dot{n}(2,C)\ \text{kmol/h})((90\ \text{kJ/kmol.K})(300\ \text{K} - 298\ \text{K}) + (\text{3E4 kJ/kmol})) = ?\ \text{kJ/h}$$

$$\dot{H}*(3) = (\dot{n}(3,A)\ \text{kmol/h})((50\ \text{kJ/kmol.K})(350\ \text{K} - 298\ \text{K}) + (\text{-2E5 kJ/kmol})) + (\dot{n}(3,B)\ \text{kmol/h})((70\ \text{kJ/kmol.K})(350\ \text{K} - 298\ \text{K}) + (\text{-1E5 kJ/kmol})) + (\dot{n}(3,C)\ \text{kmol/h})((90\ \text{kJ/kmol.K})(350\ \text{K} - 298\ \text{K}) + (\text{3E4 kJ/kmol})) = ?\ \text{kJ/h}$$

Stream Table				*Mixer solution*
Species	M	Stream		
		1	*2*	*3*
	kg/kmol	kmol/h		
A (liq)	20	4.0	3.0	7.0
B (liq)	30	8.0	5.0	13.0
C (liq)	40	12.0	14.9	26.9
Total	kg/h	800	806	1606
Phase		L	L	L
Temp.	K	400	300	***350***
Press.	kPa(abs)	150	150	140
Enthalpy*	kJ/h	***-1.05E+06***	***-6.50E+05***	***-1.70E+06***
Mass Balance Check				
Mass IN	kg/h	1606	Closure %	
Mass OUT	kg/h	1606	100.0	
Energy Balance Check				
Energy IN	kJ/h	-1.70E+06	Closure %	
Energy OUT	kJ/h	-1.70E+06	100.0	

* *Enthalpy values include heats of formation*

The combined material and energy balance has four independent equations and four unknowns, i.e. FULLY-SPECIFIED. Solve equations [1 to 4] to get the stream table.

B. Separator (adiabatic flash split)

This example shows how a known pressure is used to find an unknown adiabatic temperature and phase split.

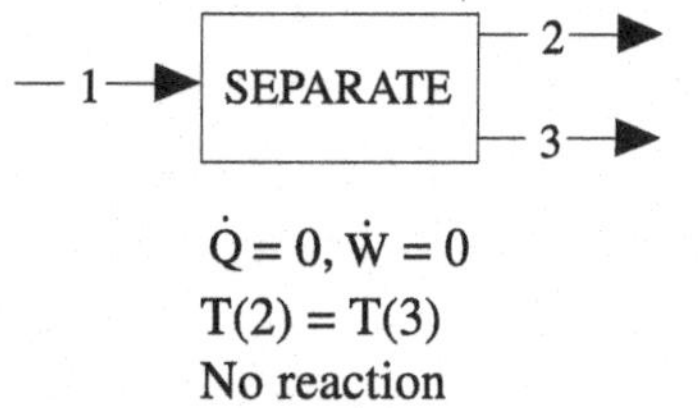

Stream Table		*Separator*		*Problem*
Species	M	Stream		
		1	*2*	*3*
	kg/kmol	kmol/h		
A	20	4	*?*	*?*
B	30	8	*?*	*?*
Total	kg/h	320	*?*	*?*
Phase		L	*G*	*L*
Temp.	K	600	*?*	*?*
Press.	kPa(abs)	25000	***140***	***140***
Enthalpy	kJ/h	*?*	*?*	*?*

Stream 2 is in thermal and V – L equilibrium with stream 3.

$$p^*(A) = \exp[17 - 3300/(T - 39)]\ \text{kPa}$$
$$p^*(B) = \exp[14 - 4000/(T - 55)]\ \text{kPa}$$

Properties	Species	A(liq)	A(gas)	B(liq)	B(gas)
$C_{p,m}$(ref.298K)	kJ/kmol.K	50	35	70	45
$h_{f,298K}$	kJ/kmol	-2.0E+05	-1.8E+05	-1.0E+05	-7.0E+04

This is the classic **adiabatic flash split**.

Solution: **Material balance**

Mole balance on A $\quad 0 = 4 - \dot{n}(2,A) - \dot{n}(3,A)$ kmol/h [1]

Mole balance on B $\quad 0 = 8 - \dot{n}(2,B) - \dot{n}(3,B)$ kmol/h [2]

Two independent equations. Four unknowns. The material balance alone is UNDER-SPECIFIED.

Equilibrium compositions:

y(2, A) = k(A) x(3, A) where k(A) = p*(A)/P(2) Raoult's law (see *Equation 2.33*) [3]
y(2, B) = k(B) x(3, B) where k(B) = p*(B)/P(2) [4]
y(i, j) = m.f. j in gas i 1 = y(2, A) + y(2, B) [5]
x(i, j) = m.f. j in liquid i 1 = x(3, A) + x(3, B) [6]

Re-arrange equations [1 to 6] by the method of *Example 4.07B* to get:

$$1 = 4/[\bar{n}(2)(k(A)-1)+12] + 8/[\bar{n}(2)(k(B)-1)+12] \quad [7]$$

Energy balance

Reference condition = *elements* at standard state, 298K

$$0 = \dot{H}^*(1) - \dot{H}^*(2) - \dot{H}^*(3) + \dot{Q} - \dot{W} \quad \text{kJ/h} \quad [8]$$

where:

$\dot{H}^*(i) = \Sigma[\dot{n}(i,j)(C_{p,m}(j)(T(i) - T_{ref}) + h^o_{f,Tref}(j))]$ kJ/h [Respect the phase]

$\dot{H}^*(1) =$ (4 kmol/h)((50 kJ/kmol.K)(300 K – 298 K)
+ (-2E5 kJ/kmol)) kJ/h
+ (8 kmol/h)((70 kJ/kmol.K)(300 K – 298 K)
+ (-1E5 kJ/kmol))

$\dot{H}^*(2) =$ ($\dot{n}$(2, A) kmol/h)((35 kJ/kmol.K)(T(2) K – 298 K)
+ (-1.8E5 kJ/kmol)) kJ/h
+ ($\dot{n}$(2, B) kmol/h)((45 kJ/kmol.K)(T(2) K – 298 K)
+ (-7E4 kJ/kmol))

$\dot{H}^*(3) =$ ($\dot{n}$(3, A) kmol/h)((50 kJ/kmol.K)(T(3) K – 298 K)
+ (-2E5 kJ/kmol)) kJ/h
+ ($\dot{n}$(3, B) kmol/h)((70 kJ/kmol.K)(T(3) K – 298 K)
+ (-1E5 kJ/kmol))

RESPECTING THE PHASE

Thermal equlibrium T(2) = T(3) [9]

The combined material and energy balance has six independent equations and six unknowns, i.e. FULLY-SPECIFIED.

The non-linear equations [7 and 8] must be solved simultaneously to find T(2) = T(3) and $\dot{n}(2)$.

The spreadsheet method is as follows:

Fix T(2) → Solve equation [7] for $\bar{n}(2)$ by the Solver → → Calculate the material balance and stream enthalpies.

Bisect T(2) manually to close the energy balance.

EXAMPLE:

T(2)	$\bar{n}(2)$	Closure %
450	7.96	95.6
400	3.89	105.5
425	5.13	102.1
437	6.09	99.7
431	5.55	101
433	5.72	100.6
435	5.89	100.2
436	**5.99**	**99.9**

Bisection solution of energy balance

k(A) = 29.873
k(B) = 0.362
x(3, A) = 0.022

Stream Table		*Separator*		*Solution*
Species	M	Stream		
		1	*2*	*3*
	kg/kmol	kmol/h		
A	20	4	***3.87***	***0.13***
B	30	8	***2.12***	***5.88***
Total moles	kmol/h	12	***5.99***	***6.01***
Total mass	kg/h	320	***141.0***	***179.0***
Phase		L	G	L
Temp.	K	600	***436***	***436***
Press.	kPa(abs)	25000	140	140
Enthalpy*	kJ/h	-1.37E+06	***-8.13E+05***	***-5.56E+05***
Mass Balance Check				
Mass IN	kg/h	320	Closure %	
Mass OUT	kg/h	320	100.0	
Energy Balance Check				
Energy IN	kJ/h	-1.37E+06	Closure %	
Energy OUT	kJ/h	-1.37E+06	99.9	

** Enthalpy values include heats of formation*

C. Heat exchanger (indirect contact)

This example shows how the known output temperature is used to find an unknown input and output flow.

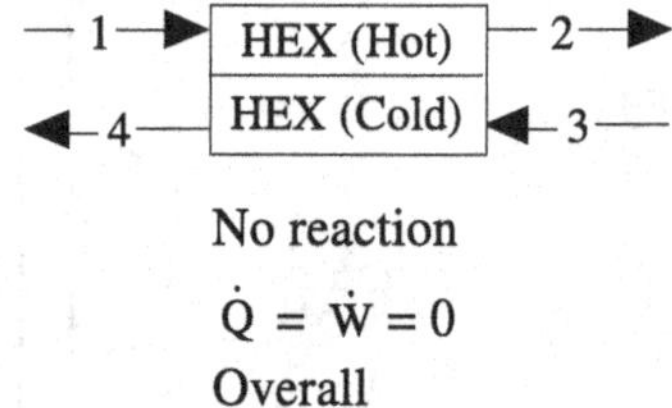

Stream Table		*Indirect Heat Exchanger*			*Problem*
Species	M	Stream			
		1	*2*	*3*	*4*
	kg/kmol	kmol/h			
A	20	4	?	3	?
B	30	?	?	7	?
Total	kg/h	?	?	270	?
Phase		G	L	L	G
Temp.	K	500	420	300	440
Press.	kPa(abs)	600	570	200	170
Enthalpy	kJ/h	?	?	?	?

Properties	Species	A(liq)	A(gas)	B(liq)	B(gas)
$C_{p,m}$(ref.298K)	kJ/kmol.K	50	35	70	45
$h_{f,298K}$	kJ/kmol	-2.0E+05	-1.8E+05	-1.0E+05	-7.0E+04
$h_{v,298K}$	kJ/kmol		2.00E+04		3.00E+04

***Solution:* Material balance**

[System = HEX hot side]

Mole balance on A

$$0 = 4 - \dot{n}(2, A) \quad \text{kmol/h} \quad [1]$$

Mole balance on B

$$0 = \dot{n}(1, B) - \dot{n}(2, B) \quad \text{kmol/h} \quad [2]$$

[System = HEX cold side]

Mole balance on A $\quad 0 = 3 - \dot{n}(4, A)$ $\quad$ kmol/h $\quad$ [3]

Mole balance on B $\quad 0 = 7 - \dot{n}(4, B)$ $\quad$ kmol/h $\quad$ [4]

Four independent equations. Five unknowns. The material balance alone is UNDER-SPECIFIED.

Energy balance [Overall HEX balance]

Reference condition = *elements* at standard state, 298K

$$0 = \dot{H}^*(1) - \dot{H}^*(2) + \dot{H}^*(3) - \dot{H}^*(4) + 0 - 0 \quad [5]$$

For a heat exchanger $\dot{Q} = \dot{W} = 0$ $\quad$ [Overall]

where:

$$\dot{H}^*(i) = \Sigma[\dot{n}(i,j)(C_{p,m}(j)(T(i) - T_{ref}) + h^{o}_{f,Tref}(j))] \text{ [Respect the phase]}$$

$$\dot{H}^*(1) = (4\ \text{kmol/h})((35\ \text{kJ/kmol.K})(500\ \text{K} - 298\ \text{K}) + (-1.8\text{E}5\ \text{kJ/kmol}))$$
$$+ (\dot{n}(1, B)\ \text{kmol/h})((45\ \text{kJ/kmol.K})(500\ \text{K} - 298\ \text{K}) + (-7\text{E}4\ \text{kJ/kmol}))$$
$$= ?\ \text{kJ/h}$$

$$\dot{H}^*(2) = (\dot{n}(2, A)\ \text{kmol/h})((50\ \text{kJ/kmol.K})(400\ \text{K} - 298\ \text{K}) + (-2\text{E}5\ \text{kJ/kmol}))$$
$$+ (\dot{n}(2, B)\ \text{kmol/h})((70\ \text{kJ/kmol.K})(400\ \text{K} - 298\ \text{K}) + (-1\text{E}5\ \text{kJ/kmol}))$$
$$= ?\ \text{kJ/h}$$

$$\dot{H}^*(3) = (2\ \text{kmol/h})((50\ \text{kJ/kmol.K})(350\ \text{K} - 298\ \text{K}) + (-2\text{E}5\ \text{kJ/kmol}))$$
$$+ (7\ \text{kmol/h})((70\ \text{kJ/kmol.K})(350\ \text{K} - 298\text{K}) + (-1\text{E}5\ \text{kJ/kmol}))$$
$$= ?\ \text{kJ/h}$$

$$\dot{H}^*(4) = (\dot{n}(4, A)\ \text{kmol/h})((35\ \text{kJ/kmol.K})(440\ \text{K} - 298\ \text{K})$$
$$+ (-1.8\text{E}5\ \text{kJ/kmol}))$$
$$+ (\dot{n}(4, B)\ \text{kmol/h})((45\ \text{kJ/kmol.K})(440\ \text{K} - 298\text{K})$$
$$+ (-7\text{E}4\ \text{kJ/kmol}))$$
$$= ?\ \text{kJ/h}$$

The combined M&E balance has five independent equations and five unknowns, i.e. FULLY-SPECIFIED.

Solve equations [1, 3 and 4] for $\dot{n}(2, A)$, $\dot{n}(4, A)$, $\dot{n}(2, B)$; then simultaneous equations [2 + 5] for $\dot{n}(1, B)$ and $\dot{n}(2, B)$. Thermal duty of HEX

$$= \dot{H}^*(1) - \dot{H}^*(2) = \dot{H}^*(4) - \dot{H}^*(3)$$
$$= 3.28\text{E}+05\ \text{kJ/h} = 91\ \text{kW}$$

Under-Specified Problem

Stream Table		*Indirect Heat Exchanger*			*Solution*
Species	M	Stream			
		1	2	3	4
	kg/kmol	kmol/h			
A	20	4	***4***	3	***3***
B	30	8	***8***	7	***7***
Total	kg/h	***320***	***320***	270	***270***
Phase		G	L	L	G
Temp.	K	500	420	300	440
Press.	kPa(abs)	600	570	200	170
Enthalpy*	kJ/h	***-1.18E+06***	***-1.51E+06***	***-1.30E+06***	***-9.70E+05***
Mass Balance Check					
Mass IN	kg/h	5.90E+02	Closure %		
Mass OUT	kg/h	5.90E+02	100.0		
Energy Balance Check					
Energy IN	kJ/h	-2.48E+06	Closure %		
Energy OUT	kJ/h	-2.48E+06	100.0		

** Enthalpy values include heats of formation*

D. Heat exchanger (direct contact)

This example shows how known output pressure and input flows are used to find an unknown output temperature and flows.

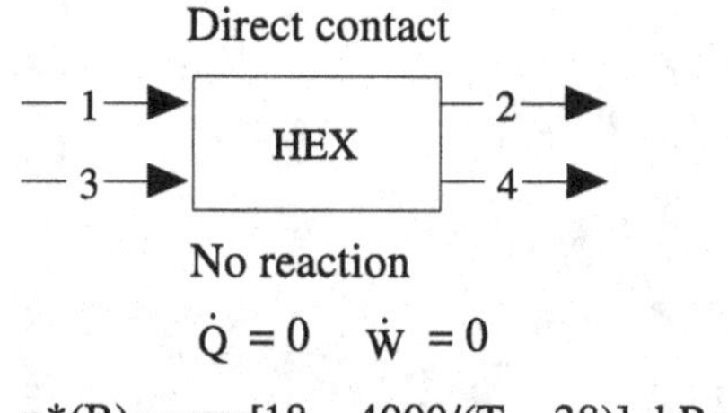

$p^*(B) = \exp[18 - 4000/(T - 38)]$ kPa

Stream 2 is in thermal and V – L equilibrium with stream 4. [This unit can also function as a humidifier]

Stream Table		Direct Contact HEX			Problem
Species	M	Stream			
		1	2	3	4
	kg/kmol	kmol/h			
A	20	100	100	0	0
B	30	0	?	800	?
Total	kg/h	2000	?	24000	?
Phase		G	G	L	L
Temp.	K	300	?	350	?
Press.	kPa(abs)	130	120	130	120
Enthalpy	kJ/h	?	?	?	?

Properties	Species	A(liq)	A(gas)	B(liq)	B(gas)
$C_{p,m}$(ref.298K)	kJ/kmol.K	50	35	70	45
$h_{f,298K}$	kJ/kmol	-2.0E+05	-1.8E+05	-1.0E+05	-7.0E+04
$h_{v,298K}$	kJ/kmol	-	2.00E+04	-	3.00E+04

Material balance

Mole balance on B $\quad 0 = 0 + 800 - \dot{n}(2,B) - \dot{n}(4,B)$ [1]

One independent equation. Two unknowns. The material balance alone is UNDER-SPECIFIED.

Energy balance Reference condition = *elements* at standard state, 298K

$$0 = \dot{H}^*(1) + \dot{H}^*(3) - \dot{H}^*(2) - \dot{H}^*(4) + \dot{Q} - \dot{W} \qquad \dot{Q} = \dot{W} = 0 \qquad [2]$$

where:

$$\dot{H}^*(i) = \Sigma[\dot{n}(i,j)(C_{p,m}(j)(T(i) - T_{ref}) + h^o_{f,Tref}(j))] \qquad \text{[Respect the phase]}$$

$\dot{H}^*(1)$ = (100 kmol/h) ((35 kJ/kmol.K) (300 K – 298 K) + (-1.8E5 kJ/kmol))
+ (0 kmol/h) ((45 kJ/kmol.K) (300 K – 298 K) + (-7E4 kJ/kmol)) = -1.80E+07 kJ/h

$\dot{H}^*(2)$ = (100 kmol/h) ((35 kJ/kmol.K) (T(2) K – 298 K) + (-1.8E5 kJ/kmol))
+ ($\dot{n}$(2, B) kmol/h) ((45 kJ/kmol.K) (T(2) K – 298 K) + (-7E4 kJ/kmol)) = ? kJ/h

$\dot{H}^*(3)$ = (0 kmol/h) ((50 kJ/kmol.K)(350 K – 298 K) + (-2.0E5 kJ/kmol))
+ (800 kmol/h)((70 kJ/kmol.K)(350 K – 298 K) + (-1.0E5 kJ/kmol)) = -7.71E+07 kJ/h

$\dot{H}^*(4)$ = (0 kmol/h)((70 kJ/kmol.K)(T(4) K – 298 K) + (-2.0E5 kJ/kmol))
+ ($\dot{n}$ (4, B) kmol/h)((50 kJ/kmol.K)(T(4) K – 298 K) + (-1.0E5 kJ/kmol)) = ? kJ/h

Equilibria.

Stream temperatures: T(2) =T(4) [3]

Stream compositions: Stream 2

$\dot{n}(2,B)/\bar{n}(2) = p^*(B)/P(2) = f(T(2))/120$ [4]

The combined M&E balance has four independent equations and four unknowns: $\dot{n}(2,B)$, $\dot{n}(3,B)$, T(2) and T(4), i.e. FULLY-SPECIFIED.

Stream Table		Direct Contact HEX			Solution
Species	M	Stream			
		1	2	3	4
	kg/kmol	kmol/h			
A	20	100	100	0	0
B	30	0	**56**	800	**744**
Total	kg/h	2000	3686	24000	22314
Phase		G	G	L	L
Temp.	K	300	**319**	350	**319**
Press.	kPa(abs)	130	120	130	120
Enthalpy*	kJ/h	-1.80E+07	-2.18E+07	-7.71E+07	-7.33E+07
Mass Balance Check					
Mass IN	kg/h	26000	Closure %		
Mass OUT	kg/h	26000	100.0		
Energy Balance Check					
Energy IN	kJ/h	-9.51E+07	Closure %		
Energy OUT	kJ/h	-9.51E+07	100.0		

** Enthalpy values include heats of formation*

E. Pump

This example shows how a known pressure range and work are used to find an unknown flow and output temperature.

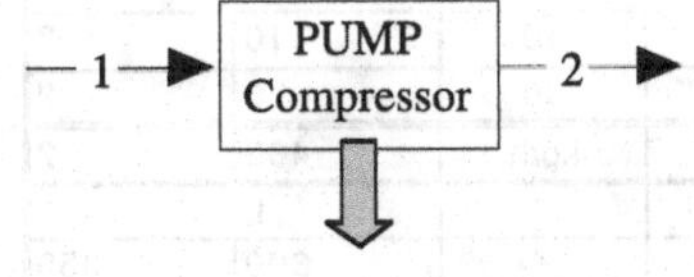

$\dot{W}$ = -3.60E + 05 kJ/h $\dot{Q} = 0$

adiabatic compression, ideal gas

Stream Table Compressor			Problem
Species	M	Stream	
		1	2
	kg/kmol	kmol/h	
A	20	?	?
Total	kg/h	?	?
Phase		G	G
Temp.	K	300	?
Press.	kPa(abs)	150	500
Volume	m^3/h	?	?
Enthalpy	kJ/h	?	?

Properties	Species	A(liq)	A(gas)
$C_{p,m}$(ref.298K)	kJ/kmol.K	50	35
$h_{f,298K}$	kJ/kmol	-2.0E+05	-1.8E+05
Density	-	-	Ideal gas

***Solution:* Material balance**

Mole balance on A $0 = \dot{n}(1,A) - \dot{n}(2,A)$ [1]

One independent equation. Two unknowns.
Material balance is UNDER-SPECIFIED.

Energy balance

Reference condition = *elements* at standard state, 298K

$$0 = \dot{H}^*(1) - \dot{H}^*(2) + \dot{Q} - \dot{W} \quad [2]$$

where: $\dot{H}^*(i) = \Sigma[\dot{n}(i,j)(C_{p,m}(j)(T(i) - T_{ref}) + h^o_{f,Tref}(j))]$ [Respect the phase]

Stream Table Compressor			*Solution*
Species	M	Stream	
		1	*2*
	kg/kmol	kmol/h	
A	20	***103.6***	***103.6***
Total	kg/h	***2071***	***2071***
Phase		G	G
Temp.	K	300	***399***
Press.	kPa(abs)	150	500
Volume	m^3/h	***1721***	***687***
Enthalpy*	kJ/h	***-1.86E+07***	***-1.83E+07***
Mass Balance Check			
Mass IN	kg/h	2071	Closure %
Mass OUT	kg/h	2071	100.0
Energy Balance Check			
Energy IN	kJ/h	-1.86E+07	Closure %
Energy OUT	kJ/h	-1.86E+07	100.0

** Enthalpy values include heats of formation*

$$\dot{H}^*(1) = (\dot{n}(1,A)(\text{kmol/h})(35\ \text{kJ/kmol.K}\ (300\ \text{K} - 298\ \text{K}) + (-1.8\text{E}5\ \text{kJ/kmol}) = ?\ \text{kJ/h}$$

$$\dot{H}^*(2) = (\dot{n}(2,A)(\text{kmol/h})(35\ \text{kJ/(kmol.K)})(T(2) - 298) + (-1.8\text{E}5\ \text{kJ/kmol}) = ?\ \text{kJ/h}$$

$$\dot{W} = -\dot{n}(1,A)RT(1)(r/(r-1))[(P(2)/P(1))^{(r-1)/r} - 1] = -3.60\text{E}+05\ \text{kJ/h} \qquad [3]$$

$$r = C_p/C_v = C_p/(C_p - R) = 35/(35 - 8.314) = 1.312$$

The M&E balance has three independent equations and three unknowns $\dot{n}(1,A)$, $\dot{n}(2,A)$ and T(2) i.e. FULLY-SPECIFIED.
$\dot{n}(1,A) = (\dot{n}(2,A) = 103.6$ kmol/h   T (2) = 399 K

F. Reactor

This example shows how a known reactor output temperature is used to find an unknown conversion.

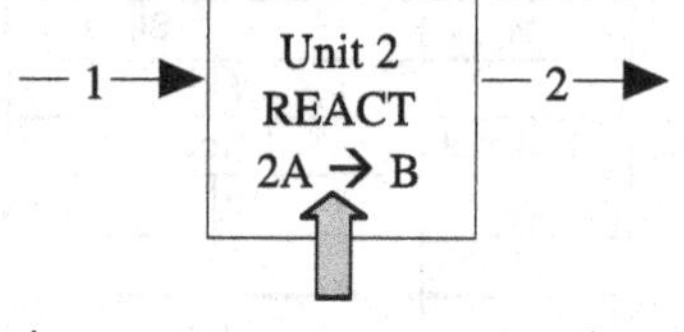

$\dot{Q} = 1.00\text{E} + 06$ kJ/h $\dot{W} = 0$

Conversion of A = X(A) unspecified

Stream Table Reactor			*Problem*
Species	M	Stream	
		1	*2*
	kg/kmol	kmol/h	
A	40	10	*?*
B	80	0	*?*
Total	kg/h	400	*?*
Phase		L	L
Temp.	K	300	350
Press.	kPa(abs)	180	150
Enthalpy	kJ/h	*?*	*?*

Properties	Species	A (liq)	A (gas)	B (liq)	B (gas)
$C_{p,m}$(ref.298K)	kJ/kmol.K	50	35	70	45
$h_{f,298K}$	kJ/kmol	-2.0E+05	-1.8E+05	-1.0E+05	-7.0E+04

Solution: **Material balance**

Mole balance on A $0 = 10 - \dot{n}(2,A) + 0 - X(A)\dot{n}(1,A)$ [1]

Mole balance on B $0 = 0 - \dot{n}(2,B) + (1/2)X(A)\dot{n}(1,A) - 0$ [2]

Two independent equations. Three unknowns $\dot{n}(2,A), \dot{n}(2,B)$ and X(A).
Material balance is UNDER-SPECIFIED.

Energy Balance

Reference condition = *elements* at standard state, 298 K

$$0 = \dot{H}*(1) - \dot{H}*(2) + \dot{Q} - \dot{W} \quad [3]$$

where:

$$\dot{H}*(i) = \Sigma[\dot{n}(i,j)(C_{p,m}(j)(T(i) - T_{ref}) + h^o_{f,Tref}(j))]$$

[Respect the phase]

$$\dot{H}*(1) = (10 \text{ kmol/h})((50 \text{ kJ/kmol.K})(300 \text{ K} - 298 \text{ K}) + (-2.0E5 \text{ kJ/kmol})) + (0 \text{ kmol/h})((70 \text{ kJ/kmol.K})(300 \text{ K} - 298 \text{ K}) + (-1.0E5 \text{ kJ/kmol})) = -2.00E+06 \text{ kJ/h}$$

$$\dot{H}*(2) = (\dot{n}(2,A) \text{ kmol/h})((50 \text{ kJ/kmol.K})(350 \text{ K} - 298 \text{ K}) + (-2.0E5 \text{ kJ/kmol})) + (\dot{n}(2,B) \text{ kmol/h})((70 \text{ kJ/kmol.K})(350 \text{ K} - 298 \text{ K}) + (-1.0E5 \text{ kJ/kmol})) = ? \text{ kJ/h}$$

Stream Table Reactor			Solution
Species	M	Stream	
		1	2
	kg/kmol	kmol/h	
A	40	10.00	***4.58***
B	80	0.00	***2.71***
Total	kg/h	400	400
Phase		L	L
Temp.	K	300	350
Press.	kPa(abs)	180	150
Enthalpy*	kJ/h	***-2.00E+06***	***-9.99E+05***
Mass balance check			
Mass IN	kg/h	400	Closure %
Mass OUT	kg/h	400	100.0
Energy balance check			
Energy IN	kJ/h	-9.99E+05	Closure %
Energy OUT	kJ/h	-9.99E+05	100.0

** Enthalpy values include heats of formation*

The M&E balance has three independent equations and three unknowns $\dot{n}(2,A), \dot{n}(2,B)$ and X(A), i.e. FULLY-SPECIFIED. Solve for X(A) = 0.54.

EXAMPLE 6.04 Simultaneous M&E balance by the overall balance approach (synthesis of ammonia).

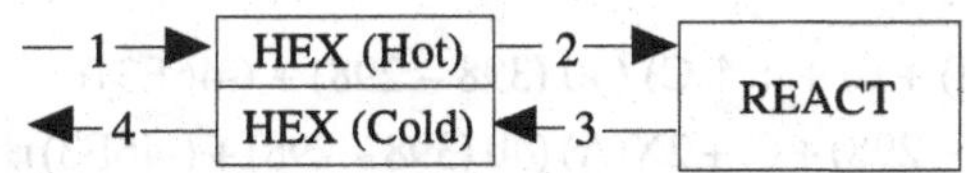

Continuous adiabatic process at steady-state.
Zero work done.

This figure shows a flowsheet of an adiabatic process for production of ammonia by the reaction:

$N_2(g) + 3H_2(g) \rightarrow 2NH_3(g)$ *Reaction 1*

Stream 1:
Mixture of N_2 and H_2 in stoichiometric proportions for *Reaction 1* GAS 2.2E4 kPa(abs) 298 K
Stream 4:
Product mixture of N_2, H_2 and NH_3 GAS 2.0E4 kPa(abs) 398 K

Problem:
Calculate the conversion of nitrogen in this process.

Data		N_2(g)	H_2(g)	NH_3(g)
$h^o_{f,298K}$	kJ/kmol	0	0	-46E3
$C_{p,m}$	kJ/(kmol.K)	30	31	40

Solution: Basis = 1 kmol/h N_2 in stream 1
Check the specifications.

	Unit 1 (HEX)	Unit 2 (REACT)	OVERALL
Material balance			
Number of variables [$\dot{n}(i, j)$ + reactions]	IJ = (4)(3) = 12	IJ + L = (2)(3) + 1 = 7	IJ + L = (2)(3) + 1 = 7
Number of independent equations*	6	6	6
Energy balance			
Number of variables [P(i),T(i),$\dot{Q}$(k),$\dot{W}$(k)]	2I + 2 = (2)(4) + 2 = 10	2I + 2 = (2)(2) + 2=6	2I + 2 = (2)(2) + 2 = 6
Number of independent equations*	7	3	7
Net M&E variables = 25, independent equations = 25			

** Includes the basis. Note that the specifications of streams <u>between</u> units are counted at each unit.*

Define the system = overall process
Component: A = N_2 B = H_2 C = NH_3
Specify the quantities = moles A, B and C X(A) = conversion of N_2

Material balance Rate ACC = Rate IN – Rate OUT + Rate GEN – Rate CON

Mole balance on N_2 $0 = 1 - \dot{n}(4, A) + 0 - X(A)$ [1]

Mole balance on H_2 $0 = 3 - \dot{n}(4, B) + 0 - 3X(A)$ [2]

Mole balance on NH_3 $0 = 0 - \dot{n}(4, C) - 2X(A) - 0$ [3]

Energy balance $0 = \dot{H}*(1) - \dot{H}*(4) + \dot{Q} - \dot{W}$ $\dot{Q} = 0, \dot{W} = 0$ [4]

$\dot{H}*(1) = 0$ [Elements at 298 K, assumed ideal gases]

$$\dot{H}*(4) = \dot{n}(4,A)\,(30\,(398-298)+0) + \dot{n}(4,B)\,(31(398-298)+0) + \dot{n}(4,C)\,(40\,(398-298)+(-46E3))$$
$$= (1-X(A))\,(30\,(398-298)+0) + 3\,(1-X(A))\,(31(398-298)+0) + 2X(A)\,(40\,(398-298)+(-46E3))$$

Four independent equations, four unknowns, i.e. FULLY-SPECIFIED.

Solve for X(A) = <u>**0.13**</u>

This problem could also be solved by calculating each process unit in sequence to get the complete M&E balance stream table, but that much detail is not needed to find the desired conversion of nitrogen.

EXAMPLE 6.05 Simultaneous M&E balance by the iterative sequential modular method.

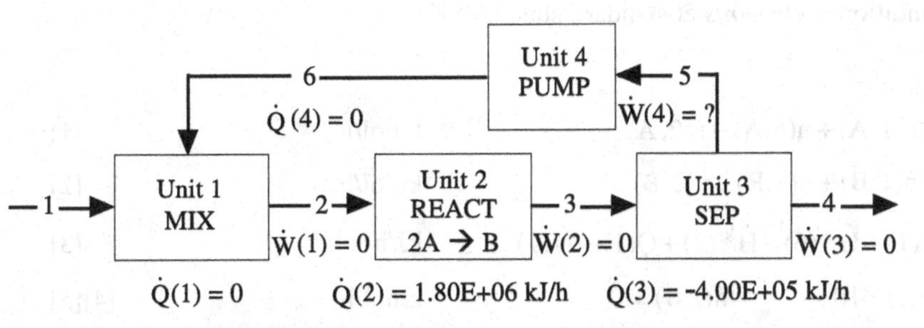

The figure shows a generic process in which a feed of "A" is partially converted a product "B". Unconverted "A" is recovered from the reaction product and recycled to the reactor to increase the overall yield of "B" from "A".

Properties	Species	A (liq)	A (gas)	B (liq)	B (gas)
$C_{p,m}$ (ref.298K)	kJ/kmol.K	50	35	70	45
$h_{f,298K}$	kJ/kmol	-2.0E+05	-1.8E+05	-1.0E+05	-7.0E+04
Density	kg/m^3	900	Ideal gas	900	Ideal gas

Specifications:

Stream 1:	10 kmol/h pure A			2000	kPa(abs)	300 K
Stream 2:	Composition unspecified	Flow unspecified	Liquid	1980	kPa(abs)	T(2)
Stream 3:	Composition unspecified	Flow unspecified	Gas	1950	kPa(abs)	T(3)
Stream 4:	Composition unspecified	Flow unspecified	Liquid	200	kPa(abs)	320 K
Stream 5:	Composition unspecified	Flow unspecified	Liquid	200	kPa(abs)	400 K
Stream 6:	Composition unspecified	Flow unspecified	Liquid	2000	kPa(abs)	T(6)

Unit 2: Conversion of A = X(A) = exp[-300/T(3)] (Due to the dependence of reaction rate on temperature)

Unit 3: Separation efficiency of A from stream 3 to stream 5 = 90%

Problem:

A. Make a degrees of freedom analysis of the problem.

B. Use a spreadsheet to calculate the stream table for the steady-state material and energy balance.

Solution: In this example the material balance and the energy balance are coupled and must be solved together.

A. Check the specifications.

	Unit 1	Unit 2	Unit 3	Unit 4	OVERALL
Material balance					
Number of variables [n(i, j) + reactions]	IJ = 6	IJ + L = 5	IJ = 6	IJ = 4	IJ + L = 5
Number of independent equations*	4	3	3	2	4
Energy balance					
Number of variables [(P(i),T(i),Q̇(k),Ẇ(k)]	2I + 2 = 8	2I + 2 = 6	2I + 2 = 8	2I + 2 = 6	2I + 2 = 6
Number of independent equations*	7	4	7	3	5
Net M&E variables = 33, independent equations = 33					

** Note that the specifications of streams between units are counted at each unit.*

B. Write the material and energy balances for each process unit in sequence.

Energy balance reference condition = *elements* at standard state, 298K

UNIT 1 (Mixer)

Mole balance on A:	$0 = \dot{n}(1,A) + \dot{n}(6,A) - \dot{n}(2,A)$		kmol/h	[1]
Mole balance on B:	$0 = \dot{n}(1,B) + \dot{n}(6,B) - \dot{n}(2,B)$		kmol/h	[2]
Energy balance:	$\dot{H}*(1) + \dot{H}*(6) - \dot{H}*(2) + \dot{Q}(1) - \dot{W}(1)$		kJ/h	[3]
Stream compositions:	$\dot{n}(1,A) = 10$	$\dot{n}(1,B) = 0$	kmol/h	[4][5]
Stream conditions:	P(1) = 2000	T(1) = 300	kPa, K	[6][7]
	P(2) = 1980		kPa	[8]
	P(6) = 2000		kPa	[9]
Unit loads:	$\dot{Q}(1) = 0$	$\dot{W}(1) = 0$	kJ/h	[10][11]

UNIT 2 (Reactor)

Mole balance on A:	$0 = \dot{n}(2,A) - \dot{n}(3,A) + 0 - X(A)\dot{n}(2,A)$		kmol/h	[12]
Mole balance on B:	$0 = \dot{n}(2,B) - \dot{n}(3,B) + (1/2)X(A)\dot{n}(2,A) - 0$		kmol/h	[13]
Energy balance:	$0 = \dot{H}*(2) - \dot{H}*(3) + \dot{Q}(2) - \dot{W}(2)$		kJ/h	[14]
Conversion:	X(A) = exp[-300/T(3)]		–	[15]
Stream conditions:	P(3) = 1950		kPa	[16]
Unit loads:	$\dot{Q}(2) = 1.8E6$	$\dot{W}(2) = 0$	kJ/h	[17][18]

UNIT 3 (Separator)

Mole balance on A:	$0 = \dot{n}(3,A) - \dot{n}(4,A) - \dot{n}(5,A)$		kmol/h	[19]
Mole balance on B:	$0 = \dot{n}(3,B) - \dot{n}(4,B) - \dot{n}(5,B)$		kmol/h	[20]
Energy balance:	$0 = \dot{H}*(3) - \dot{H}*(4) - \dot{H}*(5) + \dot{Q}(3) - \dot{W}(3)$		kJ/h	[21]
Split fraction:	$s(5, A) = \dot{n}(5, A)/\dot{n}(3, A) = 0.9$			[22]
Stream conditions:	P(4) = 150	T(4) = 320	kPa, K	[23][24]
	P(5) = 150	T(5) = 400	kPa, K	[25][26]
Unit loads:	$\dot{Q}(3) = -4.0E5$	$\dot{W}(3) = 0$	kJ/h	[27][28]

UNIT 4 (Pump)

Mole balance on A: $0 = \dot{n}(5,A) - \dot{n}(6,A)$ kmol/h [29]

Mole balance on B: $0 = \dot{n}(5,B) - \dot{n}(6,B)$ kmol/h [30]

Energy balance: $0 = \dot{H}^*(5) - \dot{H}^*(6) + \dot{Q}(4) - \dot{W}(4)$ kJ/h [31]

Unit loads: $\dot{Q}(4) = 0 \quad \dot{W}(4) = -V(5)[P(6) - P(5)]$ kJ/h [32][33]

Here there are 33 independent equations for six streams, two species, one reaction and four process units. The problem is FULLY-SPECIFIED.

The balances are calculated by the iterative sequential modular method, using the spreadsheet "Solver" to solve the non-linear equations for the reactor to get T(3) and the corresponding conversions X(A).

Everything depends – on everything!!

Stream Table		**Mixer-reactor-separator + recycle**				**Solution**	
Species	M	Stream					
		1	2	3	4	5	6
	kg/kmol	kmol/h					
A	40	10.00	*16.39*	*7.10*	*0.71*	*6.39*	*6.39*
B	80	0.00	*1.63*	*6.27*	*4.64*	*1.63*	*1.63*
Total	kg/h	400	*786*	*786*	*400*	*386*	*386*
Phase		L	L	G	L	L	L
Temp.	K	300	*347*	*528*	320	400	*402.0*
Press.	kPa(abs)	2000	1980.00	1950	200	200	2000
Volume	m^3/h	0.44	*0.87*	*30*	*0.44*	*0.43*	*0.43*
Enthalpy*	kJ/h	-2.00E+06	*-3.40E+06*	*-1.60E+06*	*-5.99E+05*	*-1.40E+06*	*-1.40E+06*
Mass Balance Check			Unit 1	Unit 2	Unit 3	Unit 4	Overall
Mass IN	kg/h		786	786	786	386	400
Mass OUT	kg/h		786	786	786	386	400
Closure %			100.0	100.0	100.0	100.0	100.0
Energy Balance Check							
Q	kJ/h		0.00	1.80E+06	-4.00E+05	0.00	1.40E+06
W	kJ/h		0.00	0.00	0.00	-774	-774
Energy IN	kJ/h		-3.40E+06	-1.60E+06	-2.00E+06	-1.40E+06	-5.99E+05
Energy OUT	kJ/h		-3.40E+06	-1.60E+06	-2.00E+06	-1.40E+06	-5.99E+05
Closure %			100.0	100.0	100.0	100.0	100.1

* *Enthalpy values include heats of formation*

EXAMPLE 6.06 Simultaneous M&E balance for sustainable development (hydrogen economy).

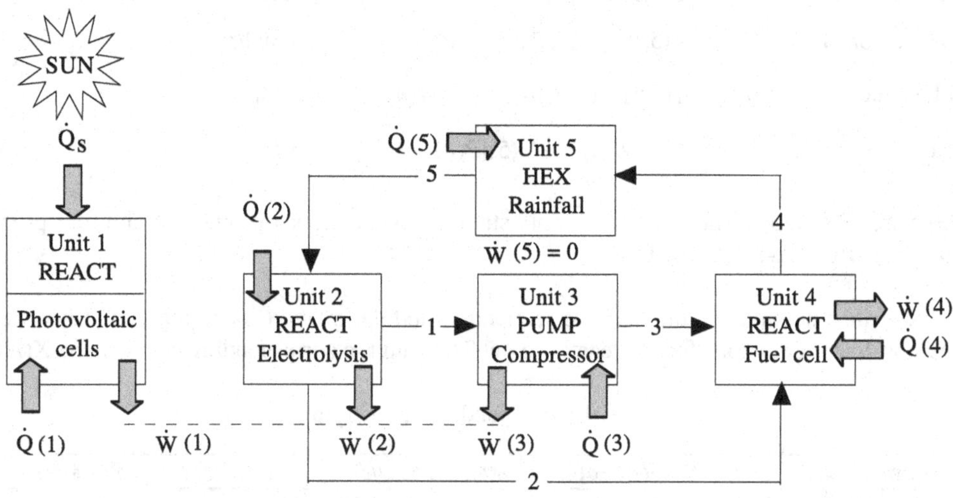

The figure shows a simplified version of the hydrogen economy, in which solar radiation ($\dot{Q}_s$) drives photovoltaic cells (*Unit 1*) to produce DC electricity ($\dot{W}(1)$) which is used to split water (*Stream 5*) to H_2 (*Stream 1*) and O_2 (*Stream 2*) by electrolysis (*Unit 2, Reaction 1*) and to drive compressors (*Unit 3*) that supply compressed hydrogen (*Stream 3*) to H_2/Air fuel cells on board vehicles (*Unit 4*). The water vapour (*Stream 4*) produced by the fuel cell (*Unit 4, Reaction 2*) is dispersed into the atmosphere then condensed as rainfall (*Unit 5*), collected and recycled to the electrolysis step. Oxygen from the electrolysis (*Stream 2*) is discharged into the air and subsequently consumed in the fuel cells.

$$2H_2O \xrightarrow[4F]{} 2H_2 + O_2 \quad \textit{Reaction 1}$$

$$2H_2 + O_2 \rightarrow 2H_2O \quad \textit{Reaction 2}$$

Properties	Species	H_2O(l)	H_2O(g)	H_2(g)	O_2(g)
$C_{p,m}$ (ref.298K)	kJ/kmol.K	75	34	29	30
$h_{f,298K}$	kJ/kmol	-2.86E+05	-2.42E+05	0.00	0.00

The process streams and units are defined as follows:

Stream 1.	Pure hydrogen	Flow unspecified	Gas	110	kPa(abs)	350	K
Stream 2.	Pure oxygen	Flow unspecified	Gas	110	kPa(abs)	350	K
Stream 3.	Unspecified	Flow unspecified	Gas	3.00E+04	kPa(abs)	350	K
Stream 4.	Unspecified	Flow unspecified	Gas	110	kPa(abs)	350	K
Stream 5.	Unspecified	Flow unspecified	Liquid	110	kPa(abs)	300	K

Unit 1.	Photovoltaic cells	Solar flux input	1 kW/m^2	6h/day		Energy efficiency = 20%
Unit 2.	Electro-chem. reactor	Input voltage		2 Volt/cell		
Unit 3.	Compressor					Energy efficiency = 70%
Unit 4.	Fuel cells	No. autos	700E+06	100 kW/auto	2 h/day	Energy efficiency = 40%
						H_2 conversion = 100%

Problem:

A. Make a degrees of freedom analysis of the problem.

B. Use a spreadsheet to calculate the stream table for the continuous **steady-state** material and energy balance.

C. Calculate the area of photovoltaic cells and the corresponding fraction of the Earth's surface required for the cells.

D. Find the increased rate of rainfall and net thermal load on the Earth required for the energy needs of 700E6 automobiles.

Solution:

A. Check the specifications:

	Unit 1	Unit 2	Unit 3	Unit 4	Unit 5	OVERALL
Material balance						
Number of variables [$\dot{n}(i, j)$ + reactions]	IJ = 0	IJ + L = 10	IJ = 6	IJ + L = 10	IJ = 6	IJ + L = 2
Number of independent equations*	0	7	5	7	3	0
Energy balance						
Number of variables [P(i), T(i), $\dot{Q}$(k), $\dot{W}$(k), $\dot{Q}_s$]	2 + 3* = 3	2I + 2 = 8	2I + 2 = 6	2I + 2 = 8	2I + 2 = 6	2I + 3* = 3
Number of independent equations*	3	8	6	8	6	2

* *Note that the specifications of streams* *<u>between</u>* *units are counted at each unit. Energy balance unknowns includes* $\dot{Q}_s$

B. Write the material and energy balances for each unit in sequence: H_2O = A, H_2 = B, O_2 = C

Energy balance reference condition = *elements* at standard state, 298 K

<u>UNIT 1 (Reactor)</u> [Zero material flows]

Energy balance: $0 = \dot{Q}_S + \dot{Q}(1) - \dot{W}(1)$ kW [1]

Efficiency: $\dot{W}(1) = 0.2\,\dot{Q}_S$ kW [2]

Unit loads: $\dot{W}(1) = -[\dot{W}(2) + \dot{W}(3)]$ kW [3]

<u>UNIT 2 (Reactor)</u>

Mole balance on H_2O: $0 = \dot{n}(5,A) - \dot{n}(1,A) - \dot{n}(2,A) + 0 - X(A)\dot{n}(5,A)$ kmol/s [4]

Mole balance on H_2: $0 = \dot{n}(5,B) - \dot{n}(1,B) - \dot{n}(2,B) + 0 - X(A)\dot{n}(5,A) - 0$ kmol/s [5]

Mole balance on O_2:	$0 = \dot{n}(5,C) - \dot{n}(1,C) - \dot{n}(2,C) + (1/2)X(A)\dot{n}(5,B) - \dot{n}(1,B) - \dot{n}(2,B)$	kmol/s	[6]
Energy balance:	$0 = \dot{H}^*(5) - \dot{H}^*(1) - \dot{H}^*(2) + \dot{Q}(2) - \dot{W}(2)$	kW	[7]
Stream compositions:	$\dot{n}(1, A) = 0$ $\quad\quad$ $\dot{n}(1, C) = 0$	kmol/s	[8][9]
	$\dot{n}(2, A) = 0$ $\quad\quad$ $\dot{n}(2, B) = 0$	kmol/s	[10][11]
Stream conditions:	P(1) = 110 $\quad\quad$ T(1) = 350	kPa, K	[12][13]
	P(2) = 110 $\quad\quad$ T(2) = 350	kPa, K	[14][15]
	P(5) = 110 $\quad\quad$ T(5) = 300	kPa, K	[16][17]
Unit loads:	$\dot{W}(2) = -E_V I' = (2)(2F(\dot{n}(5,A) - \dot{n}(1,A) - \dot{n}(2,A)))$	kW	[18]
Faraday's law:	F = 96480 kC/kmol $\quad$ I′ = current $\quad$ E_v = Voltage		

UNIT 3 (Pump)

Mole balance on H_2O:	$0 = \dot{n}(1,A) - \dot{n}(3,A)$	kmol/s	[19]
Mole balance on H_2:	$0 = \dot{n}(1, B) - \dot{n}(3, B)$	kmol/s	[20]
Mole balance on O_2:	$0 = \dot{n}(1,C) - \dot{n}(3,C)$	kmol/s	[21]
Energy balance:	$0 = \dot{H}^*(1) - \dot{H}^*(3) + \dot{Q}(3) - \dot{W}(3)$	kW	[22]
Stream compositions:	Partially specified at Unit 2		
Stream conditions:	P(3) = 30E3 $\quad\quad$ T(3) = 350	kPa, K	[23][24]
Unit loads:	$\dot{W}(3) = -P(1)\dot{V}(1)\ln[(P(3)/P(1)]/0.7$	kW	[25]

UNIT 4 (Reactor)

Mole balance on H_2O	$0 = \dot{n}(2,A) + \dot{n}(3,A) - \dot{n}(4,A) + X(B)\dot{n}(3,B) - 0$ $\quad$ X(B) = 1	kmol/s	[26][27]
Mole balance on H_2	$0 = \dot{n}(2,B) + \dot{n}(3,B) - \dot{n}(4,B) + 0 - X(B)\dot{n}(3,B)$	kmol/s	[28]
Mole balance on O_2	$0 = \dot{n}(2,C) + \dot{n}(3,C) - \dot{n}(4,C) + 0 - (1/2)X(B)\dot{n}(3,B)$	kmol/s	[29]
Energy balance.	$0 = \dot{H}^*(2) + \dot{H}^*(3) - \dot{H}^*(4) + \dot{Q}(4) - \dot{W}(4)$	kW	[30]
Stream compositions	Partially specified at Unit 2		
Stream conditions	P(4) = 110 $\quad\quad$ T(4) = 350	kPa, K	[31][32]
Unit loads	$\dot{W}(4)$ = (700E6 autos)(100 kW/auto)(2 h/24 h) = 5.83E+09	kW	[33]
Energy efficiency	$0.4 = \dot{W}(4)/(286E3(\dot{n}(3,B) - \dot{n}(4,B)))$		[34]

UNIT 5 (Heat exchanger)

Mole balance on H_2O	$0 = \dot{n}(4, A) - \dot{n}(5, A)$	kmol/s	[35]
Mole balance on H_2	$0 = \dot{n}(4, B) - \dot{n}(5, B)$	kmol/s	[36]
Mole balance on O_2	$0 = \dot{n}(4, C) - \dot{n}(5, C)$	kmol/s	[37]

Energy balance: $0 = \dot{H}*(4) - \dot{H}*(5) + \dot{Q}(5) - \dot{W}(5)$ kW [38]

Stream compositions: Unspecified

Unit loads: $\dot{W}(5) = 0$ kW [39]

NOTE: **Since H_2 and O_2 are in the same ratio (2/1) in streams 4 and 5, equations [36 and 37] give only one independent relation.**

Here there are 38 independent equations for five streams, three species, two reactions and five process units. 38 stream and unit variables. The problem is FULLY-SPECIFIED.

Stream Table		*M&E Balance for the Hydrogen Economy*		*Solution*			
Species	M	Stream					
	kg/kmol	*1*	*2*	*3*	*4*	*5*	
		kmol/s					
[A] H_2O	18	0.00E+00	0.00E+00	0.00E+00	5.10E+04	5.10E+04	
[B] H_2	2	5.10E+04	0.00E+00	5.10E+04	0.00E+00	0.00E+00	
[C] O_2	32	0.00E+00	2.55E+04	0.00E+00	0.00E+00	0.00E+00	
Total	kg/s	101981	815851	101981	917832	917832	
Phase		G	G	G	G	L	
Temp.	K	350	350	350	350	300	
Press.	kPa	110	110	3.00E+04	110	110	
Volume	m^3/s	1.35E+06	6.74E+05	4.94E+03	1.35E+06	918	
Enthalpy*	kW	7.69E+07	3.98E+07	7.69E+07	-1.22E+10	-1.46E+10	
Mass Balance Check		Unit 1	Unit 2	Unit 3	Unit 4	Unit 5	Overall
Mass IN	kg/s	0	917832	101981	917832	917832	0
Mass OUT	kg/s	0	917832	101981	917832	917832	0
Closure %		100.0	100.0	100.0	100.0	100.0	100.0
Energy Balance Check							
Energy IN	kW	9.85E+10	1.17E+08	8.88E+07	1.17E+08	-1.22E+10	9.85E+10
Energy OUT	kW	9.85E+10	1.17E+08	8.88E+07	1.17E+08	-1.22E+10	9.85E+10
Closure %		100.0	100.0	100.0	100.0	100.0	100.0

** Enthalpy values include heats of formation*

$\dot{Q}_s$ = + 9.85E+10 kW

$\dot{Q}(1)$ = - 7.88E+10 kW

$\dot{Q}(2)$ = - 4.99E+09 kW

$\dot{Q}(3)$ = - 1.19E+07 kW

$\dot{Q}(4)$ = - 6.53E+09 kW

$\dot{Q}(5)$ = - 2.33E+09 kW

$\dot{W}(1)$ = + 1.97E+10 kW

$\dot{W}(2)$ = - 1.97E+10 kW

$\dot{W}(3)$ = - 1.19E+07 kW

$\dot{W}(4)$ = + 5.83E+09 kW

$\dot{W}(5)$ = 0 kW

Surface area of Earth = 5.14 E+08 km^2 Land area of Earth = 1.48 E+08 km^2

C. Area of photovoltaic cells = 3.94 E+11 m^2 = **3.94 E+05 km^2 = 0.27% of Earth's land area**[4]

D. Increased rainfall = 917832 kg/s = **7.93E+07 tonne/day = 0.06 mm/year (average)**

Net thermal load on Earth = **9.26E+10 kW** = **0.04% of net radiant flux from Sun to Earth**

4 This is area of the active surface only. The total area needed would probably be 2 or 3 times this value, approaching 1% of Earth's land area.

SUMMARY

[1] Simultaneous M&E balances arise for systems in which the material balance is coupled with the energy balance in such a way that the two balances must be solved simultaneously to find the distribution of material and energy across the system.

[2] In simultaneous M&E balances the unknowns are material quantities together with energy variables such as pressure, temperature, heat and work. When the complete simultaneous M&E balance is fully-specified, either the material balance alone is under-specified or the material balance and the energy balance (taken separately) are both under-specified.

[3] Simultaneous M&E balances are more difficult to identify and to specify than separate material and energy balances. In addition, the solution of simultaneous M&E balances is more likely to be complicated by simultaneous non-linear equations.

[4] Simultaneous M&E balances can arise in each of the generic process units: MIX, SEPARATE, HEAT EXCHANGE, PUMP and REACT. A characteristic of these cases is that the outlet stream flows and phase splits are interdependent with the pressure and/or temperature in the unit.

[5] Simultaneous M&E balances for multi-unit processes can be solved by the sequential modular method (outlined in Chapters 4 and 5), with iteration of the recycle loops. The spreadsheet solution may require manual iteration and/or the use of macros to close the simultaneous balances on individual process units.

[6] Simultaneous M&E balances often involve complex non-linear interactions that defy an intuitive knowledge of a system's behavior. Such interactions are characteristic of chemical processes as well as of natural (ecological) systems that couple the flow of material and energy, and are a main source of the "ingenuity gap" facing society in the 21st century (see *Ref. 7*).

FURTHER READING

[1] R. M. Felder and R.W. Rousseau, *"Elementary Principles of Chemical Proceses"*, John Wiley & Sons, New York, 2000.

[2] G. V. Reklaitis, *"Introduction to Material and Energy Balances"*, John Wiley & Sons, New York, 1983.

[3] D. M. Himmelblau, "*Basic Principles of Calculations in Chemical Engineering*", Prentice Hall, Englewood Cliffs, 1989.

[4] J. M. Douglas, *"Conceptual Design of Chemical Processes"*, McGraw-Hill, New York, 1988.

[5] L. T. Biegler, I. E. Grossmann and A.W. Westerberg, *"Systematic Methods of Chemical Process Design"*, Prentice Hall, Upper Saddle River, 1997.

[6] W. D. Seider, J. D. Seader and D. R. Lewin, *"Process Design Principles"*, John Wiley & Sons, New York, 1999.

[7] T. Homer-Dixon, *"The Ingenuity Gap"*, Alfred A. Knopf, New York, 2000.

CHAPTER SEVEN

UNSTEADY-STATE MATERIAL AND ENERGY BALANCES

DIFFERENTIAL MATERIAL AND ENERGY BALANCES WITH ACCUMULATION

Differential material and energy balances can be used to follow the behaviour of systems in which conditions change over time.

When one or more condition in a system changes over time, the rate of accumulation term in the differential material balance and/or energy balance is not zero. Conditions in the system are *transient* and the system is said to be operating in the *"unsteady-state"* mode. Unsteady-state operation is common in chemical processes and is the rule in natural systems. Examples of unsteady-state systems are as follows:

- Batch chemical process units.
- Continuous chemical process units during "start-up" and "shut-down", or after disturbance to a process variable.
- Biological systems such as living cells, organs and the human body.
- The planet Earth, with respect to its environment, population and resources.

Differential balances are used to predict the temporal behavior of unsteady-state systems in areas such as process control and bio-reactor engineering, as well as for modelling the Earth's ecology and environment (e.g. predicting the global climate).

Unsteady-state differential balances are powerful tools for process modelling, but they are also relatively difficult in conception and execution.

The calculation of unsteady-state differential M&E balances begins with the general balance equation from Chapter 1, repeated here as *Equation 7.01*, with its corresponding differential form, *Equation 7.02*.

For a defined *system* and a specified *quantity*:

ACC = IN – OUT + GEN – CON *Equation 7.01*

Rate ACC = Rate IN – Rate OUT + Rate GEN – Rate CON *Equation 7.02*

where:

Rate ACC	=	Rate of accumulation of specified quantity in the system, with respect to time
Rate IN	=	Rate of input of specified quantity to the system, with respect to time (input)
Rate OUT	=	Rate of output of specified quantity from the system, with respect to time (output)
Rate GEN	=	Rate of generation of specified quantity in the system , with respect to time (source)
Rate CON	=	Rate of consumption of specified quantity in the system, with respect to time (sink)

In the case of unsteady-state systems: **Rate ACC $\neq 0$**

The "Rate ACC" term in *Equation 7.02* is expressed mathematically as a differential with respect to time.

The differential material balance for an unsteady-state system then appears as *Equation 7.03* or *7.04*.

$$\mathbf{d(m)/dt = (\dot{m})_{in} - (\dot{m})_{out} + (\dot{m})_{gen} - (\dot{m})_{con} \neq 0} \text{ (mass balance)} \qquad \textit{Equation 7.03}$$

$$\mathbf{d(n)/dt = (\dot{n})_{in} - (\dot{n})_{out} + (\dot{n})_{gen} - (\dot{n})_{con} \neq 0} \text{ (mole balance)} \qquad \textit{Equation 7.04}$$

where:

		Typical units
(m), (n)	= mass, moles of specified material in the system	kg, kmol
$(\dot{m})_{in}$, $(\dot{n})_{in}$	= mass, mole flow of specified material into the system	$kg.s^{-1}$, $kmol.s^{-1}$
$(\dot{m})_{out}$, $(\dot{n})_{out}$	= mass, mole flow of specified material out of the system	$kg.s^{-1}$, $kmol.s^{-1}$
$(\dot{m})_{gen}$, $(\dot{n})_{gen}$	= rate of generation of specified material in the system	$kg.s^{-1}$, $kmol.s^{-1}$
$(\dot{m})_{con}$, $(\dot{n})_{con}$	= rate of consumption of specified material in the system	$kg.s^{-1}$, $kmol.s^{-1}$
t	= time	s

In *Equations 7.03* and *7.04* the differential material accumulation terms "d(m)/dt" and "d(n)/dt" are the rates of change of the amount of the specified material in the system (i.e. inside the system envelope) with respect to time. The "in" and "out" terms are the rates of transfer of the specified material across the system envelope, while the "gen" and "con" terms are respectively the rates of generation and consumption of the specified material inside the system envelope. When the quantity being balanced is the total mass, which is conserved in non-nuclear processes, both "gen" and "con" terms are zero. Similarly, if *no chemical reaction* occurs in the system the "gen" and "con" terms are both zero (in non-nuclear processes) when the quantity balanced is either the amount of a component "j" or the total moles. When a chemical reaction occurs in the system, the "gen" and "con" terms for individual reactant and product species are not zero, though the "gen" and "con" terms for total moles may or may not be zero, depending on the reaction stoichiometry.

The differential energy balance on an unsteady-state system appears as *Equation 7.05*. Recall that the generation and consumption terms do not appear in the energy balance because energy is a conserved quantity (in non-nuclear processes).

$$dE/dt = \dot{E}_{in} - \dot{E}_{out} + \dot{Q} - \dot{W} - (P\dot{V}_{out} - P\dot{V}_{in})$$ *Equation 7.05*

where:

Symbol	Definition	Typical units
E	= energy content of the system (associated with material)[1]	kJ
$\dot{E}_{in}$	= rate of energy flow into the system (associated with material)	kW
$\dot{E}_{out}$	= rate of energy flow out of the system (associated with material)	kW
$(P\dot{V}_{out} - P\dot{V}_{in})$	= net rate of work to move material in and out of the system	kW
$\dot{Q}$	= net rate of heat transfer <u>into</u> the system	kW
$\dot{W}$	= net rate of work transfer <u>out</u> of the system (a.k.a. shaft power)	kW
t	= time	s

In typical chemical processes where the kinetic, potential and exotic energy terms are considered negligible (see Chapter 5), *Equation 7.05* simplifies to *Equation 7.06*.

$$dU/dt = \dot{H}_{in} - \dot{H}_{out} + \dot{Q} - \dot{W}$$ *Equation 7.06*

where:

Symbol	Definition	Typical units
U	= internal energy in the system,	kJ
$\dot{H}_{in}$	= rate of enthalpy flow into the system,	kW
$\dot{H}_{out}$	= rate of enthalpy flow out of the system,	kW
$\dot{Q}$	= net rate of heat transfer <u>into</u> the system	kW
$\dot{W}$	= net rate of work transfer <u>out</u> of the system (a.k.a. shaft power)	kW
t	= time	s

In *Equations 7.05* and *7.06* the differential energy accumulation terms "dE/dt" and "dU/dt" are the rates of change of the amount of energy <u>in the system</u> (i.e. inside the system envelope) with respect to time. The $\dot{H}_{in}$ and $\dot{H}_{out}$ terms are the rates of enthalpy flow associated with the transfer of material across the system envelope. As outlined in Chapter 5, the $\dot{Q}$ and $\dot{W}$ terms are transfers across the system envelope, respectively in the form of thermal and usually mechanical or electrical energy.

The reference state for internal energy and enthalpy in *Equation 7.06* may be either the *elements* or the *compounds* at specified conditions. As discussed in Chapter 5 the *element* reference state is preferred in processes involving chemical reaction, whereas either the *element* or *compound* reference state may be used, as is convenient, in processes without chemical reaction. Also, the energy reference states can be mixed so long as they each cancel from the energy balance.

---

[1] Remember that the energy terms E, H and U are state functions, that must be defined relative to specified reference conditions.

Equations 7.03 to *7.06* are ordinary differential equations (ODEs). In more sophisticated models of unsteady-state systems, you may encounter partial differential equations (PDEs), but PDEs are outside the scope of this text. Differential equations can be solved by analytic calculus or by numerical methods such as *Euler, Runge–Kutta, finite differencing* and *finite element.* Unsteady-state M&E balances generally involve initial-value problems whose numerial solution (e.g. by Runge–Kutta) requires the specification of an initial condition and then proceeds to step through time to obtain a temporal profile of the process variables.

This text contains example problems with ODEs that are solved by simple analytical calculus, plus some examples (*7.03, 7.06* and *7.07*) that are solved by a numerical method.

CLOSED SYSTEMS (BATCH PROCESSES)

The differential M&E balances for a closed system require the condition:

Material flow IN = Material flow OUT = 0 [each of $\dot{Q}$ and $\dot{W}$ may or may not be zero]

Examples 7.01, 7.02 and *7.03* illustrate respectively a differential material balance, a differential energy balance and a simultaneous differential material and energy balance on an unsteady-state closed system.

EXAMPLE 7.01 Differential material balance on an unsteady-state closed system (population of Earth).

The human population of Earth in 2000 AD was 6.035 billion (6.035E9), with birth and death rates respectively 21,000 and 9,000 per million people per year.

Problem: Assuming constant birth and death rates, with zero space travel, calculate the human population of Earth in 2100 AD.

Solution:
Define the system = planet Earth (a closed system, i.e. Rate IN = Rate OUT = 0)
Specify the quantity = number of people = N_p
Differential material balance on the number of people:

Rate ACC = Rate IN – Rate OUT + Rate GEN – Rate CON
$dN_p/dt = 0 - 0 + (21E3/1E6)N_p - (9E3/1E6)N_p = (12E\text{-}3)N_p$

Separate the variables, integrate and solve for N_p, with the boundary (initial) conditions:
$N_p = 6.035E9$ people $t = 2000$ years
$dN_p/N_p = (12E\text{-}3)dt$
$\ln(N_p/6.035E9) = (12E\text{-}3)(t - 2000)$
$N_p = (6.035E9)\exp((12E\text{-}3)(t - 2000))$ Substitute $t = 2100$ to get:
N_p = **20.04E9 people**[2] in the year 2100 AD

[2] This result probably over-estimates the population because the growth rate is expected to drop as the population increases.

EXAMPLE 7.02 Differential energy balance on an unsteady-state closed system (batch heater).

Initial condition

Tank contains 60 kmol pure liquid A at 300 K. Heat is transferred to the contents of the tank at a rate given by:

HEX

No reaction
No phase change

$\dot{Q} = 3(420 - T)$ kW
$\dot{W} = 0$

Properties	Species	A(liq)
$C_{p,m}$(ref.298K)	kJ/kmol.K	50
$h_{f,298K}$	kJ/kmol	-2.0E+05
Density	kg/m^3	900
Assume $C_{p,m} = C_{v,m}$ for liquids.		

$\dot{Q} = 3(420 - T)$ kW

where:

$\dot{Q}$ = rate of heat transfer into the contents of the tank, kW

T = temperature of liquid in tank, K

Problem: Calculate and plot the temperature in the tank as a function of time from t = 0 to t = 2 hours.

Solution: Define the system = contents of the tank
Specify the quantity = energy

Write the unsteady-state balances, i.e.

Rate ACC = Rate IN – Rate OUT + Rate GEN – Rate CON

Energy balance Reference condition = *elements* at standard state, 298K

$d[u^*n(A)]/dt = \dot{Q} - \dot{W}$ [1]

where:

u^* = specific internal energy of liquid A

$\approx C_{v,m}(T - 298) + h^o_{f,298K}$ kJ/kmol

n(A) = amount on A in the tank = 60 kmol

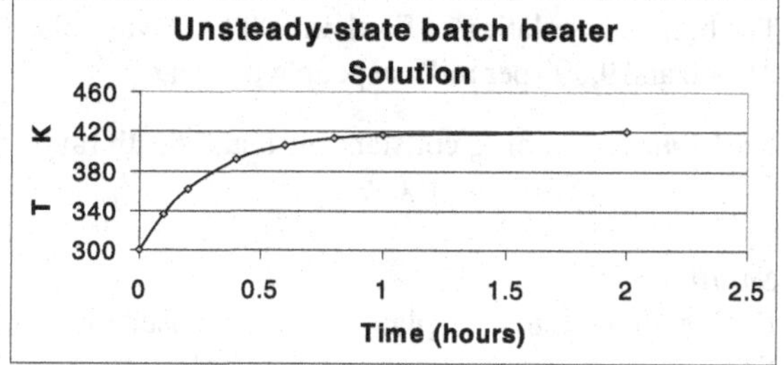

Since $C_{v,m}$ and $h^o_{f,298K}$ are constant, equation [1] simplifies to:

$(60)(50)\, dT/dt = (3600)(3)(420 - T)$ kJ/h [2]

$dT/dt = 3.6(420 - T)$ K/h [3]

Solve the differential equation [3] by separating the variables and integrating:
Initial condition T = 300, t = 0

$\mathbf{T = 420 - 120exp(-3.6t)}$ K [4]

t hours	0	0.1	0.2	0.4	0.6	0.8	1	2
T K	300	336	362	392	406	413	417	419.9

EXAMPLE 7.03 Simultaneous differential M&E balance on an unsteady-state closed system (batch reactor).

REACT
$2A(l) \rightarrow B(l)$

$\dot{W} = 0$

Component		A(l)	B(l)
MW	kg/kmol	45	90
$C_{v,m}$ ref 298K	kJ/(kmol.K)	70	80
$h^o_{f,298K}$	kJ/kmol	-50E3	-110E3
Density	kg/m³	900	900

A closed batch thermochemical reactor carries out the irreversible liquid phase reaction:

$2A(l) \rightarrow B(l)$

Where the reaction rate is a function concentration and temperature, given by the Arrhenius relation:

Reaction rate = d[A]/dt = - 2exp(-2E3/T) [A] [1]

[A] = concentration of [A] kmol/m³
T = temperature K
t = time s

The reactor initially contains 100 kmol of reactant A and zero B at 298 K, with a volume: $V_R = 5$ m³
The reactor contents are cooled by heat transfer to a water jacket at a rate given by:

$\dot{Q}$ = - 9(T – 298) = rate of heat transfer <u>into</u> reactor contents kW [note negative sign of $\dot{Q}$] [2]

Problem:
Calculate the conversion of A and the temperature in the reactor at 300 seconds after the reaction is initiated.

Solution:
Define the system = contents of the batch reactor
Specify the quantities = mole A, mole B, Energy
Differential material balances: Rate ACC = Rate IN – Rate OUT + Rate GEN – Rate CON
Mole balance on A: d(n(A))/dt = 0 – 0 + 0 – V_R(2exp(-2E3/T) [A]) = - n(A)(2exp(-2E3/T)) [3]
Mole balance on B: d(n(B))/dt = 0 – 0 + (1/2)V_R(2exp(-2E3/T) [A]) – 0 = - (1/2)d(n(A))/dt [4]
Differential energy balance:

dE/dt = $\dot{Q} - \dot{W}$ = -9(T – 298) – 0 ($\dot{W} = 0$) [5]
E = energy of reactor contents
≡ internal energy, w.r.t. *elements* at standard state, 298 K
E = (n(A) kmol)(70 kJ/(kmol.K)(T – 298) K + (-50E3 kJ/kmol))
+ (n(B) kmol)(80 kJ/(kmol.K)(T – 298) K + (-110E3 kJ/kmol))
dE/dt = 70(n(A) (T – 298))/dt – 50E3(n(A))/dt + 80(n(B)(T – 298))/dt – 110E3(n(B))/dt
= -9(T – 298) [6]

Initial conditions: n(A) = 100 kmol n(B) = 0 kmol
T = 298 K at t = 0 s [Assumes u* = h* for liquids]

Equations [1 to 6] form a set of simultaneous differential equations that embodies a complex non-linear interaction between n(A) and T. The formal solution of such simultaneous ODEs is beyond the scope of this text, however, there is a crude method, called Euler's method, that can be used to get an approximate solution by simple spreadsheet calculations. In Euler's method the calculation begins from the initial condition and marches forward in small increments of time to calculate new values of the variables (n(A) and T) using a finite difference equation, e.g:

$T_{\iota+1} = T_{\iota} + \Delta t\,(dT/dt)_{\iota}$

where:

Δt = time increment
$T_{\iota+1}$ = temperature at time $t(\iota + 1)$
T_{ι} = temperature at time $t(\iota)$
$(dT/dt)_{\iota}$ = differential of T, w.r.t. temp. at time $t(\iota)$

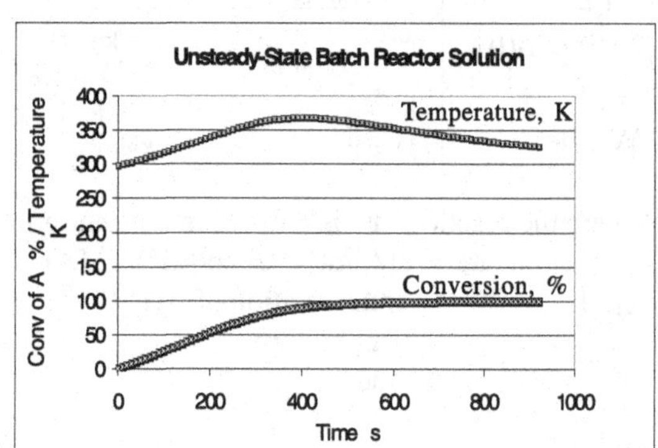

The spreadsheet solution of *Example 7.03* by Euler's method is summarised in *Figure 7.01.*

From the spreadsheet, at t = 300 s

X(A) = <u>76%</u> T = <u>361 K</u>

Component		A(l)	B(l)	k	s^{-1}	2
M	kg/kmol	45	90	Exp	K	-2.E+03
$C_{v,m}$	kJ/kmol.K	70	80	Vol	m^3	5
h^o_{298K}	kJ/kmol	-5.00E+04	-1.10E+05	$U_{hex}A$	kW/K	9

EULER   Time increment = 10 s   Energy balance converged using "Goal Seek"

t	n(A)	X(A)	n(B)	Total mass	T	$U_{hex}A(T-298)$	dn(A)/dt	dn(B)/dt	dT/dt	Energy balance closure
s	kmol	%	kmol	kg	K	kW	kmol/s	kmol/s	K/s	
0	100.0	0.0	0.0	4500	298	0	-0.243	0.122	0.174	2.E-12
100	74.2	25.8	12.9	4500	317	173	-0.271	0.136	0.215	-2.E-12
200	47.0	53.0	26.5	4500	340	379	-0.263	0.131	0.234	-2.E-12
300	23.7	76.3	38.2	4500	361	566	-0.186	0.093	0.151	3.E-12
400	9.8	90.2	45.1	4500	369	637	-0.087	0.043	-0.005	1.E-12
500	4.0	96.0	48.0	4500	364	592	-0.033	0.016	-0.089	0.E+00
600	1.8	98.2	49.1	4500	354	503	-0.013	0.006	-0.103	0.E+00
700	0.9	99.1	49.5	4500	344	413	-0.005	0.003	-0.094	0.E+00
800	0.5	99.5	49.7	4500	335	334	-0.003	0.001	-0.079	0.E+00
900	0.3	99.7	49.8	4500	328	269	-0.001	0.001	-0.065	0.E+00

Figure 7.01. Spreadsheet solution of Example 7.03.

OPEN SYSTEMS (CONTINUOUS PROCESSES)

Examples 7.04, 7.05 and *7.06* illustrate respectively a material balance, an energy balance and a simultaneous material and energy balance on unsteady-state open systems. All of these examples assume perfect mixing, for which the composition and temperature of the outlet stream are the same as those in the mixing vessel.

EXAMPLE 7.04 Differential material balance on an unsteady-state open system (mixer).

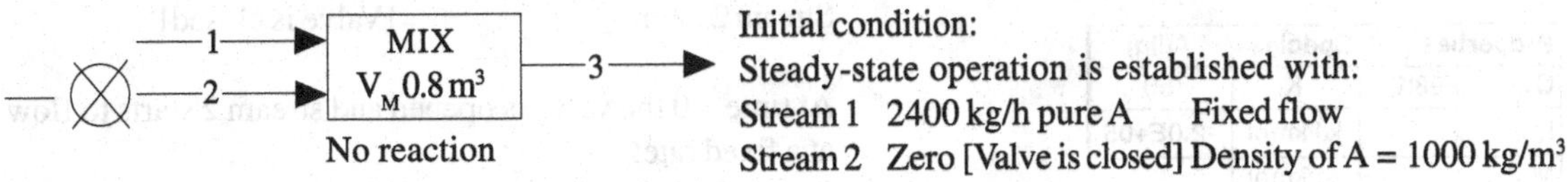

At time = 0 the valve is opened and stream 2 starts to flow at a fixed rate:
Stream 2 800 kg/h pure B Fixed flow Density of B = 1000 kg/m³
Both stream 1 and stream 2 continue to flow at the fixed rates specified above
The volume of material inside the mixing vessel remains fixed at $V_M = 0.8\ m^3$

Problem: Calculate and plot the mass fraction of B in str. 3 as a function of time from t = 0 to t = 2 hours

Solution: Define the system = mixing vessel
Specify the quantities = total mass, mass of B

Let w(3, B) = mass fraction of B in stream 3 ρ = density of stream 3, kg/m³

Write the unsteady-state balances, i.e. Rate ACC = Rate IN – Rate OUT + Rate GEN – Rate CON

Total mass balance:

$$d(V_M\rho)/dt = 2400 + 800 - \overline{m}(3) \quad kg/h \quad [1]$$

Mass balance on B

$$d[w(3,B)V_M\rho]/dt = 800 - w(3,B)\overline{m}(3) \quad kg/h \quad [2]$$

Since

$$V_M\rho = \text{constant} = (0.8\ m^3)(1000\ kg/m^3) = 800\ kg$$

From equation [1]

$$d(V_M\rho)/dt\ t = 0 = 2400 + 800 - \overline{m}(3)$$

$$\overline{m}(3) = 3200\ kg/h$$

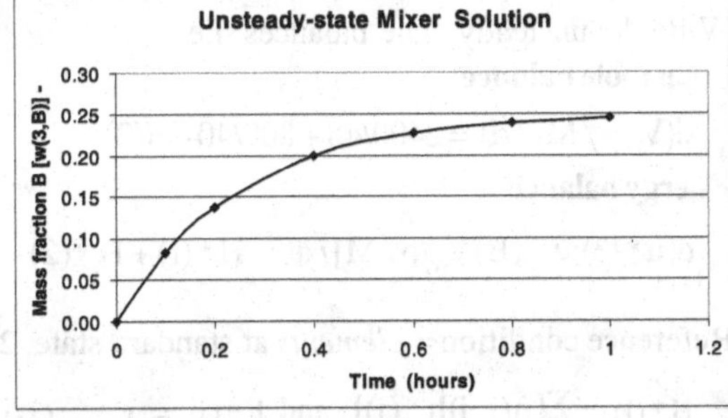

Substitute to equation [2]:

$$800\ d[w(3,B)]/dt = 800 - 3200w(3,B) \qquad \mathbf{d[w(3,B)]/dt = 1 - 4w(3,B)} \quad [3]$$

Solve the differential equation [3] by separating the variables and integrating:

Initial condition:

$$w(3,B) = 0, t = 0 \quad [4]$$

$$\mathbf{w(3,B) = (1/4)(1 - exp(-4t))} \quad [5]$$

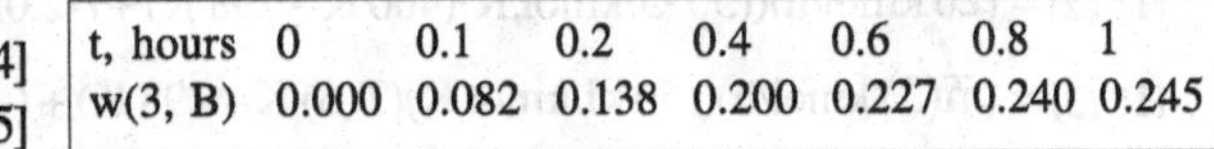

t, hours	0	0.1	0.2	0.4	0.6	0.8	1
w(3, B)	0.000	0.082	0.138	0.200	0.227	0.240	0.245

EXAMPLE 7.05 Differential energy balance on an unsteady-state open system (mixer).

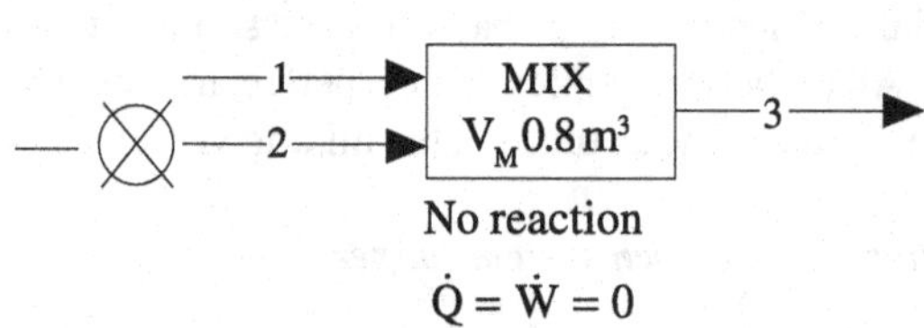

Properties	Species	A(liq)
$C_{p,m}$(ref.298K)	K	50
$h^o_{f,\ 298\ K}$	kJ/kmol	-2.0E+05
M	kg/kmol	40
Density	kg/m³	900
Assume $C_{p,m} = C_{v,m}$ for liquids		

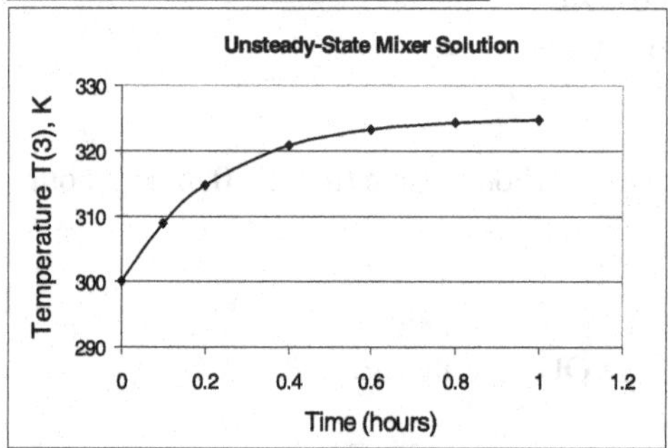

Initial condition:
Steady-state operation is established with:

Stream 1	2400 kg/h pure A	Fixed flow, P and T
	Liquid	200 kPa(abs), 300 K
Stream 2	Zero	[Valve is closed]

At time = 0 the valve is opened and stream 2 starts to flow at a fixed rate:

Stream 2	800 kg/h pure A	Fixed flow, P and T
	Liquid	200 kPa(abs), 400 K

Both *stream 1* and *stream 2* continue to flow at the fixed rates specified above. The volume of material inside the mixing vessel remains fixed at $V_M = 0.8$ m³.

Problem: Calculate and plot the temperature of stream 3 as a function of time from t = 0 to t = 2 hours.

Solution:
Define the system = mixing vessel
Specify the quantities = total moles, energy

Write the unsteady-state balances, i.e.
Total mole balance

$$d(V_M\rho/M)/dt = 2400/40 + 800/40 - \bar{n}(3) \qquad \text{kmol/h} \qquad [1]$$

Energy balance

$$d[u^*(3)w(3,B)V_M\rho/M]/dt = \dot{H}^*(1) + \dot{H}^*(2) - \dot{H}^*(3) + 0 - 0 \qquad \text{kJ/h} \qquad [2]$$

Reference condition = *elements* at standard state, 298K

$$\dot{H}^*(1) = \Sigma[\dot{n}(i,j)h^*(j)] \text{ and } h^*(j) = C_{p,m}(T(i) - 298) + h^o_{f,298K} \qquad \text{[Respect the phase]}$$

$$u^*(3) \approx C_{v,m}(T(3) - 298) + h^o_{f,298K} \qquad \text{[Assume } u^* = h^* \text{ for liquids]}$$

$$\dot{H}^*(1) = (60\ \text{kmol/h})((50\ \text{kJ/kmol.K})(300\ \text{K} - 298\ \text{K}) + (\text{-2.0E5 kJ/kmol}))$$

$$\dot{H}^*(2) = (20\ \text{kmol/h})((50\ \text{kJ/kmol.K})(400\ \text{K} - 298\ \text{K}) + (\text{-2.0E5 kJ/kmol}))$$

$$\dot{H}^*(3) = (\bar{n}(3)\ \text{kmol/h})((50\ \text{kJ/kmol.K})(T(3)\ \text{K} - 298\ \text{K}) + (\text{-2.0E5 kJ/kmol}))$$

From equation [1] $d(V_M\rho/M)/dt = 0 = 2400/40 + 800/40 - \bar{n}(3) = 80$ then $\bar{n}(3) = 80$ kmol/h

Substitute to equation [2] and collect terms, note that the heat of formation term (-2.0E5 kJ/kmol) cancels out.

$$C_{v,m}(V_M\rho/M)d(T(3))/dt = 6000 + 102000 - 4000(T(3) - 298) \quad \text{kJ/h} \quad [3]$$

$$900\, d(T(3))/dt = 1.300E6 - 4000(T(3))$$

$$\mathbf{d(T(3))/dt = 1444 - 4.444(T(3))} \quad \text{kJ/h} \quad [4]$$

Solve the differential equation [4] by separating the variables and integrating:
Initial condition $T(3) = 300, t = 0$ **<u>T(3) = 325 – 5(1 – exp(- 4.44t))</u>** K [5]

t hours	0	0.1	0.2	0.4	0.6	0.8	1
T(3) K	300.0	309.0	314.7	320.8	323.3	324.3	324.7

EXAMPLE 7.06 Simultaneous differential M&E balance on an unsteady-state open system (mixed and heated tank).

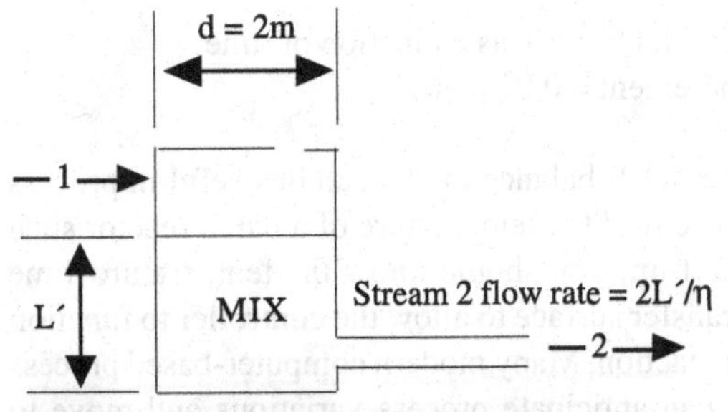

Properties	Species	A(liq)
C_{pm}(ref.298K)	kJ/(kg.K)	3
Density	kg/m^3	1200
Viscosity	kg/(m.s)	(1E-3)exp(2000/T - 6)
Assume $C_{pm} = C_{vm}$ for liquids		

Initial condition – steady-state operation with:

Stream 1 2000 kg/h A Liquid 120 kPa(abs), 300 K

Heat transfer rate: $\dot{Q} = 0$ kJ/h

At time t = 0 the heater is turned on and delivers heat to the tank at a rate: $\dot{Q} = 1.0E{+}04\,(400 - T)$ kJ/h

Stream 1 flow remains at 2000 kg/h A. Assume no phase change, plus liquid density and heat capacity are independent of temperature.
η = liquid viscosity

Problem: Calculate and plot the level (L´) and the temperature (T) in the tank as a function of time for: t = 0 to t = 10 hours.

Solution: Define the system = contents of the tank Specify the quantities = mass of A, energy

Material balance on A:

$$d(V\rho)/dt = \dot{m}(1) - \dot{m}(2) \quad [1]$$

$$V\rho = 3.14((2\,m)/2)^2(L'm)(1200\,kg/m^3) = 3768L' \quad \text{kg}$$

$$\overline{m}(1) = 2000 \quad \text{kg/h}$$

$$\overline{m}(2) = 2\,L'/\eta \quad \text{kg/h}$$

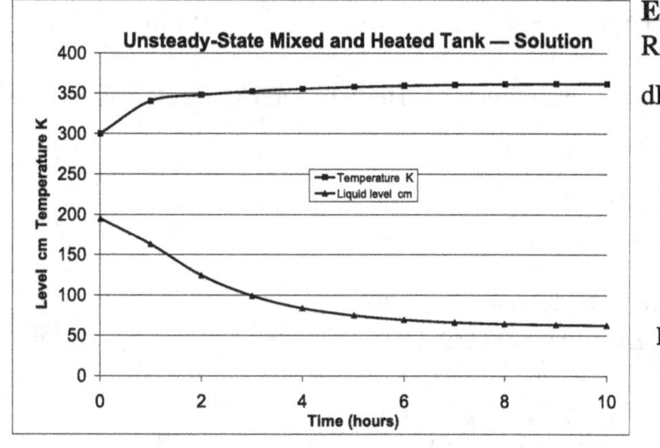

Energy balance:

Reference condition = *liquid compound* A at 298 K

$$dE/dt = \dot{H}(1) - \dot{H}(2) + \dot{Q} \quad \dot{W} = 0 \qquad [2]$$

$$E = V\rho\,(C_{v,m}(T-298))$$

$$= (3768L'\,kg)(3\,kJ/(kg.K))(T-298)\,K$$

$$= 11304\,L'\,(T-298) \qquad kJ$$

$$\dot{H}(1) = (2000\,kg/h)(3\,kJ/(kg.K))(300-298)\,K$$

$$= 12000 \qquad kJ/h$$

$$\dot{H}(2) = \dot{m}(2)(3kJ/(kg.K))(T-298)\,K$$

$$= 3\dot{m}(2)(T-298) \qquad kJ/h$$

$$\dot{Q} = 10000(400-T) \qquad kJ/h$$

Initial conditions: $\dot{m}(2) = 2000$ kg/h $T = T(2) = 300$ K

$L' = 1.95$ m $t = 0$ hours

Solve the simultaneous differential equations [1 and 2] to obtain L′ and T as a function of time. The spreadsheet solution uses Euler's method with a time increment = 0.01 hour.

In *Examples 7.02* to *7.06* you can see how an unsteady-state M&E balance model can be useful in process design, process control and safety analysis. For instance, to control the temperature of a batch reactor such as that in *Example 7.03* (or a continuous reactor during start-up) you should know the temperature–time profile and ensure the unit is designed with sufficient heat transfer surface to allow the controller to function properly and prevent a thermal "runaway" of the exothermic reaction. Many modern computer-based process controllers include differential M&E balance models that can anticipate process variations and move to prevent them (see *Ref. 8*).

Differential M&E balances can be used for a large range of problems of which only a small sampling is given above. However, practical unsteady-state problems often involve complexities beyond the scope of this text, including non-linear differential equations that must be solved by numerical methods (for which special software is available), that are more sophisticated than the Euler's method used above in *Example 7.03*.

A final example of the range and versatility of differential material and energy balances is shown in *Example 7.07*. This example presents a highly simplified differential M&E balance model of the atmosphere of planet Earth and shows how such a model can be used to predict aspects of the global climate. In common with most natural systems, the atmosphere of planet Earth is an open system subject to inputs and outputs of material and energy at rates that vary over time. The planet's climate is tied to these material and energy transfers by a set of non-linear interactions that are embodied in the simultaneous differential material and energy balance on the Earth and its atmosphere.

The complexity of *Example 7.03* is magnified many times in *Example 7.07*, but the principles and the crude solution by Euler's method are the same. The material inputs/outputs of carbon dioxide and water vapour are coupled to the input/outputs of radiant energy with non-linear interactions that result in the phenomenon known to climatologists as the *"runaway greenhouse effect"*,[3] associated with global warming (see *Refs. 6, 10 and 11*) that is looming as the most critical problem yet to face humanity. The kind of complexity shown in *Example 7.07* makes human intuition an unreliable guide in formulating policy for a sustainable future of planet Earth (see *Ref. 9*).

EXAMPLE 7.07 Simultaneous differential M&E balance on an unsteady-state open system (atmosphere of Earth).

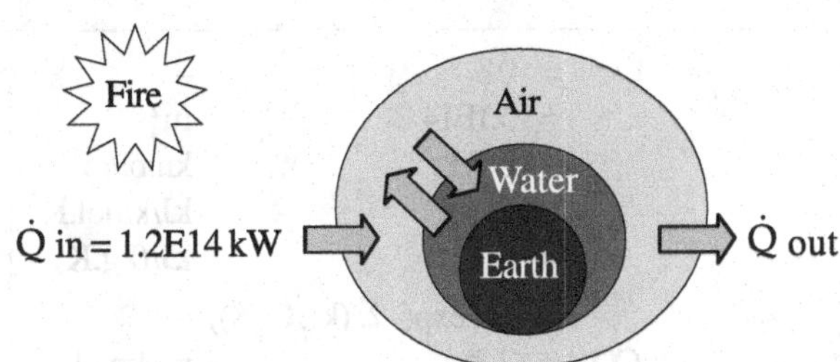

This example presents a highly simplified model of the thermal "greenhouse effect" in the environment of planet Earth caused by the accumulation of CO_2 from combustion of fossil fuels and its interaction with the increasing humidity of the atmosphere.

In this problem the atmosphere is considered as an open system that exchanges material (CO_2 and H_2O) and energy (IR radiation) with the surface of the Earth. By the *carbon cycle*[4], CO_2 enters the system from the combustion of fossil fuels and the decomposition of biomass and exits the system to support photosynthesis in forests and oceans, as well as by absorption into the oceans to form carbonates. H_2O enters the system by combustion of fossil fuels, decomposition of biomass and evaporation from the oceans, and exits the system as precipitation.

Sunlight (solar radiation) penetrates the atmosphere and is partly reflected and partly absorbed by the surface of the Earth (ice, land and sea). The absorbed solar energy is then emitted from Earth as infra-red radiation. Some of this infra-red radiation is absorbed by the atmosphere, at a rate dependent on the concentrations of both CO_2 and H_2O (i.e. greenhouse gases) in the atmosphere. Energy is also transferred back to Earth from the atmosphere to warm the oceans and melt the polar icecaps.

Specifications of the system:

Total rate of radiant energy input from Sun to planet Earth	= 0.343 (average over all Earth surface)	kW/m^2
Rate of solar radiant energy absorbed by Earth	$= 0.343(1-a) = 0.343(1-0.3) = 0.240$	kW/m^2
Rate of IR radiation from Earth surface to Atmosphere	$= f k_{SB} T_o^4$	kW/m^2

[3] The *runaway greenhouse effect* is the accelerating upward spiral of the Earth's temperature, initiated by anthropogenic (man-made) CO_2 emissions and promoted by the increasing water vapour content of the atmosphere through the coupled M&E balances. The pressure and temperature of the Earth's atmosphere relative to the phase diagram for water (*Figure 2.02*) should prevent Earth from becoming another waterless Venus!

[4] The *carbon cycle* is one of the primary geochemical recycle loops that regulate the supply of life-essential elements in the biosphere of Earth. Other primary cycles are: oxygen, nitrogen, water, etc.

Rate of IR radiation from Atmosphere to Earth surface	$= f k_{SB} T_1^4$	kW/m^2
Rate of IR radiation from Atmosphere to space	$= f k_{SB} T_1^4$	kW/m^2
Rate of CO_2 input from fossil fuels + deforestation	= 700E9 (1 + 0.03(t))	kmol/y
Rate of CO_2 input from decay and resp'n of land biomass	= 1500y (C)(T_o/288)	$kmol/m^2.y$
Rate of CO_2 output to photosynthesis on land	= (1550) y (C)(T_o/288)	$kmol/m^2.y$
Rate of CO_2 output to ocean biomass	= 2(1 – 0.005(t)) y (C)(T_o/288)	$kmol/m^2.y$
Rate of CO_2 output to ocean carbonates	= 1(y (C) – 370E-6)	$kmol/m^2.y$
Fraction of land area with photosynthesising vegetation	= 0.1(1 – 0.005 (t))	–

Data and nomenclature:

a	= Earth albedo[5]		= 0.3	–
A_E	= Earth surface area		= 5.1E14	m^2
C_{IR}	= concentration of IR absorbent gas		= p/RT	$kmol/m^3$
$C_{v,A,}$	= heat capacity of air		= 21	kJ/kmol.K
$C_{v,E}$	= heat capacity of Earth's surface		= 4.0	kJ/(kg.K)
f	= IR absorptivity of atmosphere		$= 1 - \exp(-\Sigma (k_{BL} C_{IR} S))$	–
k_{BL}	= molar absorption coefficient for IR radiation	CO_2	= 2.28	$m^2/kmol$
		H_2O	= 3.04	$m^2/kmol$
k_{SB}	= Stefan–Boltzmann constant		= 5.67E-11	$kW/(m^2.K^4)$
m_E	= mass of surface layer of Earth (100 m thick)		= 51E18	kg
n(A)	= total amount of air in Earth's atmosphere		= 1.8E17	kmol
p	= partial pressure of IR absorbent gas			kPa(abs)
RH	= relative humidity of atmosphere		= 0.3	–
S	= effective thickness of atmosphere for IR absorption	CO_2	= 5000	m
		H_2O	= 3000	m
T_0	= temperature of Earth surface			K
T_1	= temperature of Earth atmosphere			K
t	= time			years
y(C)	= mole fraction CO_2 in Earth's atmosphere			–
y(W)	= mole fraction H_2O vapour in Earth's atmosphere		= p(W)/101.3	–

Area of Earth's land surface	= 1.5E14	m^2
Area of Earth's ocean surface	= 3.6E14	m^2
Volume of Earth's oceans	= 1.4E18	m^3
Average temperature of Earth surface in 2000 AD	= 288	K
Temperature of Earth's atmosphere in 2000 AD	= 242	K
Average CO_2 content of Earth's atmosphere in 2000 AD	= 0.037	vol%
Partial pressure of H_2O in Earth's atmosphere = (RH) p* = (RH)exp[16.5362 – 3985.44/(T_0 – 38.9974)]		kPa(abs)

Other greenhouse gases (e.g. CH_4, N_2O, etc.) are ignored for simplicity.

[5] Albedo = Reflected radiant energy/Incoming radiant energy.

Problem: Calculate and plot the CO_2 and H_2O content of the atmosphere and the Earth surface temperature for t = 0 to 100 years.

Solution:

Define the system: 1 = Atmosphere (open system) 2 = Earth (open system)

Specify the quantities: = amounts of CO_2, H_2O Energy w.r.t. compounds at 298 K

Define components: A = Air C = CO_2 E = Earth W = H_2O

Differential material balances on Atmosphere:

Rate ACC = Rate IN – Rate OUT + Rate GEN – Rate CON

Mole balance on CO_2: GEN = CON = 0, assumes no reactions in the atmosphere produce or consume CO_2

d[y(C)n(A)]dt = [Fossil fuel] + [Biomass] – [Land photosynthesis] – [Sea photosynthesis] – [Sea carbonates]

$$= 700E9(1+0.03(t)) + 0.1(1.5E14)(1-0.005(t))[1500y(C)T_0/288]$$
$$- 0.1(1.5E14)(1-0.005(t))[1550y(C)T_0/288]$$
$$- 2(3.6E14)(1-0.005(t))[y(C)T_0/288] - 1(3.6E14)(y(C)-370E\text{-}6)] \quad [1]$$

Balance on H_2O: Assume relative humidity of atmosphere remains constant at 30%. i.e.

$$y(W) = (0.3)\exp[16.5362 - 3985.44/(T_0 - 38.9974)]/101.3 \quad [2]$$

Differential energy balance on Atmosphere:

$$d[C_{v,A}n(A)(T_1-298) + 44E3y(W)n(A)]/dt = \text{IR in from Earth} - \text{IR out to Earth} - \text{IR out to space}$$
$$= A_E(fk_{SB}T_0^4 - fk_{SB}T_1^4 - fk_{SB}T_1^4) \quad [3]$$

Differential energy balance on Earth:

$$d[C_{v,E}m_E(T_0-298)]/dt = \text{Solar in from Sun} + \text{IR in from atmosphere} - \text{Solar reflected} - \text{IR radiated from Earth}$$
$$= A_E(0.342 + fk_{SB}T_1^4 - 0.103 - k_{SB}T_0^4) \quad [4]$$

Equations [1, 2, 3 and 4] form a set of simultaneous differential equations to be solved for y(C), y(W), T_1 and T_0 as a function of time. Note the non-linear interactions that result from the effect of y(C) and y(W) on the atmospheric IR absorption coefficient (a "positive feedback") and the effect of y(C) and T on the rate of photosynthesis (a "negative feedback"). A crude solution is obtained by a spreadsheet calculation using Euler's method (see *Example 7.03*), stepping forward in time in increments of 0.05 year. The results are plotted on the next two pages.

Note that this crude model predicts a temperature increase of about 12 K from 2000 AD to 2100 AD. This temperature rise is about double the maximum value obtained from current global climate models (see *Ref. 10*), in part because the present model makes many simplifying assumptions, such as neglecting the effects of clouds, atmospheric aerosols, atmospheric composition profiles, heat of fusion of the polar icecaps, temperature profile in the oceans, volcanic erruptions, changing albedo, etc.

This example also assumes a *constant solar flux*, though some argue that variation in the solar flux has a major affect on Earth's climate.

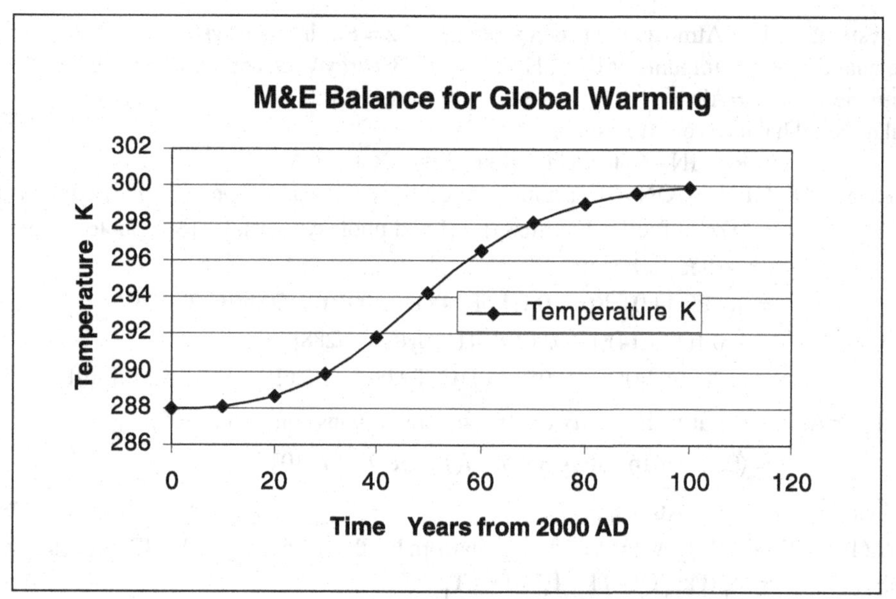
M&E Balance for Global Warming
302
300
298
296
294
292
290
288
286
Temperature K
Temperature K
0
20
40
60
80
100
120
Time Years from 2000 AD

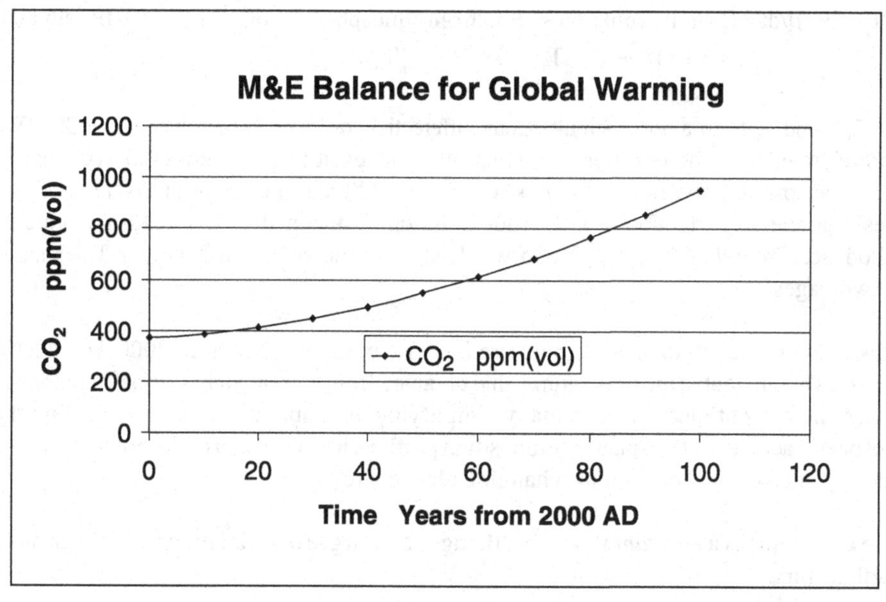
M&E Balance for Global Warming
1200
1000
800
600
400
200
0
CO_2 ppm(vol)
CO_2 ppm(vol)
0
20
40
60
80
100
120
Time Years from 2000 AD

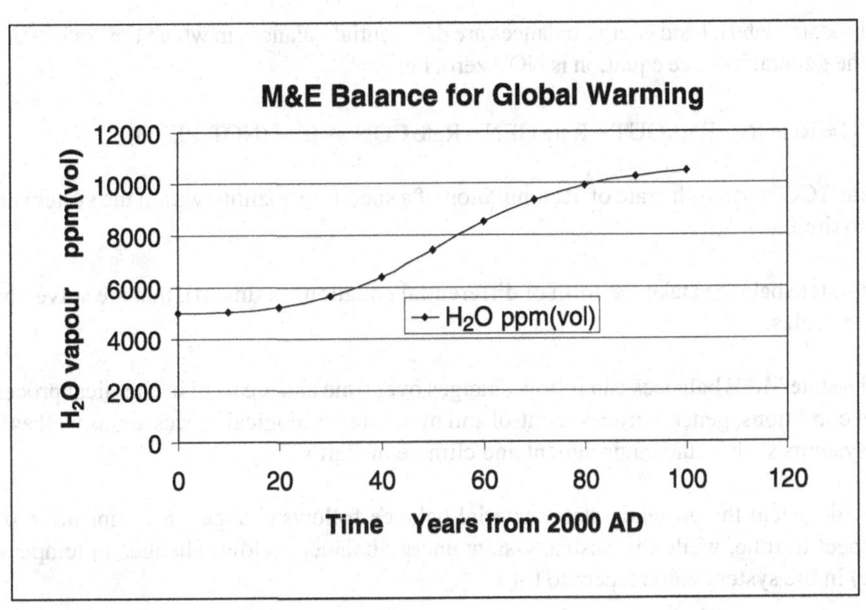
M&E Balance for Global Warming
H_2O vapour ppm(vol)
12000
10000
8000
6000
4000
2000
0
0
20
40
60
80
100
120
Time Years from 2000 AD
H_2O ppm(vol)

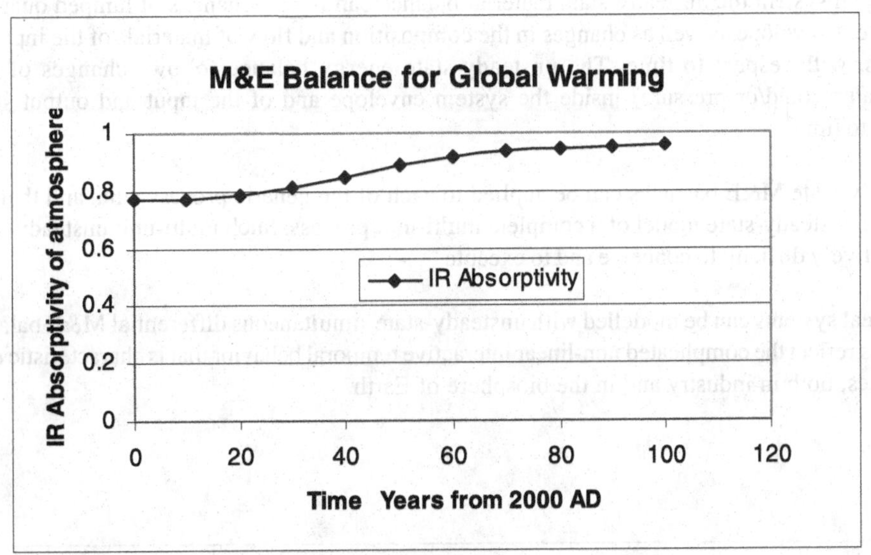
M&E Balance for Global Warming
IR Absorptivity of atmosphere
1
0.8
0.6
0.4
0.2
0
0
20
40
60
80
100
120
Time Years from 2000 AD
IR Absorptivity

SUMMARY

[1] Unsteady-state material and energy balances are differential balances in which the "rate of accumulation" term in the general balance equation is NOT zero, i.e:

Rate ACC = Rate IN – Rate OUT + Rate GEN – Rate CON $\neq 0$ (NOT ZERO)

The "Rate ACC" term is the rate of accumulation of a specified quantity within the system envelope with respect to time.

Unsteady-state balances take the form of differential equations in time (t), that are solved by analytic or numeric calculus.

[2] Unsteady-state M&E balances can follow changes over time and are used in chemical process design for transient conditions, general process control and modelling biological processes, as well as in modelling natural systems such as the environment and climate of Earth.

[3] In a closed system the unsteady-state material balance follows changes in composition of the system with respect to time, while the unsteady-state energy balance follows changes in temperature (and/or pressure) in the system with respect to time.

[4] In an open system the unsteady-state material balance can follow changes of lumped quantities inside the system envelope as well as changes in the composition and flow of materials of the input and output streams, with respect to time. The unsteady-state energy balance follows changes of the lumped temperature (and/or pressure) inside the system envelope and of the input and output streams with respect to time.

[5] Unsteady-state M&E balances can be applied to each of the generic process units and then coupled to build an unsteady-state model of a complete multi-unit process. Such multi-unit unsteady-state models are relatively difficult to conceive and to execute.

[6] Many real systems can be modelled with unsteady-state simultaneous differential M&E balances. These balances reflect the complicated non-linear interactive temporal behavior that is characteristic of chemical[6] processes, both in industry and in the biosphere of Earth.

---

[6] Throughout this text: "Chemical" process ≡ biochemical, electrochemical, photochemical, physico-chemical or thermochemical process.

FURTHER READING

[1] R. M. Felder and R.W. Rousseau, *Elementary Principles of Chemical Processes,* John Wiley & Sons, New York, 2000.

[2] D. M. Himmelblau, *Basic Principles and Calculations in Chemical Engineering,* Prentice Hall, Englewood Cliffs, 1989.

[3] P. M. Doran, *Bioprocess Engineering Principles,* Academic Press, San Diego, 1995.

[4] R. K. Sinnott, *Chemical Engineering Design,* Butterworth-Heinemann, Oxford, 1999.

[5] T. M. Duncan and J. A. Reimer, *Chemical Engineering Design and Analysis,* Cambridge University Press, Cambridge, 1998.

[6] D. Jacobs, *Introduction to Atmospheric Chemistry*, Princeton University Press, Princeton, 1999.

[7] A. Ford, *Modeling the Environment — An Introduction to System Dynamic Models of Environmental Systems*, Island Press, Washington DC, 1999.

[8] W. L. Luyben, *Process Modeling, Simulation and Control for Chemical Engineers,* McGraw-Hill, New York, 1990.

[9] T. Homer-Dixon, *The Ingenuity Gap*, Alfred A. Knof, New York, 2000.

[10] J. T. Houghton *et al.*, *Climate Change 2001: The Scientific Basis,* Intergovernmental Panel on Climate Change, Cambridge University Press, Cambridge, 2001.

[11] R. B. Stull, *Meteorology for Scientists and Engineers*, Brooks/Cole, Pacific Grove, CA, 2000.

[12] T. Jackson, *Material Concerns,* Stockholm Environmental Institute, London, 1996.

[13] Scientific American, *Energy for Planet Earth,* Scientific American, New York, 1990.

EPILOGUE

With apologies to:
"Ozymandias" by Percy Bysshe Shelley, 1792–1822

I met a traveller from a Northern land
who said: Four vast and cabless rims of chrome
stand in the desert... Near them, on the sand
half sunk, some rusted metal lies, whose frame,
and sparking tip, and gears of cast compound
tell that its owner well those fashions led
which yet survive, stamped on these lifeless things,
by glossy ad extolled and iso-octane fed.
And on the engine block these words appear:
"My name is Suv-Humongous, King of Kings:
Look on my power, ye mighty, and despair!"

Nothing beside remains.
Round the decay of that colossal wreck,
boundless and bare
the lone and level sands stretch
far away...

INDEX

N